Interactive Lecture Demonstrations

Active Learning in Introductory Physics

JOHN WILEY & SONS, INC.

Cover Photo: ©James Fraher /Getty Images

To order books or for customer service call 1-800-CALL-WILEY (225-5945).

ISBN 0-471-48774-0

Printed in the United States of America

10 9 8 7 6 5 4 3 2 1

Printed and bound by Malloy Lithography, Inc.

Acknowledgements:

We are especially grateful to Priscilla Laws of Dickinson College for her continuing collaboration that has contributed significantly to this work. We thank Jeff Marx, Shawn Kolitch and Dennis Kuhl for their contributions of ideas for *ILD* sequences, and the many college and high school faculty—including them—who have classroom tested *ILDs* over the years. The curricula that we have developed would not have been possible without the hardware and software development work of Stephen Beardslee, Lars Travers, Ronald Budworth and David Vernier. We also thank the physics faculty and students at the University of Oregon and Tufts University for participating in the *ILDs* and learning assessments.

This work was supported in part by the National Science Foundation under grant number USE-9150589, *Student Oriented Science,* grant number USE-9153725, *The Workshop Physics Laboratory Featuring Tools for Scientific Thinking,* grant number TPE-8751481, *Tools for Scientific Thinking: MBL for Teaching Science Teachers,* and grant number DUE-9950346, *Activity Based Physics Suite* and by the Fund for Improvement of Post-secondary Education (FIPSE) of the US. Department of Education under grant number G008642149, *Tools for Scientific Thinking,* and grant number P116B90692, *Interactive Physics.*

This work was supported in part by the National Science Foundation and the Fund for Improvement of Post-secondary Education (FIPSE) of the U.S. Department of Education. Opinions expressed are those of the authors and not necessarily those of these agencies.

Table of Contents

SECTION I: INTRODUCTION TO INTERACTIVE LECTURE DEMONSTRATIONS

SECTION I: INTRODUCTION TO INTERACTIVE LECTURE DEMONSTRATIONS

Why Interactive Lecture Demonstrations? Despite considerable evidence that traditional approaches are ineffective in teaching physics concepts,[1-9] most physics students in the United States continue to be taught in lectures, often in large lectures with more than 100 students. Alternative curricula such as *Workshop Physics*[10-11] that eliminate formal lectures can be used successfully, but substantial structural changes in instruction are required in large universities to implement this program.

Some attempts to increase student learning while maintaining existing structures have also been successful. A major focus of the work at the Center for Science and Mathematics Teaching (CSMT) at Tufts University and at the University of Oregon Department of Physics has been on active, discovery-based laboratory curricula supported by real-time microcomputer-based laboratory (MBL) tools. The results have been the *Tools for Scientific Thinking* [12] and *RealTime Physics*[13] laboratories. With these tools and curricula, it has been possible to bring about significant changes in the laboratory learning environment at a large number of universities, colleges and high schools without changing the lecture/laboratory structure and the traditional nature of lecture instruction.[1-5] Table I-1 compares the characteristics of our active learning environments created with these laboratory curricula to traditional, passive learning environments.

Table I-1: Passive vs. Active Learning Environments

Passive Learning Environment	Active Learning Environment
Instructor (and textbook) are the authorities--sources of all knowledge.	Students construct their knowledge from hands-on observations. Real observations of the physical world are the authority.
Students' beliefs are rarely overtly challenged.	Uses a learning cycle in which students are challenged to compare predictions (based on their beliefs) to observations of real experiments.
Students may never even recognize differences between their beliefs and what they are told in class.	Changes students' beliefs when students are confronted by differences between their observations and their beliefs.
Instructor's role is as authority.	Instructor's role is as guide in the learning process.
Collaboration with peers often discouraged.	Collaboration and shared learning with peers is encouraged.
Lectures often present the "facts" of physics with little reference to experiment..	Results from real experiments are observed in understandable ways--often in real time with the support of microcomputer-based tools.
Laboratory work, if any, is used to confirm theories "learned" in lecture.	Laboratory work is used to learn basic concepts.

While these MBL laboratory curricula do fit easily into existing structures, they also require computers, interfaces, and laboratory space. Many high school and college physics programs have only a few computers and are unable to support hands-on laboratory work for large numbers of students. Therefore, in recent years we have worked at creating successful active learning environments (like those associated with our laboratory curricula) in large (or small) lecture classes. The result of this work, primarily at the University of Oregon and at Tufts University, has been the development of a teaching and learning strategy called *Interactive Lecture Demonstrations (ILDs).*[13]

After discussing the general *ILD* procedure and guidelines for creating effective *ILDs*, we will then present research results that assess student conceptual learning gains as a result of the *ILDs,* and compare these results with learning gains as a result of traditional instruction.

The Interactive Lecture Demonstration procedure. In 1989, encouraged by our successes in fostering conceptual learning in the introductory physics laboratory,[1-5] we began to explore strategies for using the real-time data displays made possible by microcomputer-based laboratory (MBL) tools[15] to establish an active learning environment in the lecture portion of the introductory course. After several years of research, in which we tried different strategies at the University of Oregon, we formalized in 1991 a procedure for *ILDs* that is designed to engage students in the learning process and, therefore, convert the usually passive lecture environment to a more active one. The steps of the procedure are listed in Table I-2. (A copy of these steps suitable to take to lecture when doing *ILDs* is found at the end of this section, on page 13.) These steps are performed for each of the 5-8 simple demonstrations in the sequence of *ILDs*.

Table I-2: The Eight Step Interactive Lecture Demonstration Procedure

1. The instructor describes the demonstration and does it for the class without measurements displayed.
2. The students are asked to record their individual predictions on a Prediction Sheet, which will be collected, and which can be identified by each student's name written at the top. (The students are assured that these predictions will not be graded, although some course credit is usually awarded for attendance and participation at these *ILD* sessions.)
3. The students engage in small group discussions with their one or two nearest neighbors.
4. The instructor elicits common student predictions from the whole class.
5. The students record their final predictions on the Prediction Sheet.
6. The instructor carries out the demonstration with measurements (usually graphs collected with micro-computer-based laboratory tools) displayed on a suitable display (multiple monitors, LCD, or computer projector).
7. A few students describe the results and discuss them in the context of the demonstration. Students may fill out a Results Sheet, identical to the Prediction Sheet, which they may take with them for further study.
8. Students (or the instructor) discuss analogous physical situation(s) with different "surface" features. (That is, different physical situation(s) based on the same concept(s).)

Student involvement in understanding these simple conceptual demonstrations is obvious from observations in the classroom. Most students are thoughtful about the individual prediction called for in step 2, and the small group discussions (step 3) in a large lecture class are initially quite animated and "on task." In time, however, the prediction will be made and the discussions may begin to stray into extraneous matters. The instructor must observe the students carefully, and pick an appropriate time to move to the next step.

Step 4 is facilitated by using transparency made from the Prediction Sheet, and sketching student predictions using different colored pens. This is a brainstorming activity, and no commentary should be made as to whether a prediction is correct or incorrect. If no students volunteer predictions that represent the common misconceptions for the demonstrations, the instructor may want to record them, saying that "a student in my last class made this prediction." The purpose of this step is to help validate all the predictions made by students in the class. It can also be supplemented by taking a vote after all predictions are recorded. If you are running short of time, this step may be skipped.

Notice that in steps 7 and 8 it is the instructor's task to get students to give the desired answers. The instructor must have a definite "agenda," and must often guide the discussion towards the important points raised by the individual *ILDs*. The instructor should *avoid lecturing to the students*. The discussion should use the experimental results—usually the displayed graphs—as the source of knowledge about the experiment. If students have not discussed everything that is important, then the instructor may need to fill in the gaps.

Several other researchers have used a similar procedure to engage their students during lectures. While a few have used actual lecture demonstrations with real data displayed (using MBL tools),[16] most of these have not involved physical experiments but rather student reasoning or problem solving. A number of these other strategies involve the use of a student response system that collects individual student responses and feeds them into a computer system for display to the instructor and, if desired, to the class. For example, Mazur[17] has reported on his use of such a system in introductory physics lectures at Harvard University. His students are led to conclusions based primarily on reasoning processes, rather than on observations of physical phenomena. Others have made use of a similar student response strategy.[18]

Do students learn from Interactive Lecture Demonstrations? Although *ILD*s have been used in many settings, we have been able to gather the most complete data on student learning at our own institutions. We have previously reported significant learning gains in kinematics and dynamics concepts for students who completed our active learning microcomputer-based laboratory curricula.[1-5] Here we report on assessments of learning gains for introductory physics students who experienced a series of kinematics and dynamics *ILDs* at the University of Oregon during Fall, 1991, and at Tufts University during Fall, 1994 and 1995[2-3,14].

To evaluate student learning we present the results from a subset of the *Force and Motion Conceptual Evaluation* developed by the authors.[2] This test has been described in more detail elsewhere.[2-4,14] It has been developed to probe student understandings of dynamics. The choices on these carefully constructed multiple choice questions were derived from student answers on open-ended questions and from student responses in interviews.

Here we will focus on four sets of questions that investigate student views of force and motion (dynamics) concepts described by Newton's 1st and 2nd Laws, the "Force Sled," "Force Graph," "Cart on Ramp" and "Coin Toss" questions. We will present summary pre- and post-instruction results to examine how exposure to *ILDs* affects students' understanding of dynamics. We discuss the evidence for the validity of the test, and the concern that some teachers have about multiple-choice testing elsewhere.[2-4]

Both the Force Sled and the Force Graph questions explore the relationship between force and motion by asking about similar motions, but the two sets of questions are very different in a number of ways. The Force Sled questions, shown in Figure I-1, refer to a sled on ice (negligible friction) pushed by someone wearing spiked shoes. Different motions of the sled are described, and the students are asked to select the force that could cause each motion from seven different force descriptions. The Force Sled questions make no reference to graphs, make no overt reference to a coordinate system, use "natural" language as much as possible, and *explicitly* describe the force acting on the moving object. The choices are in a completely different format from the graphical displays that the students observe during the *ILDs*. We will refer to the composite of student responses on a set of these questions as the *Natural Language Evaluation* of student understanding.

A sled on ice moves in the ways described in questions 1-7 below. *Friction is so small that it can be ignored.* A person wearing spiked shoes standing on the ice can apply a force to the sled and push it along the ice. Choose the one force (**A** through **G**) which would **keep the sled moving** as described in each statement below.

You may use a choice more than once or not at all but choose only one answer for each blank. If you think that none is correct, answer choice **J**.

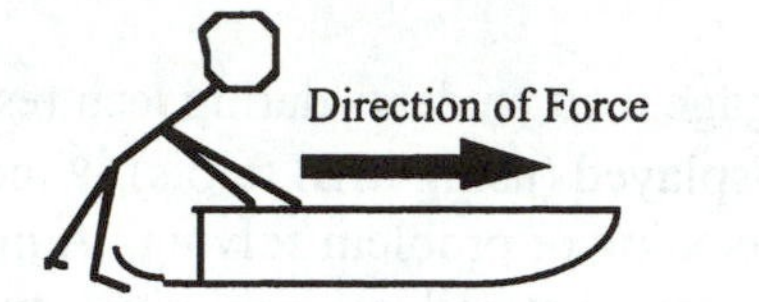

Direction of Force

A. The force is toward the **right** and is **increasing** in strength (magnitude).

B. The force is toward the **right** and is of **constant** strength (magnitude).

C. The force is toward the **right** and is **decreasing** in strength (magnitude).

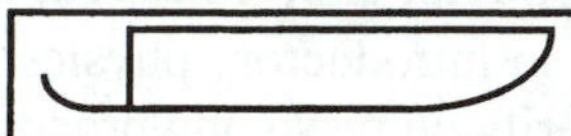

D. No applied force is needed

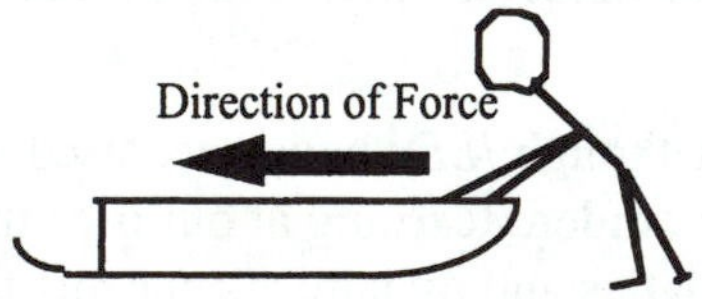

Direction of Force

E. The force is toward the **left** and is **decreasing** in strength (magnitude).

F. The force is toward the **left** and is of **constant** strength (magnitude).

G. The force is toward the **left** and is **increasing** in strength (magnitude).

___1. Which force would keep the sled moving toward the right and speeding up at a steady rate (constant acceleration)?

___2. Which force would keep the sled moving toward the right at a steady (constant) velocity?

___3. The sled is moving toward the right. Which force would slow it down at a steady rate (constant acceleration)?

___4. Which force would keep the sled moving toward the left and speeding up at a steady rate (constant acceleration)?

___5. The sled was started from rest and pushed until it reached a steady (constant) velocity toward the right. Which force would keep the sled moving at this velocity?

___6. The sled is slowing down at a steady rate and has an acceleration to the right. Which force would account for this motion?

___7. The sled is moving toward the left. Which force would slow it down at a steady rate (constant acceleration)?

Figure I-1. The Force Sled questions (*Natural Language Evaluation*) from the *Force and Motion Conceptual Evaluation.*

Unlike the Force Sled questions, the Force Graph questions use a graphical representation. Students pick the appropriate force-time graph (from 9 choices) to describe the force that could cause a toy car to move in various ways on a horizontal surface. These questions make explicit reference to a coordinate system, and do not explicitly describe the origin of the force that is acting. We will refer to the composite of student responses on a set of these questions as the *Graphical Evaluation* of student understanding. In spite of these differences in the two types of questions, student responses are very similar where there is an exact analog between a Force Sled question and a Force Graph question.

The Coin Toss and Cart on Ramp questions also probe students' understanding of Newton's first two laws, and are in general even more difficult for students to answer correctly. The Coin Toss questions are shown in Figure I-2. They refer to a coin tossed in the air, and ask the students to select among seven choices the correct description of the force acting on the coin 1) as it moves upward, 2) when it reaches its highest point and 3) as it moves downward. The Cart on Ramp questions are a coin toss analog in which a cart is given a push up an inclined ramp, and the students are asked to select (again from seven choices) the force acting on the cart during the three parts of its motion: upward, at its highest point and downward. Note that as with the Force Sled questions, the choices use a non-graphical, natural language format. For each of these sets of questions, students are considered to be understanding only if they choose *all three* forces correctly.

In the Fall of 1991, a series of kinematics and dynamics *ILDs* were used to enhance learning of Newton's First and Second Laws in the non-calculus (algebra-trigonometry based) general physics lecture class (PHYS 201) at the University of Oregon. This was a fairly standard introductory physics class except (1) there was no recitation, i.e., the class met for four lectures with approximately 200 students each week, and (2) the introductory physics laboratory was a separate course (PHYS 204), in which about half of the lecture students were simultaneously enrolled. Thus, the students in the lecture class may be divided into two groups, a NOLAB group, enrolled only in the lecture course and a LAB group, enrolled in both the lecture and laboratory courses.

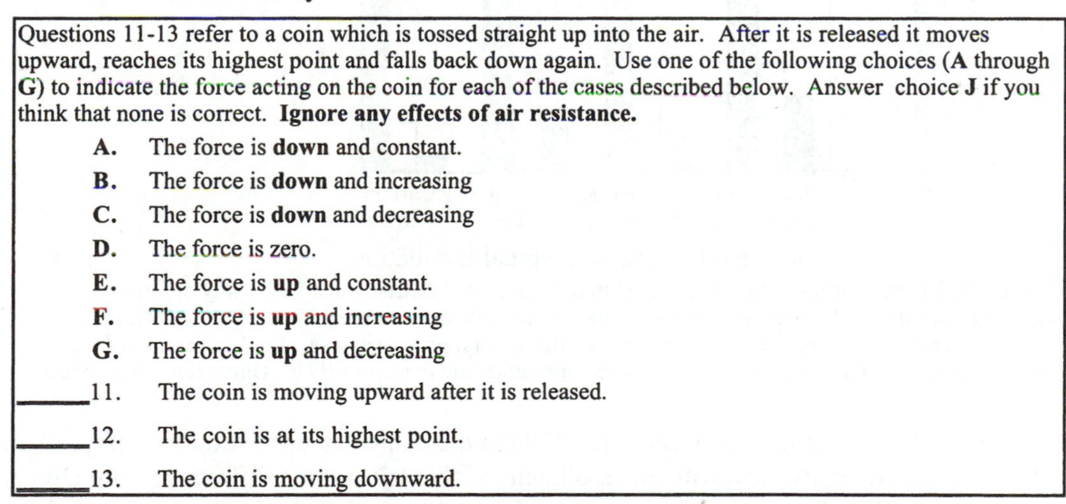
Questions 11-13 refer to a coin which is tossed straight up into the air. After it is released it moves upward, reaches its highest point and falls back down again. Use one of the following choices (**A** through **G**) to indicate the force acting on the coin for each of the cases described below. Answer choice **J** if you think that none is correct. **Ignore any effects of air resistance.**

A. The force is **down** and constant.
B. The force is **down** and increasing
C. The force is **down** and decreasing
D. The force is zero.
E. The force is **up** and constant.
F. The force is **up** and increasing
G. The force is **up** and decreasing

_____11. The coin is moving upward after it is released.
_____12. The coin is at its highest point.
_____13. The coin is moving downward.

Figure I-2: The Coin Toss questions from the *Force and Motion Conceptual Evaluation*.

Students at Oregon were first briefly introduced to kinematics with some of the *Kinematics 1: Constant Velocity Motion* sequence of *ILDs* , looking at body motions just as in our laboratory curricula.[1,5] Next, after all traditional kinematics instruction, the *Kinematics 2: Accelerated Motion ILD* sequence was completed in one 50-minute lecture. After all traditional lecture instruction on dynamics, the students experienced the *Newton's 1st & 2nd Laws ILD* sequence in another 50-minute lecture period. Students were awarded a small number of points towards their final grades for attending and handing in their

Prediction Sheets on the days when these demonstrations were carried out, but their predictions were not graded.

Figure I-3 compares student learning of dynamics concepts in traditional instruction (where students listen to lectures, do homework problems, and take quizzes and exams) to learning in the identical course where just two lectures were replaced with *ILDs*. The baseline for traditional instruction shown in the first two bars in Figure I-3 are the results for 1989-90 Oregon students before and after traditional instruction. (The pre-test results for Oregon students in 1991, and for Tufts students in 1994, shown in Figure I-4, were very similar to this combined 1989-90 group of Oregon students.) As can be seen all traditional instruction resulted in only a 7-10% overall improvement on these dynamics questions. In comparison, the last bar shows that the effect of experiencing *just two lectures* of *ILDs* was very substantial for the 1991 Oregon NOLAB group. (Recall that these students *did not* participate in the conceptual laboratories. The addition of *ILDs* also improved the scores of the LAB students, but most of these students were able to answer the questions correctly after completing just the laboratories.[2-4])

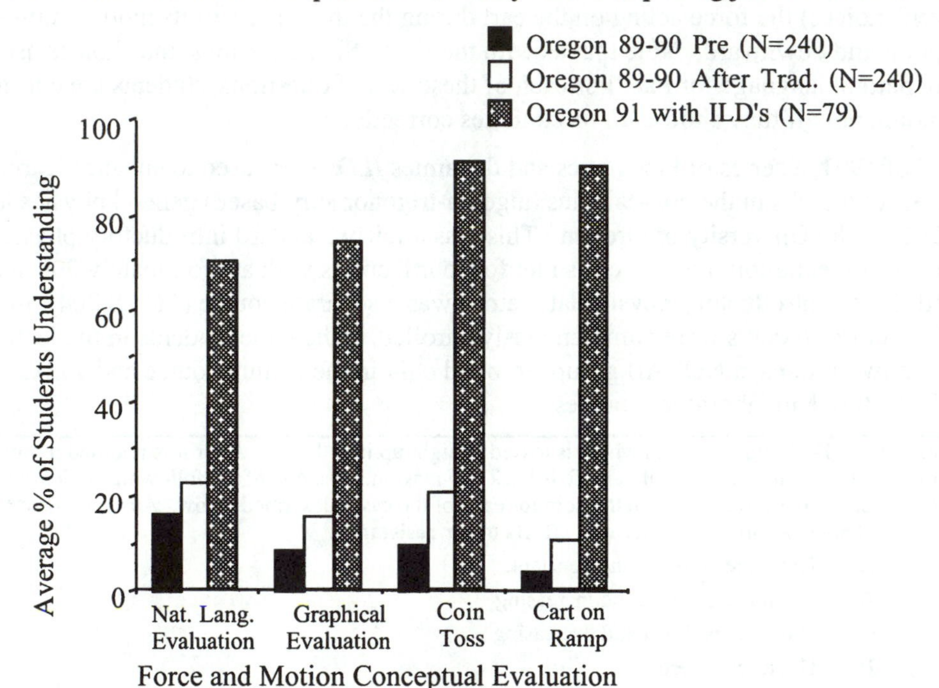

Figure I-3: Traditional Instruction Compared to *ILD*-Enhanced Instruction at University of Oregon. For each set of questions, the first two bars show student conceptual understanding of dynamics before and after traditional instruction in the Oregon non-calculus general physics course. The last bar shows the result of enhancing the introductory course with kinematics and dynamics *ILDs*. Only NOLAB students are included here.

A similar set of *ILDs* was carried out during Fall, 1994 in the non-calculus introductory physics class (Physics 1) at Tufts University, also with an enrollment of about 200. One difference from Oregon was that at Tufts in 1994 all traditional instruction in both kinematics and dynamics was completed *before any ILDs were presented.* The timelines at both Oregon and Tufts were necessitated by our desire to assess the effectiveness of the *ILDs* independently from traditional lecture instruction. All students at Tufts were offered one traditional recitation each week, and all but a few students were enrolled in the laboratory, where they completed two of our active learning (*Tools for Scientific Thinking*) <u>kinematics</u> laboratories but did <u>not</u> do any dynamics laboratories.[1,12-13]

Because most Tufts students did the two kinematics laboratories, we began with the *Accelerated Motion* Each *ILD* sequence was done in one lecture periods. As at Oregon, students were awarded a small number of points towards their final grades for attending and handing in their Prediction Sheets.

The results of two lectures of kinematics and dynamics *ILDs* on student understanding of Newton's 1st and 2nd Laws are gratifying as shown in Figure I-4. As at Oregon, our studies show less than a 10% gain for questions like these when students only experience good traditional lecture instruction.

Because of the results at Oregon and Tufts, similar *ILDs* were repeated at Tufts in the Fall of 1995, only this time the *ILDs* were more integrated into the lectures. There was a different instructor and the three *ILD* sequences were given near the beginning of the lectures on kinematics, dynamics, and the third law respectively rather than after all lectures. The results were similar to 1994.

The research data also show that the *ILD*-enhanced learning is persistent both at Oregon and at Tufts. As a test of retention, the Force Graph questions were included on the Oregon final examination. The final was given about six weeks after the dynamics *ILDs*, during which time no additional dynamics instruction took place. There was no decrease in understanding. In fact, there

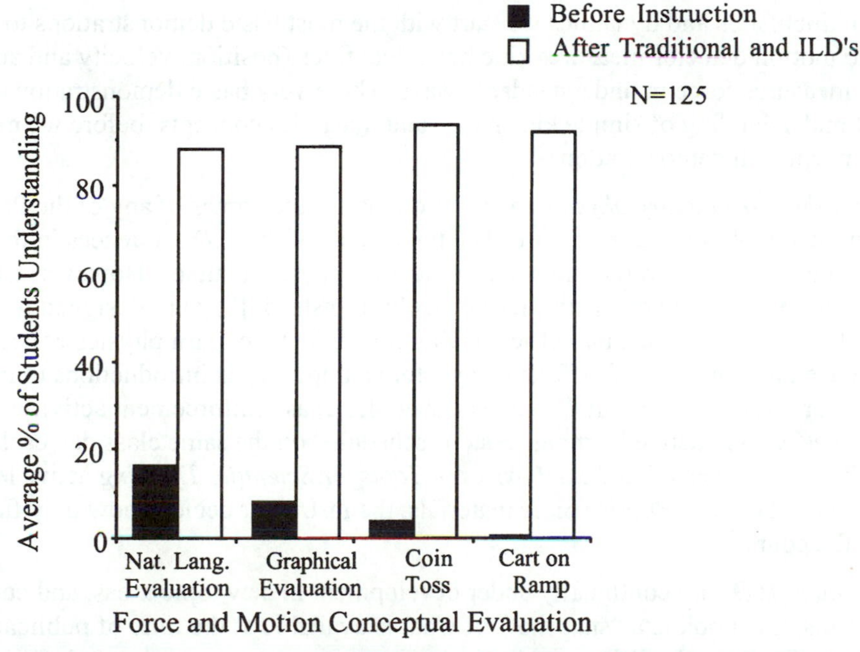

Figure I-4: *ILD*-Enhanced Instruction at Tufts University. For each set of questions, the black bar shows student conceptual understanding of dynamics before instruction in the Tufts non-calculus general physics course during Fall, 1994. The striped bar shows the result of enhancing the traditional introductory course with kinematics and dynamics *ILD* sequences.The Tufts students also experienced two *Tools for Scientific Thinking* kinematics labs.

was a 6% improvement in spite of the fact that there is little room for further gain. At Tufts a final exam was given seven weeks after dynamics instruction (including *ILDs*) had ended. There was a 7% improvement. We have seen student understanding of concepts increase after instruction has ended in many contexts were there has been conceptual learning. We ascribe the increase to assimilation of the concepts by the students. Additional different questions about dynamics were also asked on the final exam at Tufts, and more than 90% of the students were able to answer them correctly.

Why do ILDs work? The eight step *ILD* procedure is designed to engage students in the learning process. Students are asked to make predictions based on their beliefs on a sheet that will be collected. They are forced to contemplate each demonstration in terms of the models they commonly use. Students are then asked to defend their predictions to their peers. After these two steps, most students care what happens in the demonstration. They are engaged by these steps. Since the results they observe often disagree with their naïve predictions--often based on incorrect models--there is a chance for their models to be changed by the discussion that follows.

We have used two basic guidelines in designing the short, simple experiments that make up *ILD* sequences. First, the order and content of the sequences are based on the results of research in physics learning. Our experiences in developing hands-on guided discovery laboratory curricula and evaluating the learning results[1-5,10-13] have been invaluable in selecting simple but fundamental lecture demonstrations. If the sequences are to be successful, they must begin with what students know and lay the basis for additional understanding. Secondly, the *ILDs* must be presented in a manner such that students understand the experiments and "trust" the apparatus and measurement devices used. The real-time display of data gives students feedback in a way that builds confidence in the measurement devices and the resulting data. Many traditional exciting and flashy demonstrations are too complex to be effective learning experiences for students in the introductory class.

For example, in kinematics and dynamics we start with the most basic demonstrations to convince the students that the motion detector measures kinematic quantities (position, velocity and acceleration) and the force probe measures force in understandable ways. These very basic demonstrations also begin to solidify student understanding of simple kinematics and dynamics concepts, before we move on to more complex and concept-rich demonstrations.

How ILDs fit into the introductory physics course: You may make copies of any of the Prediction and Results sheets in this book for classroom use. The topic areas of the *ILD* sequences in this book are distributed over the concepts covered in most introductory physics courses. However, taken together, they do not constitute an introductory physics curriculum. Instead, they are designed to supplement the other components of the course with an efficient way for students to learn physics concepts in lecture. They have been used in a number of different ways, for example: 1) as introductions to important concepts at the appropriate moment in the course schedule, 2) as reinforcement activities for concepts already taught, 3) as weekly active learning sessions scheduled on the same class day each week and 4) as summaries and bridges between *RealTime Physics* or *Tools for Scientific Thinking* active learning laboratories. As with the other *Physics Suite* materials, the instructor decides how they fit into his/her overall plan for the course.

ILDs on other topics: ILDs are continually under development in new topic areas, and current *ILDs* are often updated. Thus, this book is a "snapshot" of what is available at the time of publication. New *ILDs* will be posted periodically at the ILD web-site, http://darkwing.uoregon.edu/~sokoloff/ILD.htm , and you are welcome to download these for your use. We are currently developing web-based *ILDs*, and information on these will also be available there.

Another way to learn about new developments in *ILDs* (and other *Physics Suite* materials) is to attend one of the workshops that we offer several times each year at national American Association of Physics Teachers meetings and at our institutions under the auspices of the NSF-sponsored national Chautauqua Short Course program. We also offer extended NSF-sponsored summer institutes on the effective use of *Physics Suite* materials. For more information on these programs visit the following web-site: http://darkwing.uoregon.edu/~sokoloff/physcourse.htm.

You are also encouraged to design your own *ILD* sequences. If you do so, please share them. Any new ideas and new *ILD* sequences are welcomed by the authors. Visit the web-site for instructions on how to submit them. Remember two important points: 1) the apparatus should be simple, easily understood and trusted by your students and 2) do all eight steps of the *ILD* procedure for each demonstration in the sequence.

In fact, many traditional, popular lecture demonstrations are amenable to the eight-step *ILD* procedure. We encourage you to present them in this way, and believe that learning will be much improved by doing so. You will need to be sure that the students actually possess the prior knowledge needed to understand the demonstration. If they don't, you can prepare a sequence of *ILDs* to prepare them.

When to use computer-supported tools. Of the *ILD* sequences in this book, roughly two thirds make use of computer-supported tools: microcomputer-based laboratory (MBL) tools, video analysis software or vector visualization software. These tools are incorporated into the demonstrations because our experiences show that the collection and display of data afforded by them make student learning more effective. Some of the *ILD* sequences in this book may be done with or without the use of computer-supported tools, e.g., the electric circuit *ILDs*. Other *ILDs* are more appropriately done without the use of technology. In all cases, the eight-step *ILD* procedure should be used for each demonstration in a sequence.

Conclusions. Our studies of student understanding using research-based conceptual evaluations with large numbers of students show that introductory physics students do not commonly understand physics concepts as a result of thorough traditional instruction. This research and that of others, along with the development of user-friendly microcomputer-based laboratory tools and our experience with computer-supported active laboratory curricula have allowed us to develop a strategy for more active learning of these concepts in lectures using *Interactive Lecture Demonstrations.* Assessments indicate that student understanding of concepts is improved when these *ILD*s are substituted.

One last crucial question: why should we care if students understand physics concepts? We believe that this is fundamental to a real understanding of our discipline. Students cannot hope to be able to do more than algorithmic solutions to simple physics problems without a sound grasp of the fundamental concepts.

References:

1 R. K. Thornton, and D. R. Sokoloff, "Learning motion concepts using real-time, microcomputer-based laboratory tools," *Am. J. Phys*. **58**, 858-867 (1990).

2 Ronald K. Thornton and David R. Sokoloff, "Assessing Student Learning of Newton's Laws: The Force and Motion Conceptual Evaluation and the Evaluation of Active Learning Laboratory and Lecture Curricula," *Am. J. Phys.* **66**, 338-352 (1998).

3 R.K.Thornton, "Learning physics concepts in the introductory course, Microcomputer-based Labs and Interactive Lecture Demonstrations" in *Proc. Conf. on the Intro. Physics Course,* (Wiley, New York, 1996), pp. 69-85.

4 R. K. Thornton, "Using large-scale classroom research to study student conceptual learning in mechanics and to develop new approaches to learning," chapter in *Microcomputer-Based Laboratories: Educational Research and Standards, Series F, Computer and Systems Sciences, Vol. 156,* Robert F. Tinker, ed., (Springer Verlag, Berlin, Heidelberg, 1996), pp. 89-114. Also available on WWW.

5 R. K. Thornton and D.R. Sokoloff, "RealTime Physics: Active Learning Laboratory," *The Changing Role of the Physics Department in Modern Universities, Rroceedings of the International Conference on Undergraduate Physics Education*1101-1118 (College Park, American Institute of Physics, 1997).

6 L.C. McDermott, "Millikan lecture 1990: What we teach and what is learned--closing the gap," *Am. J. Phys* **59**, 301-315 (1991).

7 L.C. McDermott, "Research on conceptual understanding in mechanics," *Phisics. Today* **37**, 24-32 (July, 1984)

8 D. Hestenes, M. Wells and G. Schwackhammer, "Force Concept Inventory," *The Physics Teacher* **30**:3, 141-158 (1992).

9 J. A. Halloun and D. Hestenes, "The initial knowledge state of college physics students," *Am. J. Phys.* **53**, 1043-1056 (1985).

10 P. W. Laws, "Calculus-based physics without lectures," Physics Today **44**:12, 24-31 (December, 1991).

11 P.W. Laws, *Workshop Physics Activity Guide: The Core Volume with Module 1: Mechanics*, (New York, Wiley, 1997), pp. 125-136.

12 Ronald K. Thornton and David R. Sokoloff, *Tools for Scientific Thinking--Motion and Force Curriculum and Teachers' Guide, Second edition*, (Vernier Software and technology, Portland, 1992).

13 David R. Sokoloff, Ronald K. Thornton, and Priscilla W. Laws, *RealTime Physics Module 1: Mechanics,Module 2: Heat and Thermodynamics, Module 3: Electric Circuits and Module 4: Light and Optics,* (New York, Wiley, 2004).

14 David R. Sokoloff and Ronald K. Thornton, "Using Interactive Lecture Demonstrations to Create and Active Learning Environment," *The Physics Teacher* **36**: 6, 340 (1997).

15 Computer interfaces, probes and software used in Interactive Lecture demonstrations are available from Vernier Software and Technology (www.vernier.com) and PASCO Scientific (www.pasco.com).

16 E. Sassi, Department of Physics, University of Naples, Mostra D'Oltremare pad. 20, I80125 Naples, Italy, private communication.

17 E. Mazur, *Peer Instruction: A User's Manual* (Upper Saddle River, NJ, Prentice Hall, 1997).

18 R. Defresne, W. Gerace, W. Leonard, J. Mestre and L. Wenk, "Classtalk: A classroom communication systemfor active learning," *Journ. Computing in Higher Ed.* **7**, 3-47 (1996).

The Eight Step
Interactive Lecture Demonstration Procedure

1. Describe the demonstration and do it for the class without measurements displayed.
2. Ask students to record their individual predictions on a Prediction Sheet.
3. Have the students engage in small group discussions.
4. Elicit common student predictions from the whole class.
5. Ask each student to record final prediction on the Prediction Sheet.
6. Carry out the demonstration with measurements displayed.
7. Ask a few students to describe the results and discuss them in the context of the demonstration. Students may fill out the Results Sheet.
8. Discuss analogous physical situation(s) with different "surface" features. (That is, different physical situation(s) based on the same concept(s).)

Section II: Interactive Lecture Demonstrations in Mechanics

KINEMATICS 1—HUMAN MOTION (KIN1)

Hand in this sheet **Name**____________________________

INTERACTIVE LECTURE DEMONSTRATIONS
PREDICTION SHEET—KINEMATICS 1—HUMAN MOTION

Directions: This sheet will be collected. Write your name at the top to record your presence and participation in these demonstrations. Follow your instructor's directions. You may write whatever you wish on the attached Results Sheet and take it with you.

Demonstration 1: Sketch below on the left axes your prediction of the *distance (position)-time* graph for a person moving away from the origin (the motion detector) at a steady (constant) velocity. On the other axes sketch your prediction for a person moving toward the origin at a steady (constant) velocity.

moving away — distance vs. time

moving toward — distance vs. time

Demonstration 2: Sketch on the left axes below your prediction of the *velocity-time* graph for a person moving away from the the orgin (the motion detector) at a steady (constant) velocity. On the other axes sketch your prediction for a person moving toward the orgin at a steady (constant) velocity.

moving away — velocity (+, 0, -) vs. time

moving toward — velocity (+, 0, -) vs. time

Demonstration 3: Sketch on the axes below your predictions for the *distance-time* and *velocity-time* graphs of a person moving away from the motion detector at approximately twice the speed of Demonstration 1 and Demonstration 2.

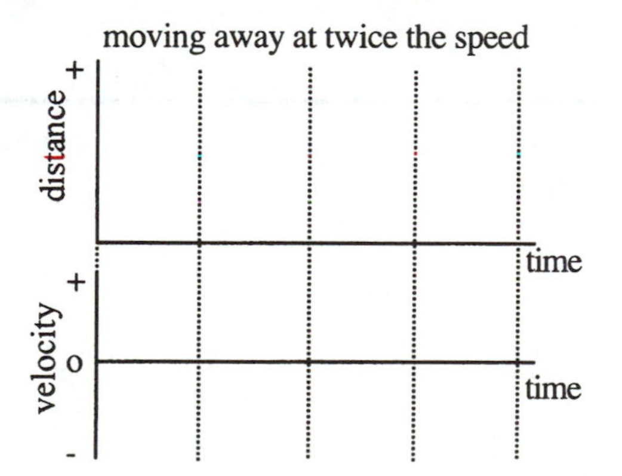

Describe in words how the *distance-time* graph changes when the speed is twice as fast.

Describe in words how the *velocity-time* graph changes when the speed is twice as fast. fast

<u>Demonstration 4:</u> Sketch on the axes below using a *dashed line* your *prediction* of the *<u>velocity-time</u>* graph produced when a person—

- walks away from the motion detector slowly and steadily for 6 seconds
- then stands still for 6 seconds
- and then walks toward the motion detector steadily about twice as fast as before

Compare your prediction with your neighbor(s), and see if you can agree. Sketch in the prediction you agreed on using a *solid line.*

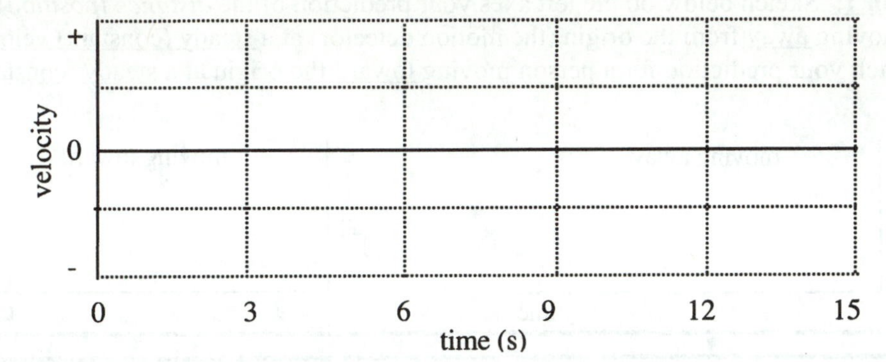

Predict the *<u>distance (position)-time</u>* graph for the motion described above with a *dashed line* on the axes below. (Align the distance and velocity graphs correctly in time.)

Again, sketch in the prediction you agreed on with your neighbor(s) using a *solid line.*

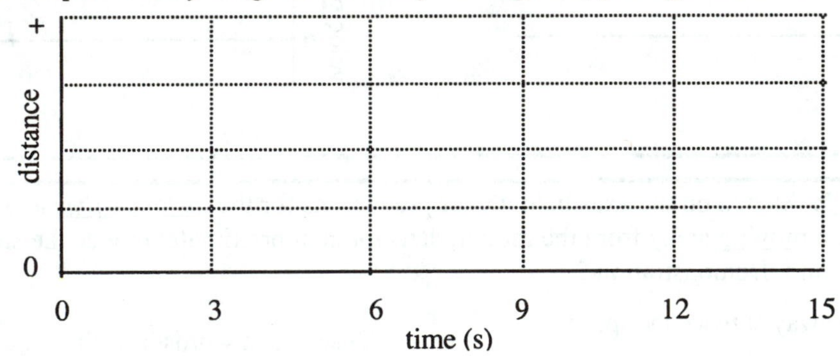

Keep this sheet

INTERACTIVE LECTURE DEMONSTRATIONS
PREDICTION SHEET—**KINEMATICS 1—HUMAN MOTION**

You may write whatever you wish on this sheet and take it with you.

Demonstration 1: Sketch below on the left axes your prediction of the *distance (position)-time* graph for a person moving away from the origin (the motion detector) at a steady (constant) velocity. On the other axes sketch your prediction for a person moving toward the origin at a steady (constant) velocity.

moving away — distance / time

moving toward — distance / time

Demonstration 2: Sketch on the left axes below your prediction of the *velocity-time* graph for a person moving away from the the orgin (the motion detector) at a steady (constant) velocity. On the other axes sketch your prediction for a person moving toward the orgin at a steady (constant) velocity.

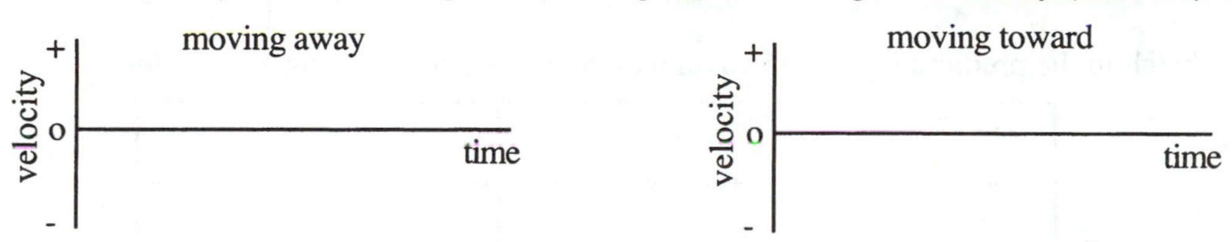

Demonstration 3: Sketch on the axes below your predictions for the *distance-time* and *velocity-time* graphs of a person moving away from the motion detector at approximately twice the speed of Demonstration 1 and Demonstration 2.

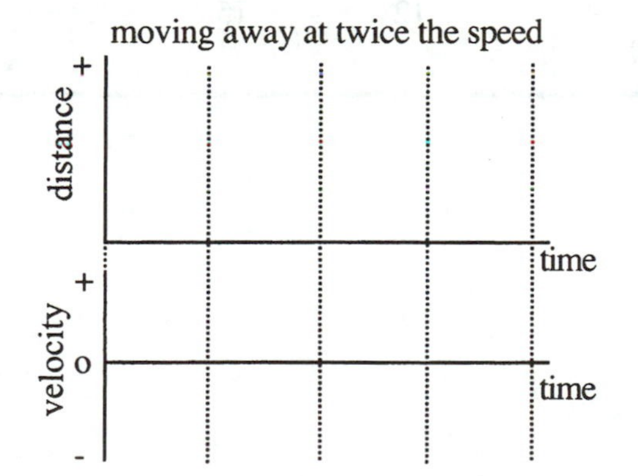

Describe in words how the *distance-time* graph changes when the speed is twice as fast.

Describe in words how the *velocity-time* graph changes when the speed is twice as fast. fast

Demonstration 4: Sketch on the axes below using a *dashed line* your *prediction* of the velocity-time graph produced when a person—

- walks away from the motion detector slowly and steadily for 6 seconds
- then stands still for 6 seconds
- and then walks toward the motion detector steadily about twice as fast as before

Compare your prediction with your neighbor(s), and see if you can agree. Sketch in the prediction you agreed on using a *solid line.*

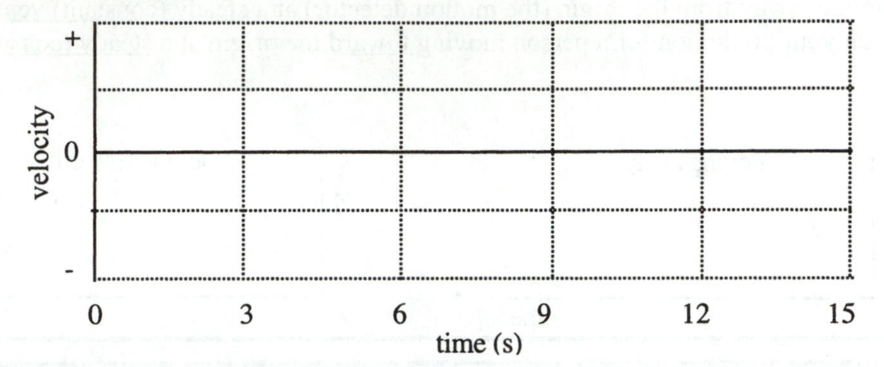

Predict the *distance (position)-time* graph for the motion described above with a *dashed line* on the axes below. (Align the distance and velocity graphs correctly in time.)

Again, sketch in the prediction you agreed on with your neighbor(s) using a *solid line.*

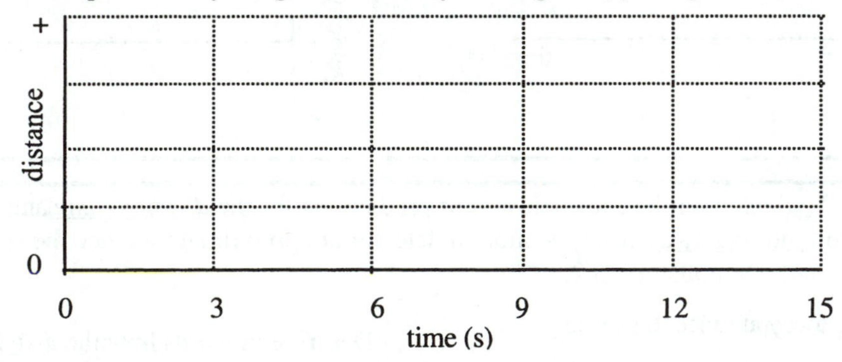

KINEMATICS I—HUMAN MOTION (KIN1)
TEACHER'S GUIDE

Prerequisites:

This sequence of *ILDs* is the first in the series and has no prerequisites. If your students have done *RealTime Physics Mechanics* Lab 1 or *Tools for Scientific Thinking Motion and Force* Lab 1, you can skip the *Human Motion ILDs,* and go directly into *Kinematics 2—Motion of Carts.*

Equipment:

computer-based laboratory system
motion detector (See below.)
ILD experiment configuration files

General Notes on Preparation and Equipment:

The motion detector and software:
Motion detectors are available from Vernier Software and Technology (www.vernier.com) (MD-BTD), and PASCO Scientific (www.pasco.com) (CI-6742 or PS-2103). These work with the appropriate computer interface and software available from these companies. The motion detector is a sonar device, sending out pulses of ultra-sound with a frequency range around 50,000 Hz. (The rate of emission of these pulses is adjustable from 10 to 50 per second. 20 per second is the default value, set in the experiment configuration files for these demonstrations.)

After a pulse is emitted, the motion detector converts from a transmitter to a receiver, and listens for echoes. By timing the delay from sending a pulse to receiving an echo, and using the speed of sound in air, the software determines the distance from the detector to the object that reflected the pulse. The software then plots these distances as a function of time. Velocities and accelerations are calculated in real time using a fit to the distance data.

Walking in front of the motion detector:
Body motions have been shown to be very effective in teaching kinematics concepts. While it may be better to use a student volunteer for reasons of class interest, the student is unlikely to move initially in ways that you wish for the demonstrations. Therefore, it is best for you to walk in front of the motion detector yourself. (If you have extra time, feel free to use students.)

Paying attention to the following should help to assure reasonable results:

1. Some motion detectors will not correctly measure objects closer than 0.5 meter from them. If you are using one of these, be sure that you never get closer than 0.5 m away. The motion detector should be able to detect objects out to at least 4 meters.
2. The motion detector sends out ultrasonic waves in a cone of about 15°. It will see the closest object to it within this cone. Extraneous objects should be cleared from your path. Be sure to leave lots of room! If the motion detector is consistently seeing something at a fixed distance from it, read this distance from the graph (or using the analysis feature in the software) and check for objects at that distance from the motion detector. Occasionally the motion detector will even see small cracks in the table or floor. Tilting it a bit upward usually solves this problem.
3. The motion detector will detect small movements of arms, bends of the waist, etc. Best results will be obtained if the motion detector is high enough to be aimed at the lower chest rather than the legs.

4. Ultrasonic waves are absorbed by some materials like fleece, some furry wool, and some course woven cotton sweaters. Most clothes are sufficiently reflective. If something is absorbing the signal, the *maximum* distance will be recorded (since no reflected pulse is received by the motion detector). The best idea is to check a sweater ahead of time, if possible. Holding a card or book against the body in front of the motion detector will generally work, but the card (or book) must be held steady while walking, not carelessly moved or waved.

One last consideration:
At a data rate of 20 per second, a data point is measured every 0.05 seconds. While some averaging is used in calculating velocities and accelerations, the graphs display a real-world look at body motions. These generally are not smooth, textbook-like graphs. Body motion is more interesting to students and gives them a chance to decide what features of a graph are important. It is worthwhile in these demonstrations to make careful efforts to move in as smooth, and non-jerky a fashion as possible. Smooth, short, shuffling steps will give the most uniform, easiest to interpret graphs. Sampling and averaging rates should be chosen appropriately for the moving object. Be aware of this if you change them from the values assigned in the experiment configuration files. If motion is looked at over too short a time, the small motions made by clothes, for example, can cause substantial bumps on velocity graphs of a moving person. If motion is looked at over too long a time, the measurements will not appear responsive to movements. If you use the default values in the experiment configuration files, you should have no trouble. Finally, because of the real jerkiness of body motions, it is not productive to look at acceleration graphs.

Nomenclature note:
We begin by labeling the position-time graph as "distance" where distance is defined as "distance from the motion detector." This is generally called "position" in physics courses. "Distance from the motion detector" immediately makes sense to most students. In later *ILDs* we use the term position.

<u>Demonstrations and Sample Graphs:</u>

Do not be tempted to skip the *Human Motion ILDs* even though they seem very simple. They offer an introduction to kinematics that gets all the students started at the same place. The motion detector and real-time graphing are novel enough that even students who know this material stay interested. The simple motions convince the students that the motion detector is measuring what we claim it is measuring.

The demonstrations are described below for your review with a few additional notes to help you prepare. There are few additional equipment notes since the demonstrations are so simple. In later *ILD* sequences, there are more notes. The *Presentation Notes* that follow are designed to guide your actual presentations. You should know how to save a data set in the software you are using so that the graphs are persistently displayed when new data are collected.

Classroom introduction to *Constant Velocity Motion ILDs*:
After demonstrating a simple motion in front of the motion detector, explain that the detector works like a bat. A bat sends out a high-pitched sound, waits for the echo, and from the roundtrip time knows how far away an object is. From multiple readings a bat can tell how fast an object is moving. The motion detector works the same way. The computer times the interval from sending the pulse to receiving the echo back and calculates the position from the speed of sound. From multiple readings it calculates velocity and acceleration.

<u>Demonstration 1:</u> Distance (position)-time graphs. (Use experiment configuration file **KIN1D1**.)
Start at least 0.5 m away from the motion detector and walk away slowly and steadily, displaying distance vs. time. Move smoothly and don't swing your arms. Figure II-1 is a typical distance (position)-

time graph. In general you should walk so that you can see the screen as you move. This means backing away from the motion detector. Save the data so that the graph is persistently displayed on the screen, and then move toward the motion detector slowly and steadily, again displaying a distance-time graph.

If you get spikes on your distance graphs, it is likely that the motion detector is not aimed correctly, and is intermittently missing your body. Be sure that you are walking in a line straight away from the motion detector. If the graph levels off at a certain distance and continues as a horizontal line even though you are still moving away, this indicates that some other object is in the way between you and the motion detector. Find the distance where it levels out from the graph--or by using the analysis feature in the software, and look for any objects that might be in the way.

Discussion after the graphs are displayed: Remind the students that distance means "distance from the motion detector." Ask the students to describe the difference between *moving away* and *moving toward* on a distance (position)-time graph. Be sure to select the relevant portion of the graph if you were not moving in the appropriate way for the entire time. It is usually best not to re-scale, however.

Demonstration 2: Velocity-time graphs. (Use experiment configuration file **KIN1D2**.) Start at least

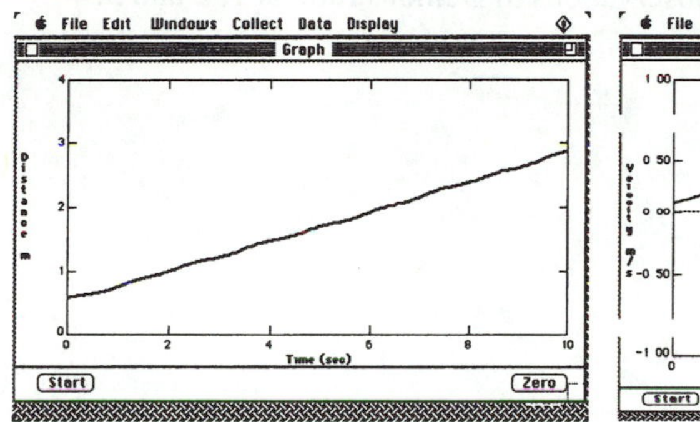

Figure II-1: Distance-time graph for walking slowly and steadily away from the motion detector.

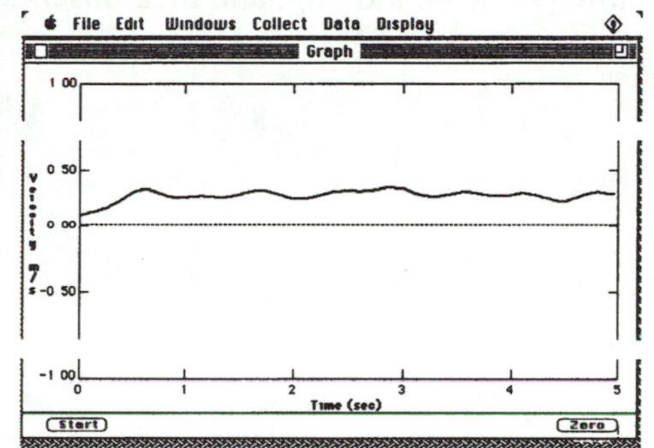

Figure II-2: Velocity-time graph for walking slowly and steadily away from the motion detector.

0.5 m away from the motion detector and walk away slowly and steadily (the same speed as in Demonstration 1), displaying velocity vs. time. Figure II-2 is a typical velocity-time graph. Note that the individual steps can be seen as bumps on the velocity-time graph. These can be minimized by taking care not to swing your arms or jerk your body as you walk. Small shuffling steps are best. Save the data so that the graph is persistently displayed on the screen, and then move toward the motion detector slowly and steadily, again displaying a velocity-time graph. Keep the graphs from Demonstrations 1 and 2 persistently displayed for comparison to Demonstration 3.

Discussion after the graphs are displayed: Ask students to ignore the bumps, and extrapolate the bumps to a horizontal line. Explain the convention that *away* is *positive*. Ask students to describe the difference between *moving away* and *moving toward* on a velocity-time graph. Point out that the sign of the velocity only gives direction. Lead into the next demonstration: how can you know the magnitude of the velocity from a velocity-time graph?

Demonstration 3: Distance and velocity graphs moving faster. (Use the same experiment configuration file as in Demonstration 2.) With the velocity vs. time graph for *moving away* from Demonstrations 2 still displayed, start at least 0.5 m away from the motion detector and walk *away* steadily and about twice as fast as in Demonstrations 1 and 2. Then display both distance and velocity in a different window that displays these on two different sets of axes.

Discussion after the graphs are displayed: Ask students to describe the difference in the velocity-time and distance-time graphs between moving away *slower* (Demonstrations 1 and 2) and then *twice as fast.* Ask how the graphs would look for moving twice as fast *toward* the motion detector.

__Demonstration 4:__ **Distance from velocity.** (Use experiment configuration file **KIN1D4**.) Display the velocity-time graph *only* as you walk away from the motion detector slowly and steadily (about the same speed as in Demonstrations 1 and 2) for 6 seconds, stand still for 6 seconds and then walk toward the motion detector steadily at about twice the speed (about the same speed as Demonstration 3). After discussing the velocity-time graph, open a new window to display both velocity-time and distance–time graphs for the same motion.

Discussion after the graphs are displayed: With the velocity graph *only* displayed, ask students to describe the different portions of the graph in terms of how you were moving, and their observations in Demonstrations 1, 2 and 3. Ask why you ran into the motion detector on the way back. Could you avoid that by starting your motion in a different place? Does a velocity-time graph tell you anything about where you started your motion? Display the distance-time graph, and ask students to describe it in terms of how you were moving, and their observations in Demonstrations 1, 2 and 3.

KINEMATICS 1—HUMAN MOTION (KIN1)
TEACHER PRESENTATION NOTES

Classroom introduction to *Constant Velocity Motion ILDs*:
Demonstrate a simple motion in front of the motion detector, and explain that the itworks like a bat.

Demonstration 1: Distance-time graphs. Move *away from* the origin (motion detector) at a *steady (constant) velocity*. Use experiment configuration file **KIN1D1**.

- After showing the graph, save the data so that they are persistently displayed on the screen.
- Then move *toward* the origin at a *steady (constant) velocity*.
- Remind students that Distance means "distance from the motion detector."
- Ask the students to describe the difference between *moving away* and *moving toward*.

Demonstration 2: Velocity-time graphs. Move *away from* the origin (motion detector) at a *steady (constant) velocity*. Use experiment configuration file **KIN1D2**.

- Save the data so that they are persistently displayed for comparison in Demonstration 3.
- Have students notice the effect of your steps and conclude that the velocity is essentially constant, i.e., a horizontal line. Explain convention that away is positive.
- Then move *toward* the origin at a *steady (constant) velocity*.
- Ask students to describe the difference between *moving away* and *moving toward*.
- Point out that the sign of velocity only gives direction.

Demonstration 3: Use the same experiment configuration file. Select a second window with distance and velocity axes, with the graphs for *moving away* from Demonstration 2 still persistently displayed. Move *away from* the detector at approximately twice the speed of Demonstration 2.

- Ask students to describe the difference between moving away slower and then twice as fast.
- Ask how graphs would look for twice as fast *toward* the motion detector.
- Lead into the next demonstration: how can you know the velocity from a distance-time graph?

Demonstration 4: Use experiment configuration file, **KIN1D4**. Walk *away* from the detector *slowly and steadily* for 6 seconds, stand still for 6 seconds, and then walk *toward* the detector *steadily about twice as fast* as before.

- After discussing velocity result, display the distance-time graph as well as velocity-time. (Select the second window to do this.) Compare differences.
- Why do you crash into motion detector on the way back? Does a velocity–time graph tell you where you started? Where should you start to get the same velocity-time graph without crashing into motion detector?

Kinematics 2—Motion of Carts (KIN2)

Hand in this sheet **Name**______________________________

INTERACTIVE LECTURE DEMONSTRATIONS
PREDICTION SHEET—**KINEMATICS 2—MOTION OF CARTS**

Directions: This sheet will be collected. Write your name at the top to record your presence and participation in these demonstrations. Follow your instructor's directions. You may write whatever you wish on the attached Results Sheet and take it with you.

Demonstration 1: On the left velocity axes below sketch your prediction of the *velocity-time* graph of the cart moving away from the motion detector at a steady (constant) velocity. On the left position axes below sketch your prediction of the *position-time* graph for the same motion.

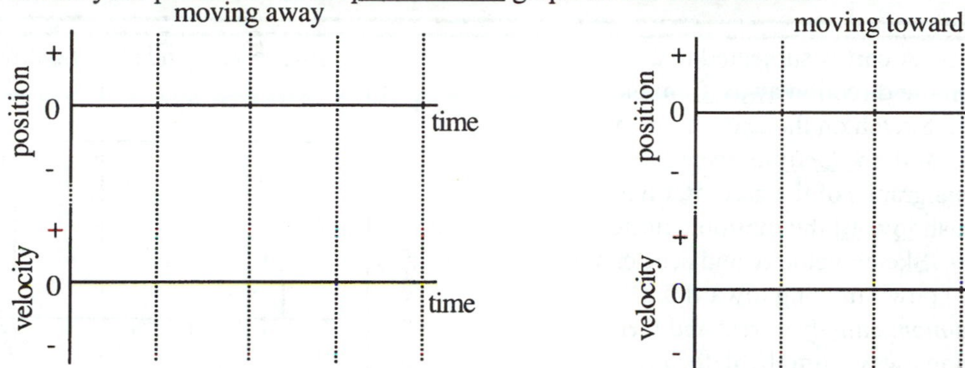

Demonstration 2: On the right velocity axes above sketch your prediction of the *velocity-time* graph for the cart moving toward the motion detector at a steady (constant) velocity. On the right position axes above sketch your prediction of the *position-time* graph for the same motion.

Demonstration 3: Sketch on the axes on the right your predictions for the *velocity-time* and *acceleration-time* graphs of the cart moving away from the motion detector and speeding up at a steady rate.

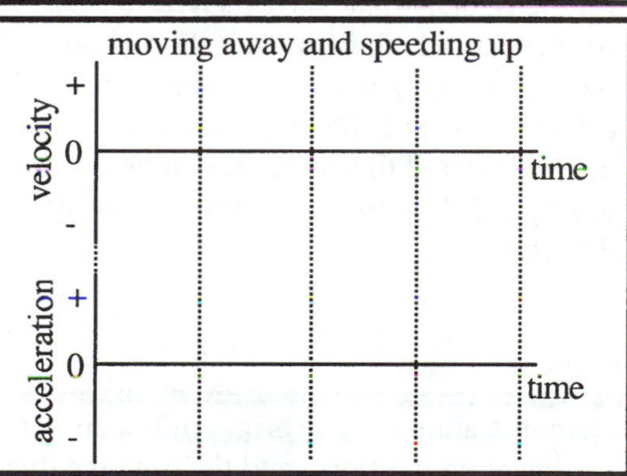

Demonstration 4: Sketch on the axes on the right your predictions for the *velocity-time* and *acceleration-time* graphs of the cart moving away from the motion detector and slowing down at a steady rate.

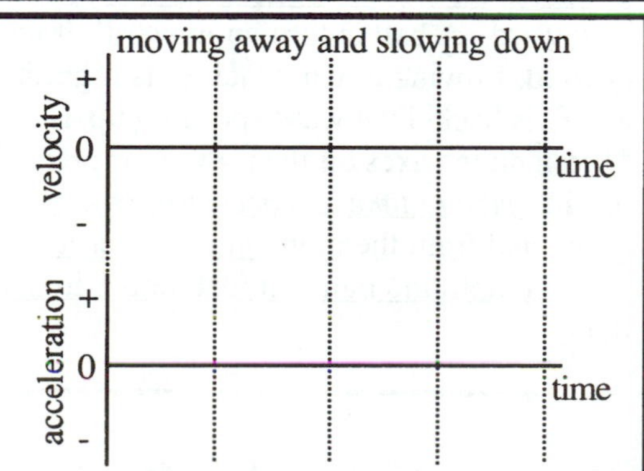

Demonstration 5: A cart is subjected to a constant force in the direction away from the motion detector. Sketch on the axes on the right your predictions for the *velocity-time* and *acceleration-time* graphs of the cart moving toward the motion detector and slowing down at a steady rate. (Start your graph after the push that gets the cart moving.)

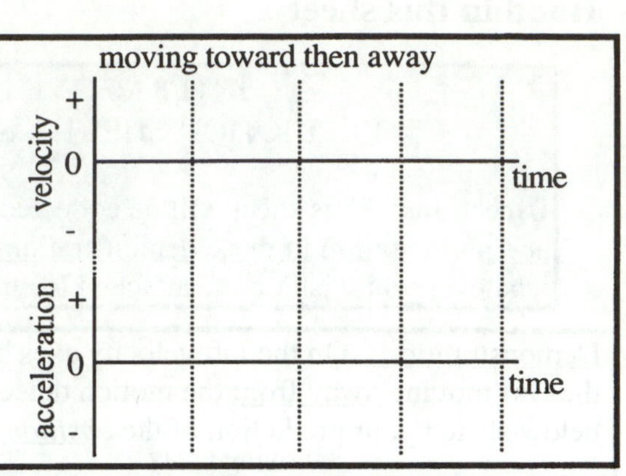

Demonstration 6: A cart is subjected to a constant force in the direction away from the motion detector. Sketch on the axes on the right your predictions of the *velocity-time* and *acceleration-time* graphs of the cart after it is given a short push toward the motion detector (and is released). Sketch velocity and acceleration as the cart slows down moving toward the detector, comes *momentarily* to rest and then speeds up moving away from the detector.

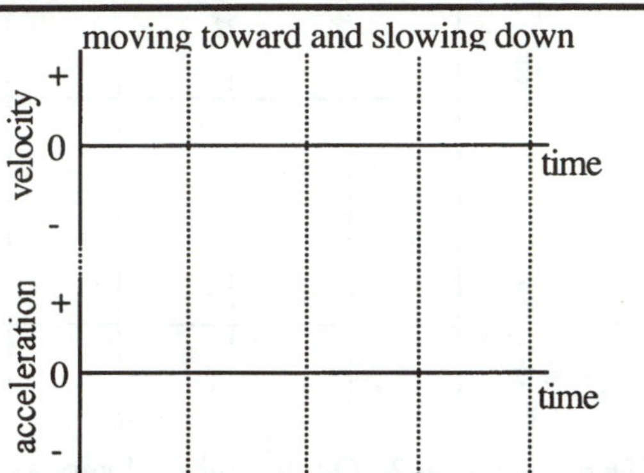

Demonstration 7: Sketch below your predictions for the *velocity-time* and *acceleration-time* graphs for the cart which is given a short push up the inclined ramp toward the motion detector (and is released) Sketch the graph as the cart slows down moving toward the detector, comes *momentarily* to rest and then speeds up moving away from the detector.

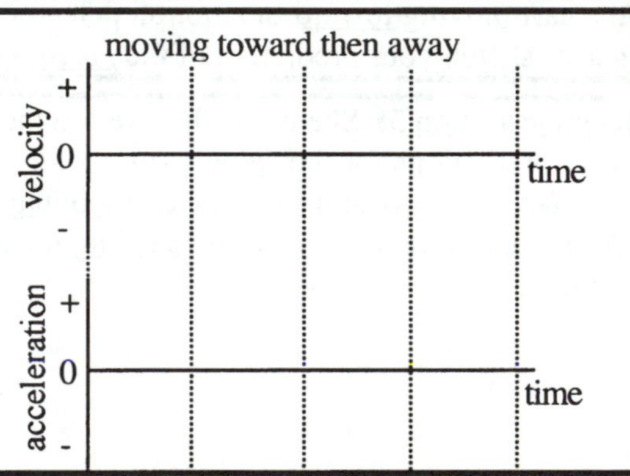

Demonstration 8: The origin of the coordinate system is on the floor, and the positive direction is upward. A ball is thrown upward. It moves upward, slowing down, reaches its highest point and falls back downward speeding up as it falls. Sketch on the axes on the right your predictions for the *velocity-time* and *acceleration-time* graphs of the ball from the moment just after it is released until the moment just before it hits the floor.

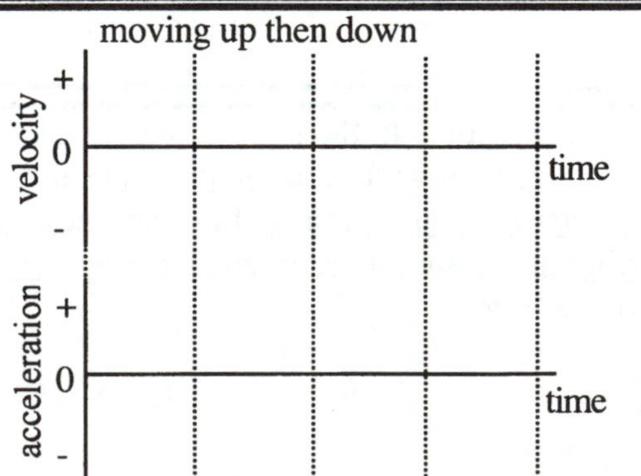

Keep this sheet

INTERACTIVE LECTURE DEMONSTRATIONS
PREDICTION SHEET—**KINEMATICS 2—MOTION OF CARTS**

You may write whatever you wish on this sheet and take it with you.

Demonstration 1: On the left velocity axes below sketch your prediction of the *velocity-time* graph of the cart moving away from the motion detector at a steady (constant) velocity. On the left position axes below sketch your prediction of the *position-time* graph for the same motion.

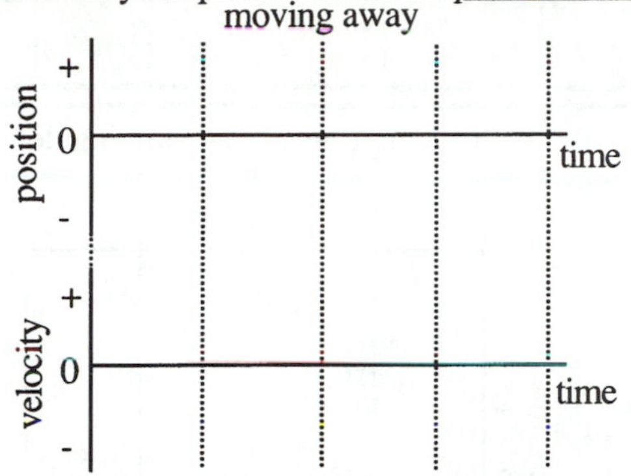

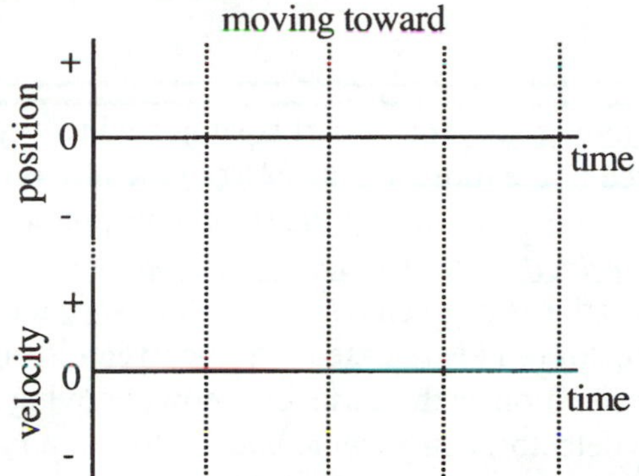

Demonstration 2: On the right velocity axes above sketch your prediction of the *velocity-time* graph for the cart moving toward the motion detector at a steady (constant) velocity. On the right position axes above sketch your prediction of the *position-time* graph for the same motion.

Demonstration 3: Sketch on the axes on the right your predictions for the *velocity-time* and *acceleration-time* graphs of the cart moving away from the motion detector and speeding up at a steady rate.

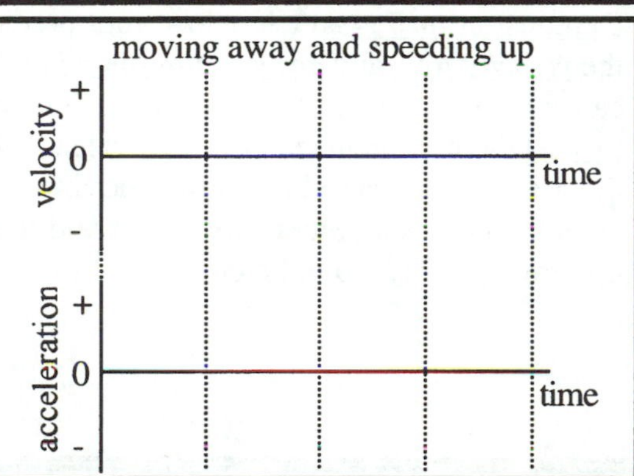

Demonstration 4: Sketch on the axes on the right your predictions for the *velocity-time* and *acceleration-time* graphs of the cart moving away from the motion detector and slowing down at a steady rate.

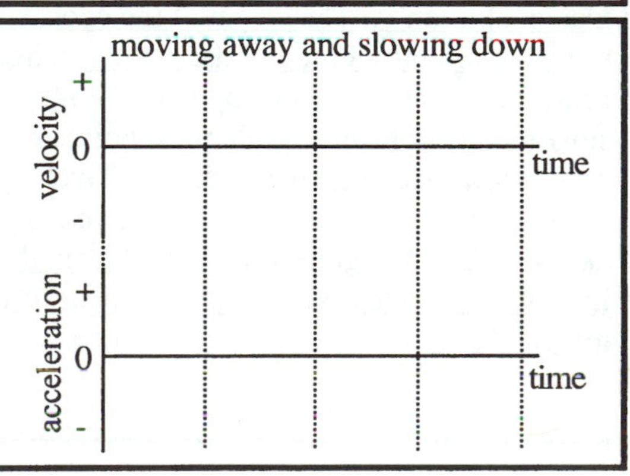

Demonstration 5: A cart is subjected to a constant force in the direction away from the motion detector. Sketch on the axes on the right your predictions for the *velocity-time* and *acceleration-time* graphs of the cart moving toward the motion detector and slowing down at a steady rate. (Start your graph after the push that gets the cart moving.)

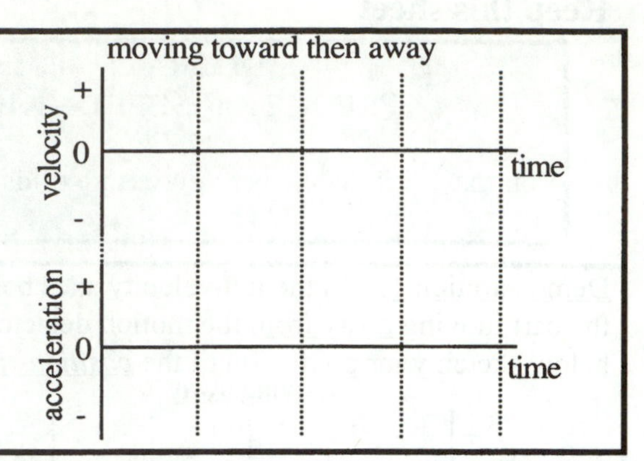

Demonstration 6: A cart is subjected to a constant force in the direction away from the motion detector. Sketch on the axes on the right your predictions of the *velocity-time* and *acceleration-time* graphs of the cart after it is given a short push toward the motion detector (and is released). Sketch velocity and acceleration as the cart slows down moving toward the detector, comes *momentarily* to rest and then speeds up moving away from the detector.

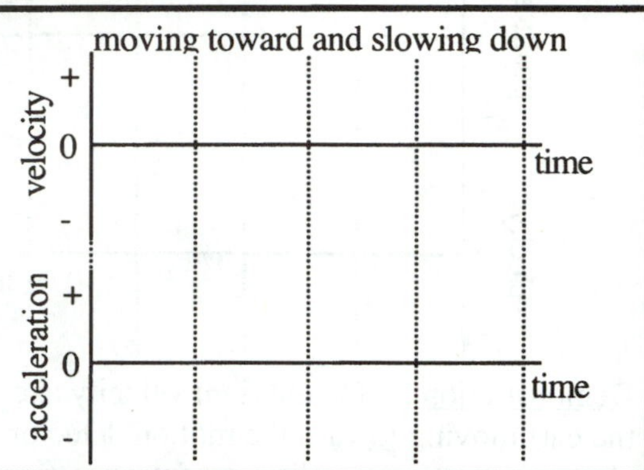

Demonstration 7: Sketch below your predictions for the *velocity-time* and *acceleration-time* graphs for the cart which is given a short push up the inclined ramp toward the motion detector (and is released) Sketch the graph as the cart slows down moving toward the detector, comes *momentarily* to rest and then speeds up moving away from the detector.

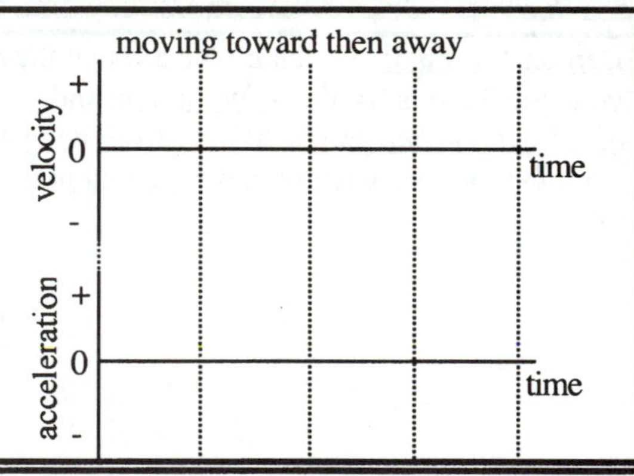

Demonstration 8: The origin of the coordinate system is on the floor, and the positive direction is upward. A ball is thrown upward. It moves upward, slowing down, reaches its highest point and falls back downward speeding up as it falls. Sketch on the axes on the right your predictions for the *velocity-time* and *acceleration-time* graphs of the ball from the moment just after it is released until the moment just before it hits the floor.

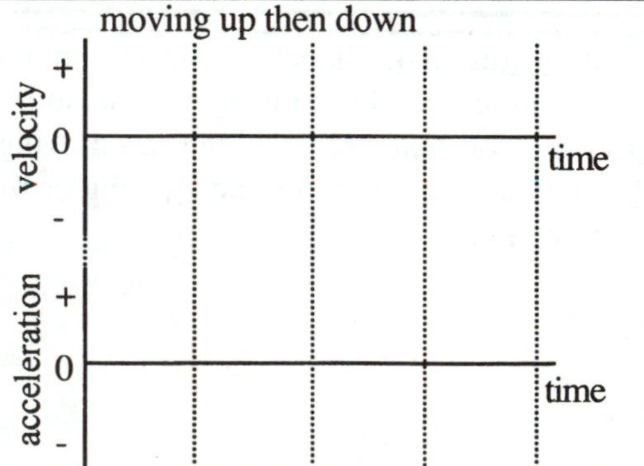

KINEMATICS 2—MOTION OF CARTS (KIN2)
TEACHER'S GUIDE

Prerequisites:

The *Kinematics 1—Human Motion ILD* sequence is the only prerequisite. If your students have done *RealTime Physics Mechanics* Lab 1 or *Tools for Scientific Thinking Motion and Force* Lab 1, you can skip the *Human Motion ILDs,* and go directly into *Motion of Carts.* We have found the *Motion of Carts ILDs* to be a useful review even for students who have completed the first two kinematics labs in the above laboratory modules.

Equipment:

computer-based laboratory system

ILD experiment configuration files

motion detector (one is okay but two are better, since one will need to be mounted on the ceiling)

two low-friction kinematics carts are better, one with the fan unit pre-mounted (If you only have one, you can mount the fan unit during the demonstrations. Any cart with very low friction will work but truly low friction carts are difficult to find.) (See below.)

fan unit (See below.)

2.2 meter aluminum track, long door threshold or a very smooth, level table or ramp (See below.)

basketball or other large round ball (Tennis balls do not work well due to the fury covering and soccer balls are not round.)

General Notes on Preparation and Equipment:

Low friction carts:
In order to get the smoothest possible acceleration graphs, considerable care must be taken in choosing a cart and a ramp. The cart must have smooth wheels that do not rub or bind. The best results are obtained by using dynamics carts with roller bearing wheels such as the PASCO ME-9430, ME-9454, ME-6950 or ME-6951. The PASCO carts have very little friction and are very sturdy. One additional advantage is that there is a Friction Cart Accessory Kit (ME-9457) available that converts the PASCO low friction cart to a cart with adjustable friction.

Track or ramp:
This laboratory does not strictly need a ramp except for Demonstration 7 since all other demonstrations are done on a level surface. (Even Demonstration 7 could be done with a suitable table that could be tilted.) A table top with a very smooth surface (e.g., Formica) and a clear distance of about 2 meters (with no cracks) will work just fine. However, most tables are too short or have too many nicks and scratches. Pushing two tables together won't work because of the crack in the middle.

An additional advantage to having a ramp is that it can be easily elevated to provide a longer falling distance for the hanging mass in the modified Atwood's machine for the *Newton's 1st & 2nd Laws ILDs.*

A 2.2 m Dynamics Track with grooves for the cart is available from PASCO (ME-9458 or ME-9453). If you choose to purchase PASCO (www.pasco.com) carts, this track is ideal but expensive. (Note that the 1.2 m Dynamics Track (ME-9435A) and the 1.2 m Force and Motion Track (ME-6858) are both too short for these demonstrations.

If instead you choose to build your own ramp, there are several options. A 2-2.5 m door threshold, available from most building supply or door and window stores, works very well, and is much less expensive than the PASCO tracks. It is not necessary for the grooves to match up with the spacing between the wheels on the cart. In fact, the wheels may bind if the grooves are narrow and exactly matched to the spacing. It is better for the cart to ride in just one groove. Since a threshold is somewhat flexible, it will need to be placed on a flat, level table or mounted on a flat board.

It is also possible to fabricate a wooden ramp. It should be about 2.2 m long and 20 - 30 cm wide, with a very smooth top surface. A design that works well uses 3/4" plywood with one side furniture grade. Cut one 12" wide strip about 8' long and two 2" strips to glue and screw edgewise to the 12" strip. Without these strips, the plywood will bow too much when supported at two points. Glue plastic coated paneling or Formica on the board to provide a smooth surface for the carts. The design is shown in Figure II-3.

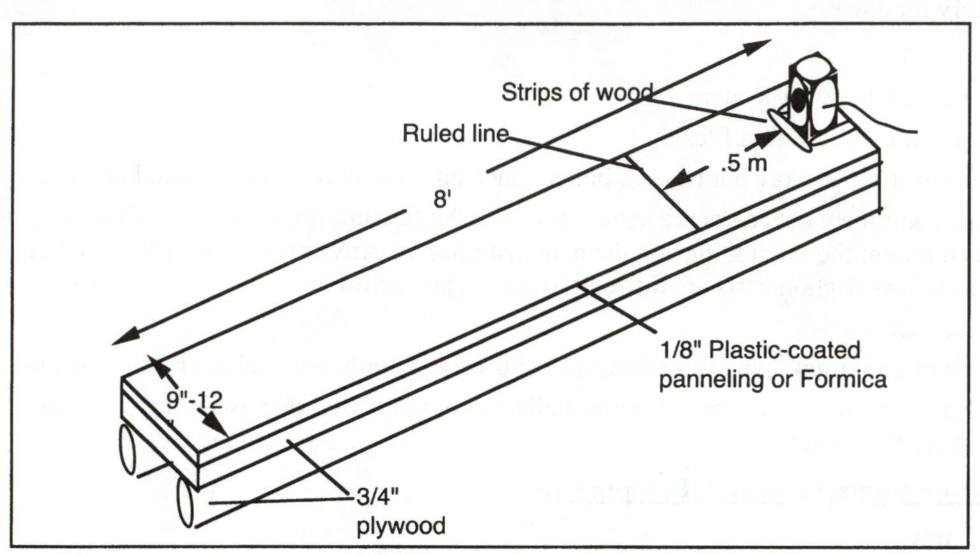

Figure II-3. A suggested ramp construction

You may also build a small holder for the motion detector at one end, and mark a line across the width of the board, 0.5 m from the detector.

Fan unit:
A fan unit is available from PASCO that mounts on any of the PASCO carts listed above (Fan Accessory—ME-9491).

It is also possible to construct fan units for use with the PASCO carts. The design described below and pictured in Figure II-4 is adapted from Robert Morse's design in his October, 1993 paper in *The Physics Teacher* (vol. 31, pp. 336-438). A later design can be found on the Workshop Physics web site at http://physics.dickinson.edu.

You will need the following major parts for each fan unit:

- 6.5 cm long piece of 6.5 cm cross-section PVC downspout
- DC motor with a no-load speed of about 8000 rpm (Radio Shack #273-223 works)
- 12.5 cm nylon propeller (e.g., #858 from Cox Hobbies, Corona, CA)
- Battery holder for 4 AA batteries (e.g., Radio Shack #270-391)
- SPST push-button switch (e.g., Radio Shack #275-1565). (Substituting a DPDT switch will make it possible to reverse the thrust of the fan unit. Adding a potentiometer will provide an adjustable acceleration.)

Figure II-4: Photograph of homemade fan unit mounted on a PASCO cart.

Cut one side off the downspout, leaving a U-shaped piece that will grasp onto the top of the PASCO cart. (When in use, the fan unit should always be taped onto the cart or held on with a rubber band. This precaution will prevent it from flying off when the cart is brought to a sudden stop, such as by colliding with a bumper.) The motor is fastened to the top of the PVC section, and the switch is inserted in a hole drilled through the top. The battery holder is fastened to the side with self-tapping sheet metal screws. Finally, the most difficult problem is fastening the propeller to the motor shaft. The best, most permanent solution is to machine a piece of metal with a hole and set screw on one end to fasten to the motor shaft and a threaded hole on the other end for fastening the propeller with a machine screw. This is labor intensive, but we have not found a satisfactory method of gluing or pressure-fitting the propeller to the motor shaft.

Rechargeable batteries are very convenient for any model fan unit.

Great care should be taken to keep fingers away from the propeller. Nicks caused by sticking a finger in the path of the blade are painful, but not particularly dangerous. It is also important to keep the fan units from falling on the floor. The impact can bend the brushes in the motor and stop it from working.

A Fan Cart is available from PASCO (ME-9485). The cart has the same low friction wheels as the PASCO dynamics cart. However, the fan cart does not have collision bumpers or a friction pad assembly, and may not be as convenient for other standard experiments (e.g., collisions) as a dynamics cart. It does not allow easy mounting of a force probe for later dynamics demonstrations.

Substitute for fan unit:
A modified Atwood's machine (pulley and falling mass) can be used in place of the fan unit, and works quite well, although acceleration by a fan unit is more transparent to most students. (See the Teacher's

Guide for the *Newton's 1st & 2nd Laws ILDs*, page 54. It is important in this case to provide at least a 1.5 m distance for the mass to fall by elevating the ramp or using two pulleys. Elevating the ramp and using one pulley is better because of the difficulties in aligning two pulleys.)

Experimental setup:

We elevate a PASCO 2.2 m track on a rolling table so it can be easily seen by the class, and check to see that the track is level when the table is in demonstration position.

As you practice the following *ILDs*, remember that for many motion detectors the cart must never be closer than 0.5 m.

For Demonstration 8, it is easiest to mount the motion detector on the ceiling and throw the ball up toward the motion detector. (The experiment configuration file is set so that the upward direction (toward the motion detector) is positive.) In this configuration you are less likely to put your hands between the ball and the motion detector, and it is easier to keep the ball in view of the motion detector. (This also eliminates the need for a screen to protect a motion detector on the floor.) It still takes some practice to get good graphs. The hardest part is keeping hands and other body parts out of the view of the motion detector. Be sure that the motion detector is seeing the ball.

Demonstrations and Sample Graphs:

We suggest you work through the demonstrations before you do them for the class making use of the suggestions below and comparing your results to those shown. Work out any difficulties and be sure the experiment configuration files display well with your equipment.

Demonstration 1: Cart moves away from motion detector at constant velocity. (Use experiment configuration file **KIN2D1**.) Prediction begins just after cart leaves hand and ends just before the cart is stopped.

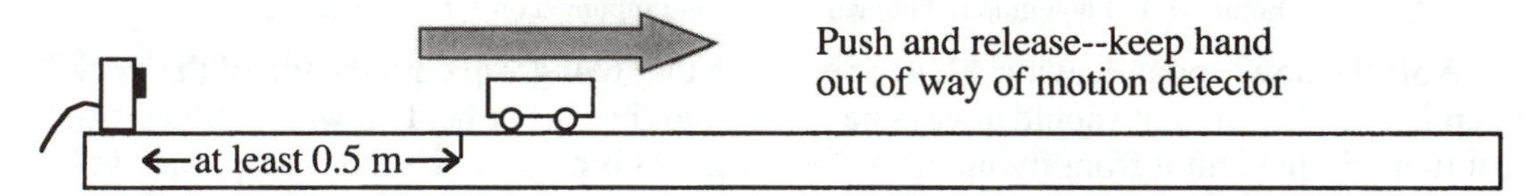

Figure II-5 shows typical position-time and velocity-time graphs. Figure II-6 shows the acceleration-

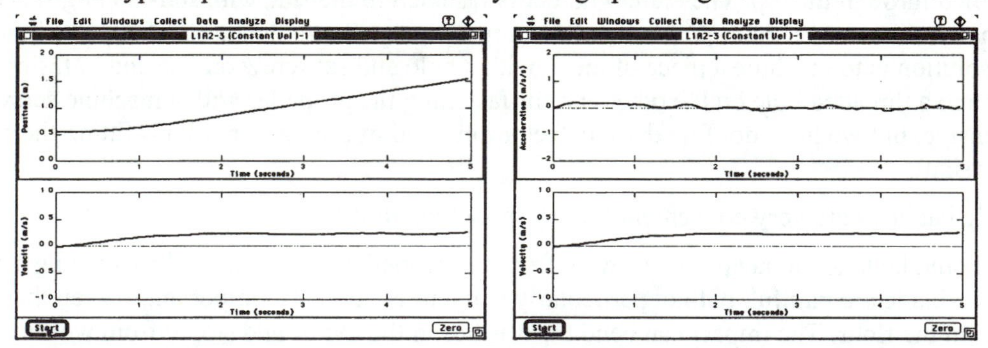

Figure II-5: Position-time and velocity-time graphsfor a cart moving away from the motion detector at a constant velocity.

Figure II-6: Acceleration-time and velocity- time graphs for the same motion as in Figure II-5.

time and velocity-time graphs for the same motion of the cart.

If your velocity and acceleration graphs are much bumpier than these, check your experimental setup. See the suggestions under Demonstration 3.

Discussion after the graphs are displayed: The velocity is in the positive direction. The position-time and velocity-time graphs are the same as for walking away from the motion detector at a constant velocity in the *Human Motion ILDs*.

Constant velocity means acceleration is essentially zero. (A small amount of friction may be evident, but students generally don't even notice it.) Discuss the slope of a position-time graph and its relationship to the velocity.

After the discussion, these graphs should be saved for persistent display on the screen and then hidden.

<u>**Demonstration 2:**</u> **Cart moves toward the motion detector at a constant velocity.** (Use experiment the same configuration file.) Prediction begins just after cart leaves hand and ends just before the cart is stopped.

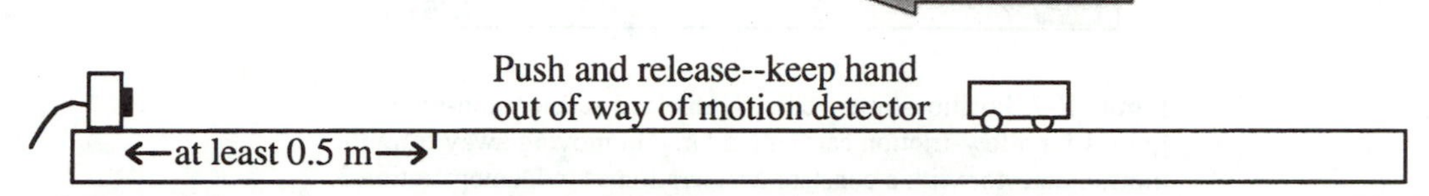

Again, if your velocity and acceleration graphs are much bumpier than those in Figures II-5 and II-6, check your experimental setup. See the suggestions under Demonstration 3.

Discussion after the graphs are displayed: The velocity is in the negative direction. Constant velocity means acceleration is essentially zero. (A small amount of friction may be evident.) Discuss slope of position-time graph and relationship to velocity. Show the graphs from Demonstration 1 and compare them.

<u>**Demonstration 3:**</u> **Cart moves away from the motion detector and speeds up at a steady rate.** (Use experiment configuration file **KIN2D3**.) Prediction begins just after cart leaves hand and ends just before the cart is stopped.

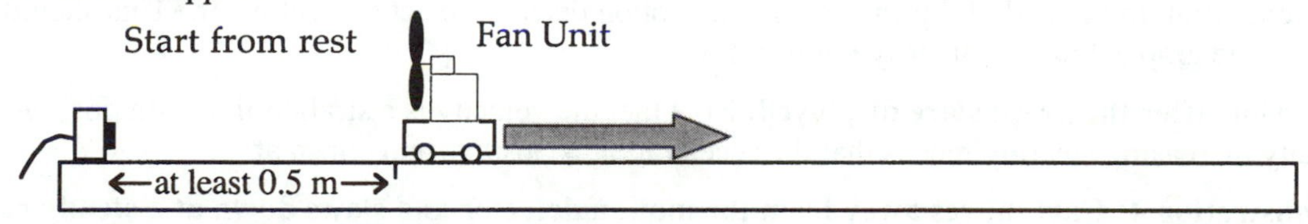

Figure II-7 shows typical graphs of position-time, velocity-time and acceleration-time graphs.

<u>Troubleshooting your graphs:</u> If your position-time graph levels off and the velocity and acceleration fall to zero at the same instant, then the motion detector is likely seeing something other than the cart. Read the position-time graph to locate the position of this object, and move it or tilt the motion detector slightly away from it.

If your velocity-time and especially your acceleration-time graphs are bumpier than those in Figure II-7, several things might be wrong:

1. The track is not smooth, and the bumps represent real characteristics of the motion of the cart.
2. The bearings of the cart's wheels are bad, and they are causing the acceleration to be non-constant.

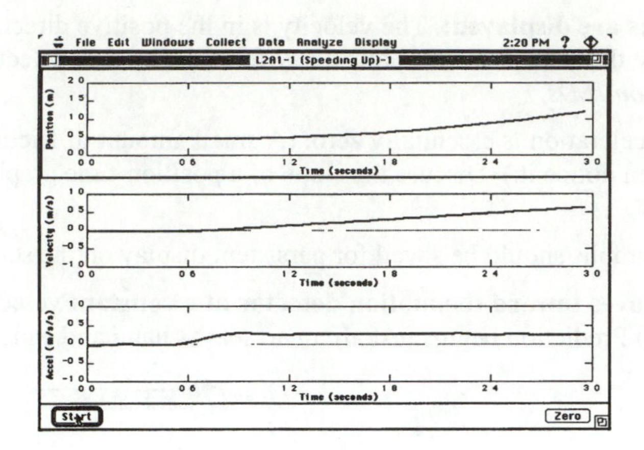

Figure II-7: Position-time, velocity-time and acceleration-time graphs for a low-friction cart with a fan unit moving away from the motion detector with a constant acceleration as in Demonstration 3.

3. The fan blade is extending beyond the end of the cart, and the motion detector is intermittently seeing the fan blade and the cart.
4. If the acceleration is okay for part of the run but varies as the cart moves away, the detector may be at an angle and shifting from one part of the cart to another. Adjusting the detector may improve things.
5. The motion detector, interface or cables may be too close to an electronically noisy monitor. Try moving the monitor further away

If you have eliminated (1) - (5), and the graphs are still bumpy, try taping a stiff piece of cardboard a few inches high to the end of the cart facing the motion detector to act as a reflector. This should improve the graphs but is ordinarily not necessary.

Discussion after the graphs are displayed: Note that the velocity is a straight line with positive slope. Velocity increasing steadily means that the acceleration is positive and constant.

Demonstration 4: Cart moves away from the motion detector and slows down at a steady rate (fan opposed to the push). (Use experiment configuration file **KIN2D4**.) Prediction begins just after cart leaves hand and ends just before the cart is stopped.

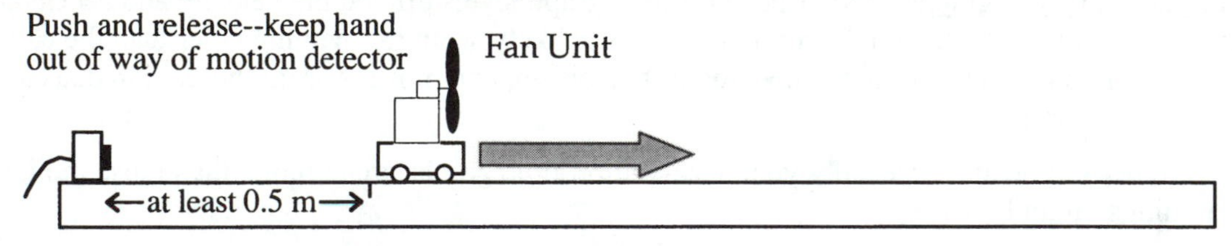

Figure II-8 shows typical velocity-time and acceleration-time.

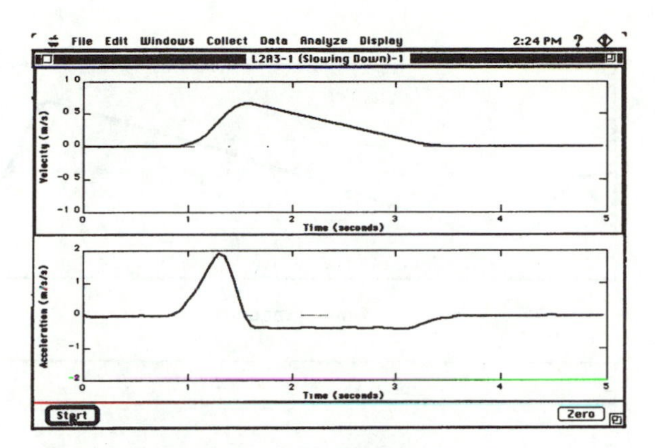

Figure II-8: Velocity-time and acceleration-time graphs for a low-friction cart with a fan unit moving away from the motion detector and slowing down at a steady rate as in Demonstration 4.

Discussion after the graphs are displayed: Select the relevant portions of the graphs. The velocity is decreasing (straight line with negative slope) although the velocity is always positive. The acceleration is in the opposite direction to the velocity so it must be negative. Since the cart is slowing down at a steady rate, the acceleration is negative and constant.

<u>Demonstration 5:</u> Cart moves toward the motion detector and slows down at a steady rate (fan opposes push). (Use the same experiment configuration file as in Demonstration 3.) Prediction begins just after cart leaves hand and ends just before the cart is stopped.

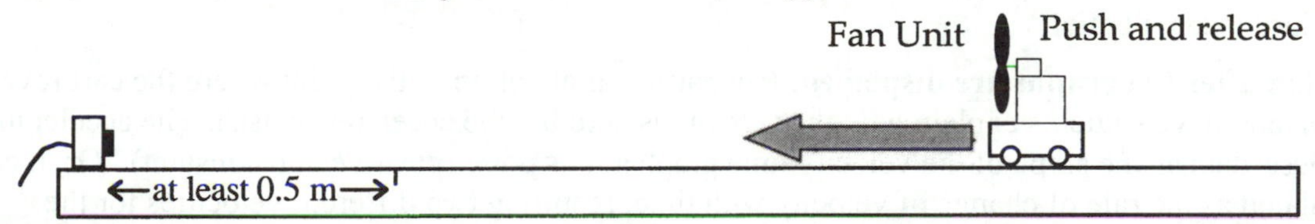

Cart is slowing down so velocity begins as large negative number and becomes smaller negative number.

Discussion after the graphs are displayed: The velocity is in the negative direction. (The cart is slowing down, so the velocity begins as large negative number and becomes smaller negative number.) Note that the acceleration is positive even though the cart is slowing down. *Deceleration* is not necessarily *negative* acceleration. Sign just shows the direction. Whenever the acceleration is in the direction opposite to velocity, the cart is slowing down. When acceleration and velocity are in the same direction, the cart speeds up. It is probably best to avoid the word *deceleration*.

<u>Demonstration 6:</u> Cart moves toward the motion detector and slows down, then reverses direction and speeds up. (Use experiment configuration file **KIN2D6**.) Prediction begins just after cart leaves hand and ends just before the cart is stopped.

Figure II-9 shows typical velocity-time and acceleration-time graphs.

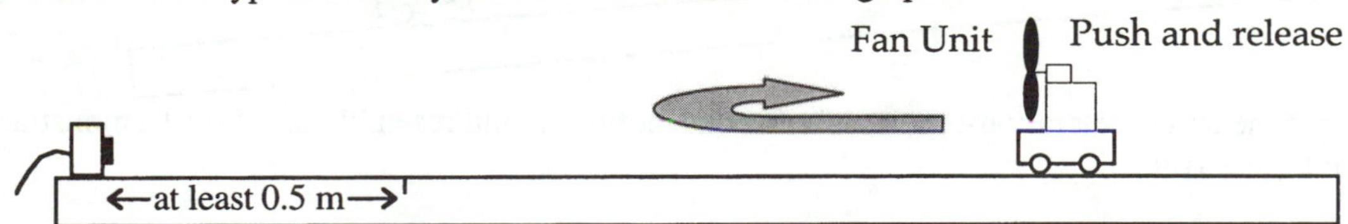

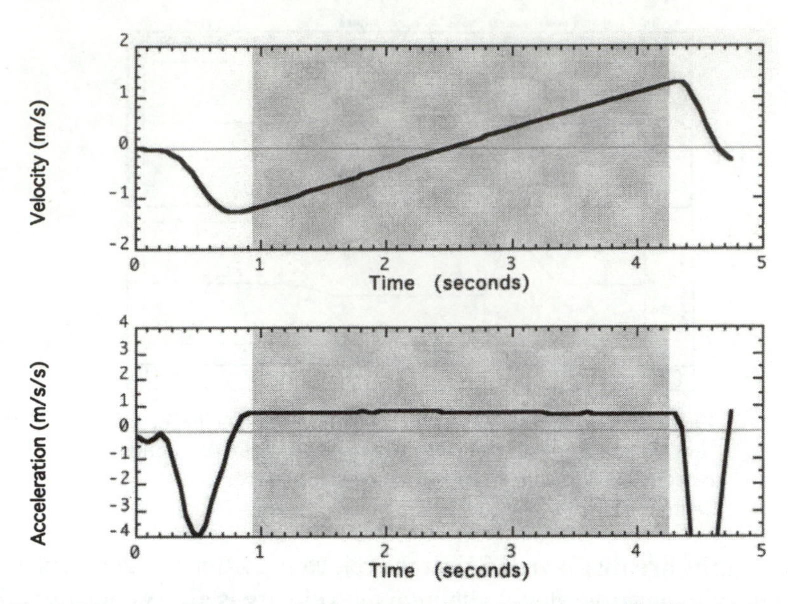

Figure II-9: Actual data from Demonstration 6 in which a cart with a fan unit opposing the initial motion was given a quick push toward the motion detector. The part of the motion that the students were asked to predict is selected just as you should do in class.

When you push the cart toward the detector, be sure it doesn't come closer than 0.5 m or you could end up with a false display of zero velocity for an extended period at the top of the motion. Since acceleration will also be displayed falsely as zero, you must avoid this at all costs or you will reinforce the standard student belief.

Discussion after the graphs are displayed: Pay particular attention to the point where the cart reverses direction, and have students explain why the velocity is zero but the acceleration isn't. The acceleration can be described as the slope of the velocity time graph at this point (positive and constant). Or, it can be calculated as the rate of change of velocity with time, requiring two different velocities for the calculation. Only one of these velocities is zero. Explain the sign and direction of the acceleration.

If the cart has substantial friction you may see different slopes on the velocity graph and also different values of the acceleration for motion of the cart toward and away from the motion detector. Try to avoid this result even though it is interesting and could be demonstrated later with substantial friction. If you discuss this now, you may fail to get the main point across which is that the acceleration is constant (does not go to zero as the cart turns around) and the velocity (consequently) is a straight line with positive slope across zero.

Demonstration 7: Cart moves up inclined ramp, reaches highest point, and rolls back down. (Use the same experiment configuration file as in Demonstration 6.) Prediction begins just after cart leaves hand and ends just before the cart is stopped.

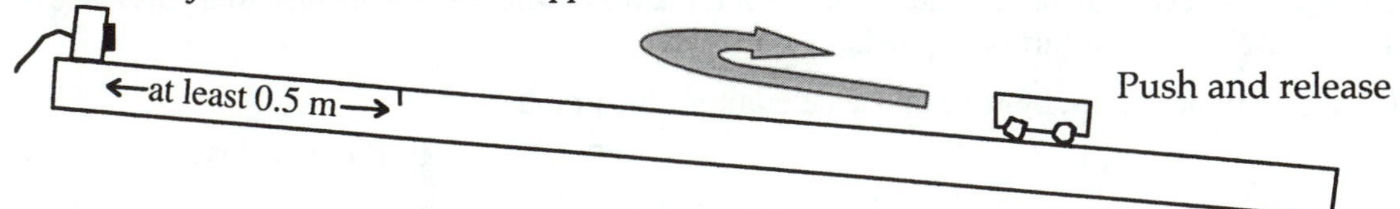

The velocity-time and acceleration-time graphs for this motion should resemble those for Demonstration 6, shown in Figure II-9.

Discussion after the graphs are displayed: Discuss analogies with previous demonstration and with the coin toss. (Again, direct student attention to the point where the cart reversed direction.) As in Demonstration 6, if the cart has substantial friction you may see different slopes on the velocity graph and also different values of the acceleration for motion of the cart on the way up and on the way down.

Demonstration 8: A ball is thrown straight upward, reaches its highest point, and comes back down. (Use experiment configuration file **KIN2D8**. The origin is set to be on the floor with the positive direction upward even though motion detector is best mounted on the ceiling. (See the discussion under **Experimental setup**.) Figure II-10 shows typical velocity-time and acceleration-time. You can see the throw and the ball hitting the floor in the acceleration graph.

The data collection rate has been set at 30 points per second. Still the number of data points collected before the ball hits the floor is small. You might try a larger data rate--50 points per second, but this can give spurious data because of multiple reflections of the ultrasound pulses between the motion detector and the floor. A carpeted floor or a cloth under the motion detector can make this effect less likely.

Discussion after the graphs are displayed: Discuss analogies with previous demonstration. Ask what is the same (constant force initially opposing the motion) and what is different (much larger force). (Again, direct student attention to the point where the ball reversed direction.)

Note: Demonstrations 6, 7 and 8 may seem repetitive, but our research shows that all of these are necessary for students to learn the concepts.

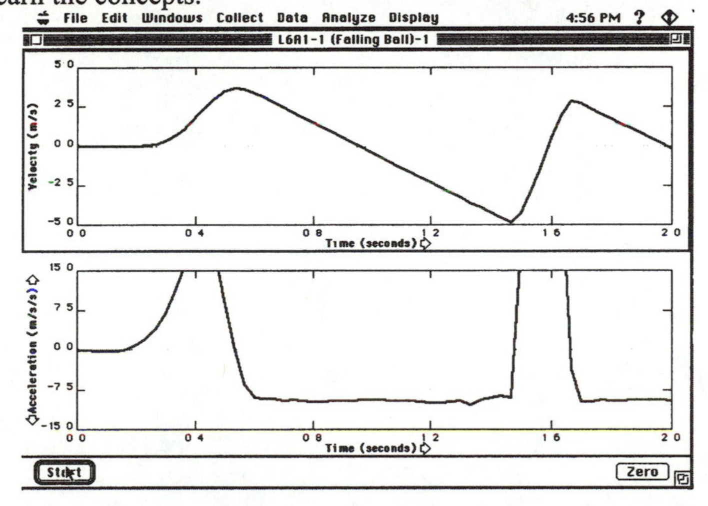

Figure II-10: Velocity-time and acceleration-time graphs for Demonstration 8 for a basketball thrown up toward the motion detector and allowed to move upward, reverse direction and fall back down again. The origin has been set to be at the floor rather than at the motion detector on the ceiling.

KINEMATICS 2—MOTION OF CARTS (KIN2)
TEACHER PRESENTATION NOTES

Classroom introduction to the *Accelerated Motion ILDs*:
Students should be familiar with the motion detector from the *Human Motion ILDs*. Push a wood block on the track or a book on the table (without measurement) to show students the motion with friction that they are accustomed to seeing where objects stop moving when they are not pushed. Emphasize the point that the cart you are using has very low friction.

Demonstration 1: Cart moves away from motion detector at a constant velocity. Use experiment configuration file **KIN2D1**. Prediction from just after cart leaves hand to just before the cart is stopped.

- Velocity is in positive direction.
- Constant velocity means acceleration is essentially zero.
- Discuss slope of position-time graph.
- After discussion save data to display graphs persistently and then hide them.

Demonstration 2: Cart moves toward the motion detector at a constant velocity. Use same experiment configuration file as in Demonstration 1. Prediction from just after cart leaves hand to just before the cart is stopped.

- Velocity is in negative direction.
- Constant velocity means acceleration is essentially zero.
- Discuss slope of position-time graph.
- Compare to Demonstration 1 by showing stored data.

Demonstration 3: Cart moves away from the motion detector and speeds up at a steady rate. Use experiment configuration file **KIN2D3**. Prediction from just after cart leaves hand to just before the cart is stopped.

- Velocity is straight line with positive slope.
- Velocity increasing steadily means acceleration is positive and constant.

Demonstration 4: Cart moves away from the motion detector and slows down at a steady rate (fan opposed to push). Use experiment configuration file **KIN2D4**. Prediction from just after cart leaves hand to just before the cart is stopped.

- Velocity is decreasing (straight line with negative slope) although velocity is always positive.
- Acceleration has opposite direction to velocity. It is negative and constant.

Demonstration 5: Cart moves toward the motion detector and slows down at a steady rate (fan opposed to push). Use same experiment configuration file as in Demonstration 3. Prediction from just after cart leaves hand to just before the cart is stopped.

- Velocity in negative direction. (Cart is slowing down so velocity begins as large negative number and becomes smaller negative number.)
- Note that acceleration is positive even though cart is slowing down. When acceleration is in direction opposite to velocity, the cart is slowing down. When acceleration and velocity are in the same direction, cart speeds up.

Demonstration 6: Cart moves toward the motion detector and slows down then reverses direction and speeds up. Use experiment configuration file **KIN2D6**. Prediction from just after cart leaves hand to just before the cart is stopped.

- Pay particular attention to the point where cart reverses direction and have students explain why velocity is zero but acceleration isn't.
- Explain direction of acceleration.

<u>Demonstration 7:</u> Cart moves up inclined ramp, reaches highest point, and rolls back down. Use same experiment configuration file as in Demonstration 6. Prediction from just after cart leaves hand to just before the cart is stopped.

- Discuss analogies with previous demonstration and with the coin toss.
- Again, direct student attention to the point where cart reversed direction.

<u>Demonstration 8:</u> A ball is thrown straight upward, reaches its highest point, and comes back down. . Use experiment configuration file **KIN2D8**. Prediction from just after ball leaves hand to just before the ball is stopped or hits floor.

- Discuss analogies with previous demonstration.
- Ask what is the same (constant force initially opposing the motion) and what is different (much larger force, ball initially moves in positive direction).
- Again, direct student attention to the point where the ball reversed direction.

NEWTON'S 1ST & 2ND LAWS (N1&2)

Hand in this sheet **Name**______________________________

INTERACTIVE LECTURE DEMONSTRATIONS
PREDICTION SHEET--NEWTON'S 1ST & 2ND LAWS

Directions: This sheet will be collected. Write your name at the top to record your presence and participation in these demonstrations. Follow your instructor's directions. You may write whatever you wish on the attached Results Sheet and take it with you.

Demonstration 1: The frictional force acting on the cart is very small (almost no friction) and can be ignored. The cart is pulled with a constant force (the applied force) so that it moves away from the motion detector speeding up at a steady rate (constant acceleration). On the axes to the right sketch your predictions of the velocity and acceleration of the cart and the applied and net force on the cart after it is released and during the time the cart is moving under the influence of the constant force. (Applied and net force are the same in this case. Why?)

Demonstration 2: The frictional force acting on the cart is now increased. The cart is pulled with the same constant force (the applied force) as in Demonstration 1 so that it moves away from the motion detector speeding up at a steady rate (constant acceleration). On the same axes to the right sketch your predictions of the velocity and acceleration of the cart and the applied and net force on the cart after it is released. (Note that the applied and net force are different now. Which determines the acceleration?) We are measuring only the applied force.

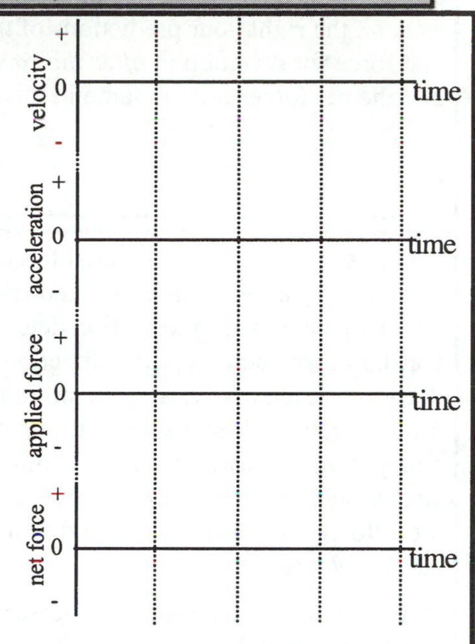

Demonstration 3: The cart has equal and opposite forces acting on it (due to two fans blowing in opposite directions). The frictional force is very small (almost no friction) and can be ignored. The cart is given a quick push away from the motion detector and released. Sketch on the right your predictions of the velocity and acceleration of the cart *after it is released*. What is the net (or resultant) force after it is released?

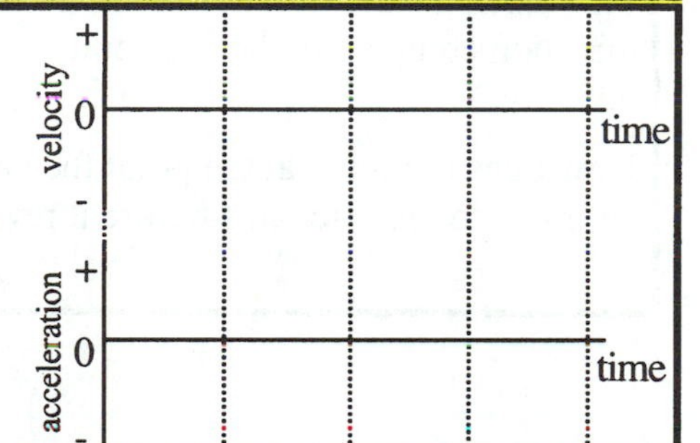

Demonstration 4: The frictional force acting on the cart remains very small (almost no friction). The cart is given a brief pull away from the motion detector and then released. Sketch on the axes on the right your predictions of the velocity and applied force for the motion, *including the time during the pull.* Is the net force the same as the applied force in this case?

What does the acceleration look like? Sketch your prediction on the acceleration-time axes on the right (below the force).

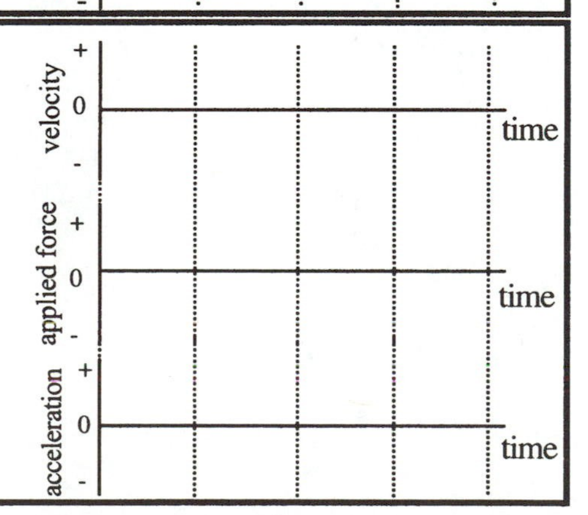

Demonstration 5: The frictional force acting on the cart remains very small (almost no friction) and can be ignored. The cart is given a push toward the motion detector and released. A constant force pulls it in the direction away from the motion detector. The cart moves toward the motion detector slowing down at a steady rate (constant acceleration). Sketch on the axes on the right your predictions of the velocity, acceleration and force for this motion *after the cart is released.* (The applied and the net forces are the same in this case.)

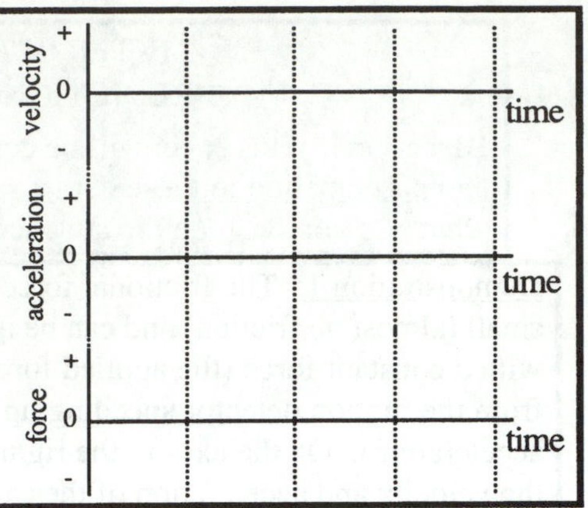

Demonstration 6: The frictional force acting on the cart remains very small (almost no friction) and can be ignored. The cart is given a push toward the motion detector and released A constant force pulls it in the direction away from the motion detector. It moves toward the motion detector slowing down at a steady rate (constant acceleration), comes to rest *momentarily* and then moves away from the motion detector speeding up at a steady rate. Sketch on the axes on the right your predictions of the velocity and acceleration and of the force on the cart *after the cart is released.*

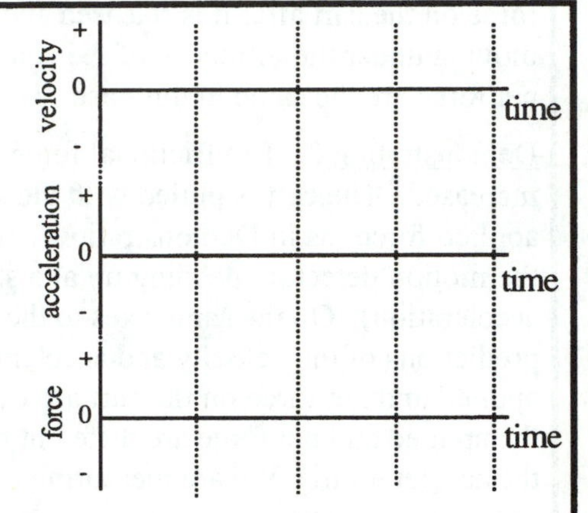

Why is the net force on the cart essentially the same as the applied force in this case?

How does the acceleration at the point the cart reverses direction compare to the acceleration just before it reverses direction?

How does the force at the point the cart reverses direction compare to the force just before it reverses direction?

Keep this sheet

INTERACTIVE LECTURE DEMONSTRATIONS
RESULTS SHEET--**NEWTON'S 1ST & 2ND LAWS**

You may write whatever you wish on this sheet and take it with you.

Demonstration 1: The frictional force acting on the cart is very small (almost no friction) and can be ignored. The cart is pulled with a constant force (the applied force) so that it moves away from the motion detector speeding up at a steady rate (constant acceleration). On the axes to the right sketch your predictions of the velocity and acceleration of the cart and the applied and net force on the cart after it is released and during the time the cart is moving under the influence of the constant force. (Applied and net force are the same in this case. Why?)

Demonstration 2: The frictional force acting on the cart is now increased. The cart is pulled with the same constant force (the applied force) as in Demonstration 1 so that it moves away from the motion detector speeding up at a steady rate (constant acceleration). On the same axes to the right sketch your predictions of the velocity and acceleration of the cart and the applied and net force on the cart after it is released. (Note that the applied and net force are different now. Which determines the acceleration?) We are measuring only the applied force.

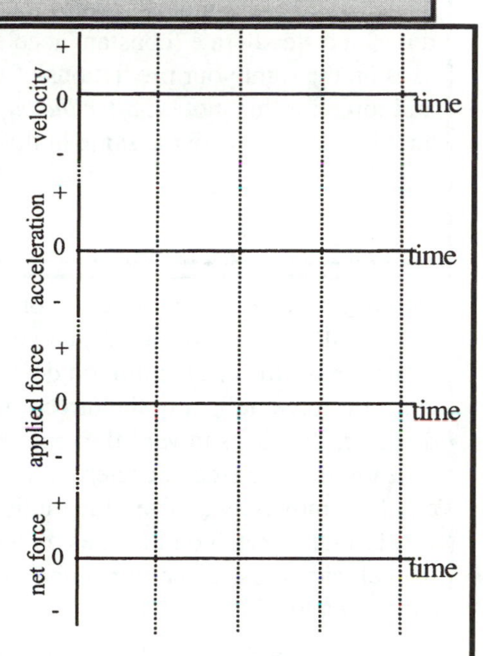

Demonstration 3: The cart has equal and opposite forces acting on it (due to two fans blowing in opposite directions). The frictional force is very small (almost no friction) and can be ignored. The cart is given a quick push away from the motion detector and released. Sketch on the right your predictions of the velocity and acceleration of the cart *after it is released*. What is the net (or resultant) force after it is released?

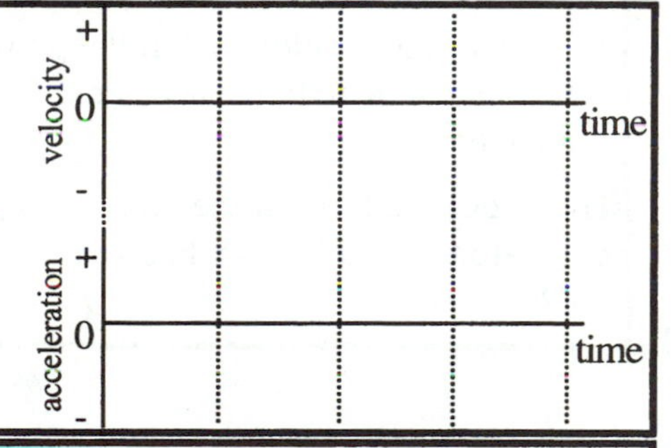

Demonstration 4: The frictional force acting on the cart remains very small (almost no friction). The cart is given a brief pull away from the motion detector and then released. Sketch on the axes on the right your predictions of the velocity and applied force for the motion, *including the time during the pull*. Is the net force the same as the applied force in this case?

What does the acceleration look like? Sketch your prediction on the acceleration-time axes on the right (below the force).

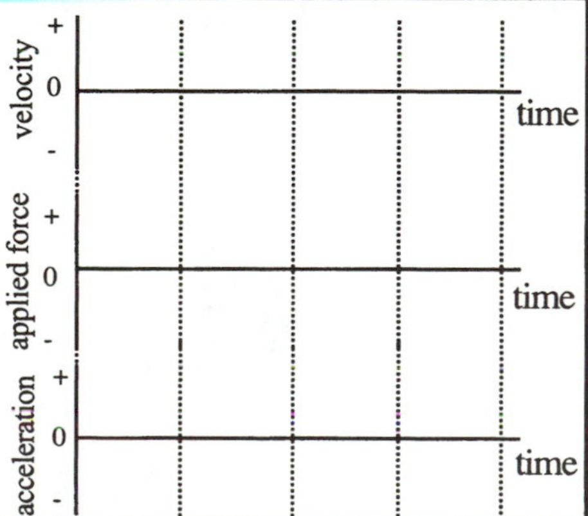

Demonstration 5: The frictional force acting on the cart remains very small (almost no friction) and can be ignored. The cart is given a push toward the motion detector and released. A constant force pulls it in the direction away from the motion detector. The cart moves toward the motion detector slowing down at a steady rate (constant acceleration). Sketch on the axes on the right your predictions of the velocity, acceleration and force for this motion *after the cart is released.* (The applied and the net forces are the same in this case.)

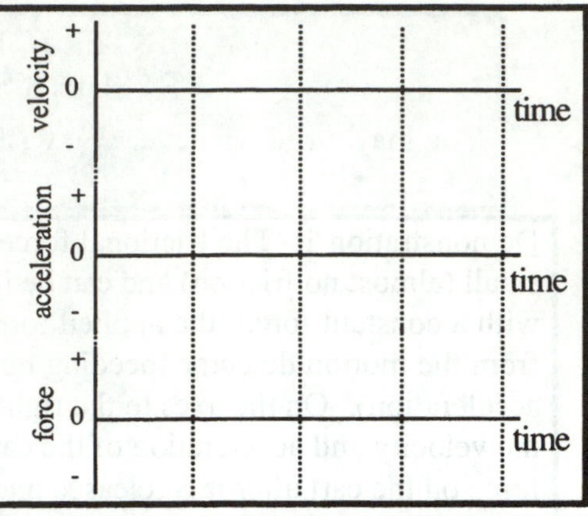

Demonstration 6: The frictional force acting on the cart remains very small (almost no friction) and can be ignored. The cart is given a push toward the motion detector and released A constant force pulls it in the direction away from the motion detector. It moves toward the motion detector slowing down at a steady rate (constant acceleration), comes to rest *momentarily* and then moves away from the motion detector speeding up at a steady rate. Sketch on the axes on the right your predictions of the velocity and acceleration and of the force on the cart *after the cart is released.*

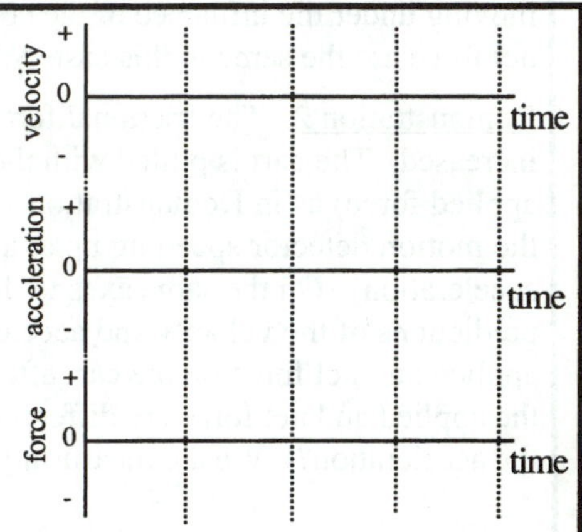

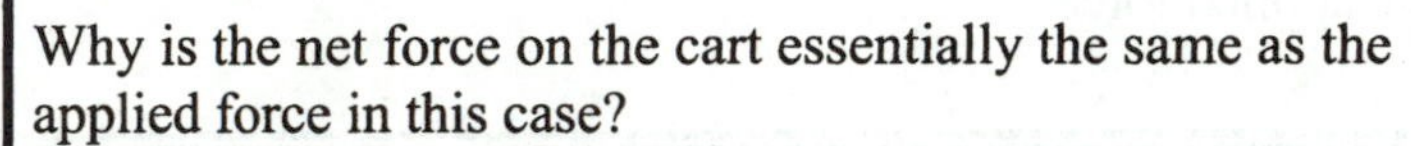

Why is the net force on the cart essentially the same as the applied force in this case?

How does the acceleration at the point the cart reverses direction compare to the acceleration just before it reverses direction?

How does the force at the point the cart reverses direction compare to the force just before it reverses direction?

NEWTON'S 1ST & 2ND LAWS (N1&2)
TEACHER'S GUIDE

Prerequisites:

The *Kinematics 1: Human Motion* and *Kinematics 2: Motion of Carts ILD* sequences are prerequisites. If students have done *RealTime Physics Mechanics* Labs 1 and 2 or *Tools for Scientific Thinking Motion and Force* Labs1 and 2, you can skip the *Human Motion ILDs*. We have found the *Motion of Carts ILDs* to be a useful review even for students who have done the suggested kinematics labs, but both sequences could be skipped if the students have done the labs.

Equipment:

computer-based laboratory system
ILD experiment configuration files
motion detector (See below.)
two or three low-friction kinematics carts are better, one with the fan unit pre-mounted and one with an adjustable friction pad pre-mounted (See below.)
sufficient mass to make the mass of the cart approximately 1 kg (See below.)
track or ramp (See below.)
two fan units (See below.)
one or two force probes (See below.)
low friction, low mass pulley and flexible light weight string (See below.)
variety of small hanging masses (10 - 50 grams)

General Notes on Preparation and Equipment:

Carts, friction pad, fan units and force probes:
See the Teacher's Guide for *Kinematics 2: Motion of Carts* for recommendations on low friction carts, friction pad and fan units. Everything can be done with one cart with increased setup time during the *ILDs*. A force probe needs to be mounted on the cart and the cart should also have an adjustable friction pad. It is convenient to have two balanced fan units pre-mounted on another cart. If you only have one cart, you can mount the fan units during the demo but it will take time.

The Vernier Dual-Range Force Sensor (DFS-BTA) is very stable and maintains its calibration, and is easily mounted on any of the PASCO low friction carts. The method of mounting on the cart, however, prevents the use of a PASCO 0.5 kg mass bar. Vernier sells a mass that fits, or a standard cylindrical mass can be used. (It is possible to do these demonstrations using just the cart, but the force needed to accelerate the cart appropriately is fairly low and harder to measure accurately.) Either of the PASCO force sensors (CI-6537 or CI-6746) should also work well for these demonstrations. <u>Always zero the force probe with nothing pushing or pulling on it before each demonstration</u>.

Motion detector, track and pulley: The motion detector is described in the Teacher's Guide for *Kinematics 1: Human Motion* the track is described in the Teacher's Guide for *Kinematics 2: Motion of Carts*. The PASCO Super Pulley (ME-9448A), that comes with the PASCO track, is an excellent low-friction, lightweight pulley for these demonstrations.

Experimental setup:
In general you should raise whatever ramp you are using to get a 1.5 m height for the weight to fall through. The lightweight PASCO ramp can easily be raised above the table. If you are not using a

PASCO ramp and you are having difficulty getting 1.5 m off the floor, you can use the arrangement shown in Figure II-11, that we do not recommend. The disadvantages of this arrangement are that it doesn't look straightforward to students, and it requires two pulleys. Misaligned pulleys can cause substantial frictional forces.

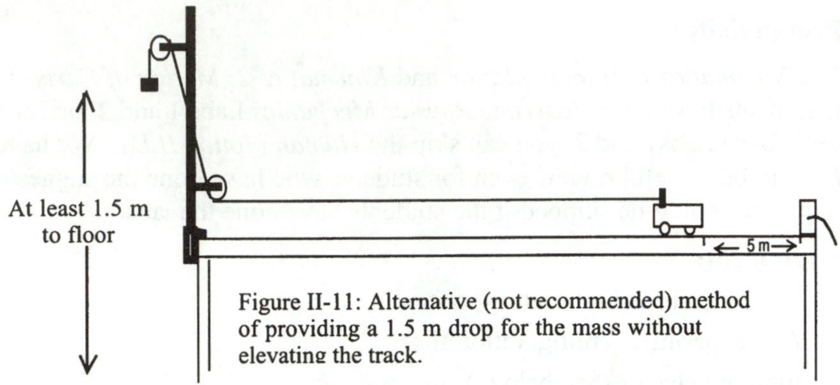

Figure II-11: Alternative (not recommended) method of providing a 1.5 m drop for the mass without elevating the track.

The activities in which the cart is accelerated using the modified Atwood's machine are also more easily done if the cart has a mass of about 1 kg. Then a reasonable hanging mass can be used, and this results in a force that can be accurately measured with the force probe. Also, it is important that the motion of the cart be on a time scale such that the students can actually see that it is accelerating as it moves across the ramp. Thus, even with a 1 kg cart, the applied force should still be a fairly small one, and it is very important that attention be paid to the following:

1. Be sure you are using the most sensitive range of the force probe.
2. It is important that the motion detector see the end of the cart and not the force probe or cable, that may be seen as intermittent targets. This means keeping the force probe cable off to the side or above the cart. If the motion detector is not seeing the cart consistently, mounting a card on the end of the cart may help. However, this card must be rigid so that it does not wave in the air as it moves.
3. Care in setting up the cable will also avoid any drag on the cart as it moves along. Hooking the force probe cable to a raised support (ring stand) in the middle of the cart's range of motion on the track can be helpful.
4. Be sure that the string attached to the force probe is parallel to the track. Adjust the pulley height to make it so.

Demonstrations and Sample Graphs:

Introduction to force probe:
If not already done, demonstrate the operation of the force probe by graphing as you pull and push on the hook. Note that a pull is positive and a push is negative. Explain experimental setup with the falling weight exerting a constant force on the force probe.

Demonstration 1: The cart (with very small friction) is pulled with a constant force so that it moves away from the motion detector, speeding up at a steady rate. (Use experiment configuration file **N1&2D1**.) Prediction from just after cart leaves hand to just before the cart is stopped. Figure II-12 shows typical velocity-time, acceleration-time and force-time graphs.

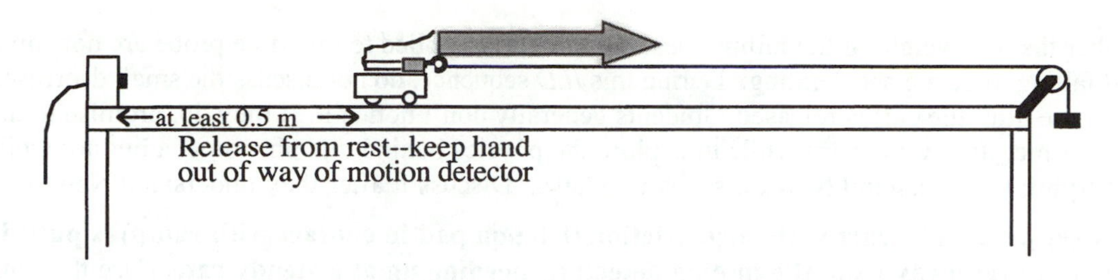

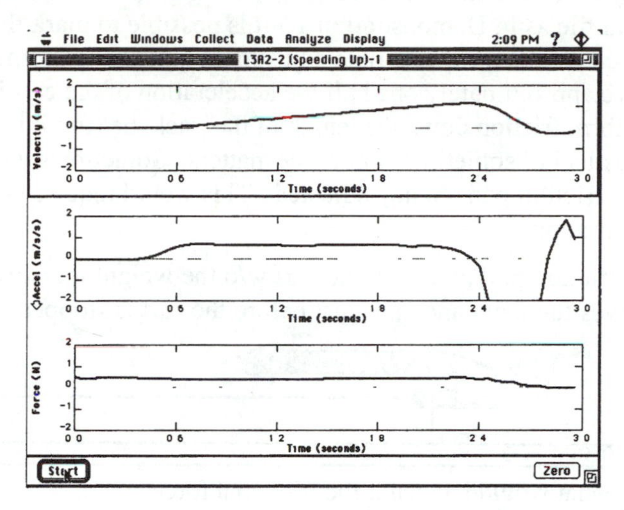

Figure II-12: Velocity, acceleration and force-time graphs for a low friction cart accelerated by a constant applied force as in Demonstration 1. The cart was stopped by catching it at the end of the track.

The constant acceleration region is roughly between 0.6 sec and 2.2 sec. The large negative dip in the acceleration after 2.4 sec is caused by the cart being stopped. This force is not seen on the force graph, since the cart was stopped without contact with the force probe. Save the graphs displayed persistently on the screen for comparison in Demonstration 2.

Troubleshooting your graphs: If your graphs are not as smooth as the ones in Figure II-12, the following suggestions may help.

1. See the Teacher's Guide for *Kinematics 2: Motion of Carts* for possible problems with the cart, track and alignment of the motion detector.
2. The motion detector may be seeing the force probe cable or the force probe and the end of the cart intermittently. Taping or holding the cable off to the side may help. Taping a stiff cardboard card to the end of the cart may also help.
3. If the acceleration changes as the cart moves along, it may be that the force probe cable is dragging. Suspend or hold the cable above the cart so that the cart can move freely without being pulled by the cable.

Discussion after the graphs are displayed: Select the region you are discussing. Explain applied and net force. Why are they the same in this demonstration? Which is constant when a constant force is applied to the low-friction cart—the velocity or the acceleration?

Save the data so that the graphs are persistently displayed on the screen, and then hide them for later comparison.

Remember that the weight of the falling mass and the force applied to the force probe are not equal if the cart and falling mass are accelerating. During this *ILD* sequence, do not discuss the small decrease in applied force after the cart is released. Students generally don't notice it. If they do, promise to discuss it at another time. It is very worthwhile to explore the physics behind this phenomena but not while they are first trying to understand Newton's first two laws. Discuss it after they understand Newton's laws.

Demonstration 2: The cart with large friction (friction pad in contact with ramp) is pulled so that it still moves away from the motion detector, speeding up at a steady rate. (Use the same experiment configuration file as in Demonstration 1.) It is possible to mark the adjustment screw for the friction pad so that you can make an appropriate adjustment quickly. Determine an adjustment ahead of the demonstration to give approximately one half the acceleration of the cart in Demonstration 1. If you have trouble getting uniform friction down the length of the track, be sure it is clean. Cleaning oil or grease off the track with alcohol sometimes improves matters. Sometimes we have found it helpful to use another cart with the friction pad already adjusted. This would require switching to another force probe with the same calibration.

First show students the friction pad and push the cart w/o the weight to show it slows down. Prediction begins just after cart leaves the hand and ends just before the cart is stopped.

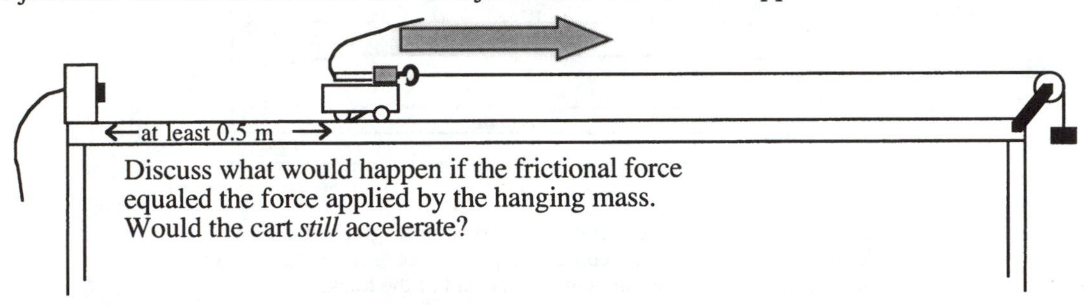

Discussion after the graphs are displayed: Make the point that *we are measuring only the applied force. Net* (or resultant) force determines acceleration. Display the graphs from Demonstration 1 and compare. Why is the acceleration smaller now?

Demonstration 3: The cart with equal and opposite forces moves away from the motion detector. (Use experiment configuration file **N1&2D3**.) First show that the cart accelerates in either direction when only one fan unit is on (as seen in previous demonstrations). With both fans balanced, the cart does not move. Now push and release and observe velocity and acceleration. Prediction from just after cart leaves hand to just before the cart is stopped.

The "dueling fan carts" in this activity will be most convincing if you have balanced the batteries in the fan units and leveled the ramp so that the cart has no tendency to move before giving it a push. You can switch batteries around or run down the stronger fan unit to achieve balance.

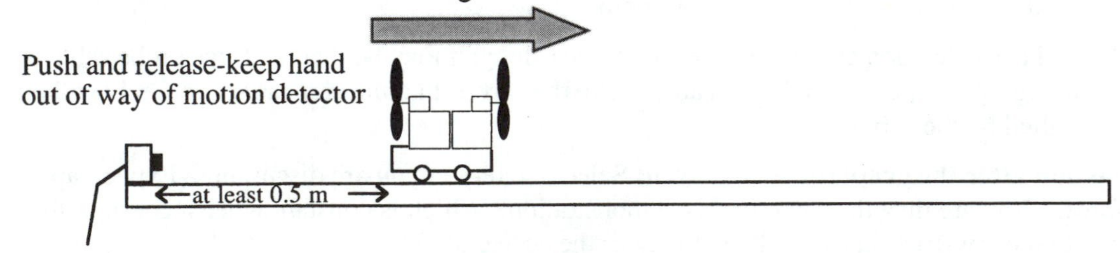

Figure II-13 shows typical velocity-time and acceleration-time graphs.

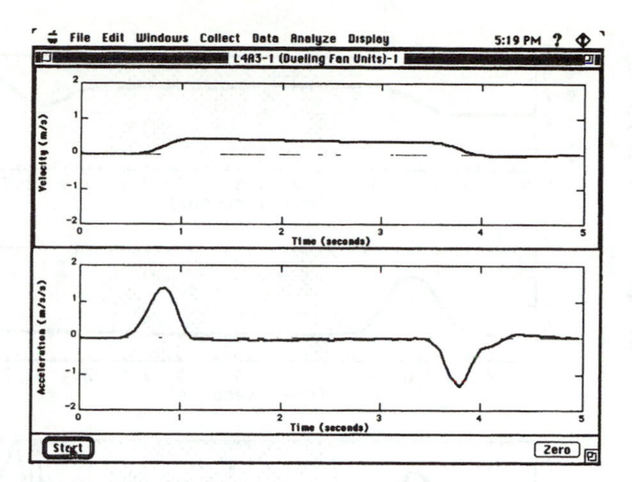

Figure II-13: Velocity-time and acceleration-time graphs for motion of a cart given a push away from the motion detector and released, with two fan units pushing with equal forces in opposite directions, as in Demonstration 3.

Discussion after the graphs are displayed: Research shows most students will agree that the net force is zero when the cart is not moving but not if it is moving at a constant velocity. Discuss in the context of a bicycle and/or a car moving down the road at a constant velocity. Why is it necessary to pedal or step on the accelerator? The combined (net) force is zero, and after the cart is pushed and released, it moves with a nearly constant velocity, and zero acceleration, as they should be according to Newton's First Law.

Demonstration 4: Cart with very small frictional force is given a pull away from the motion detector and released. (Use experiment configuration file **N1&2D4.**)

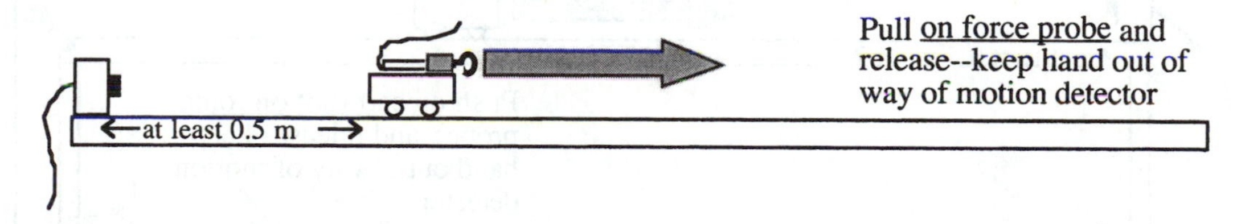

A short quick pull works best. Prediction includes pull of the hand and ends just before the cart is stopped. Figure II-14 shows typical graphs.

Discussion after the graphs are displayed: Discuss in context of previous demonstration and Newton's First Law--constant velocity motion with net force equal to zero. Many students believe force continues after the pull (or push).

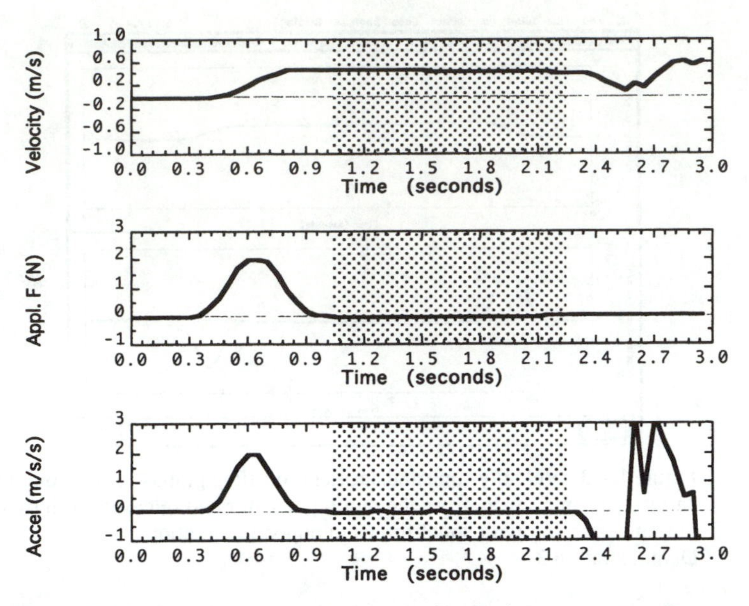

Figure II-14. Actual data from Demonstration 4 where a cart was given a quick pull away from the motion detector. The part of the motion after the pull is selected.

<u>Demonstration 5:</u> The cart (with very small friction) moves toward the motion detector slowing down at a steady rate. (Use experiment configuration file **N1&2D5**.) Prediction from just after cart leaves hand to just before the cart is stopped. Be sure to push the cart so that it comes no closer than 0.5 m from the motion detector when it starts to turn around. Catch it at this point.

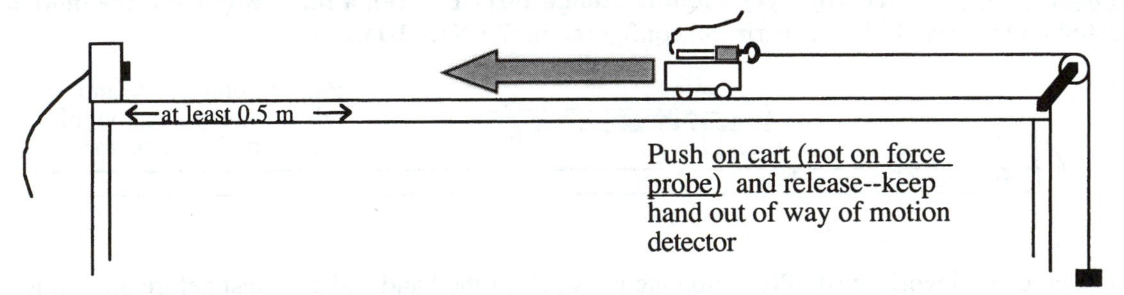

Discussion after the graphs are displayed: Note that acceleration and force are positive even though the cart is slowing down. As was seen in the *Kinematics 2: Motion of Carts ILDs*, "deceleration" is not necessarily *negative* acceleration. The sign just shows direction. When acceleration and force are in the direction opposite to the velocity, the cart slows down. When all are in the same direction, the cart speeds up. Applied and net force are the same.

<u>Demonstration 6:</u> The cart (with very small friction) moves toward the motion detector slowing down at a steady rate, comes to rest *momentarily*, and then moves away from the motion detector. (Use the same experiment configuration file as in Demonstration 5.)

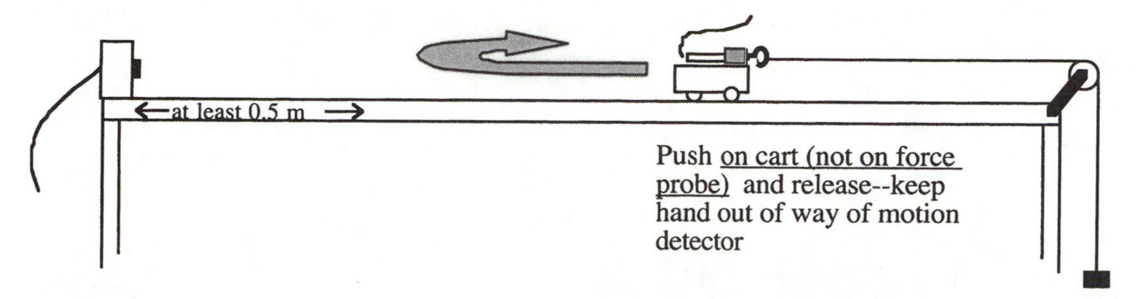

Prediction from just after cart leaves hand to just before the cart is stopped. Be sure to use a low friction cart so that the acceleration is essentially the same toward and away. Do not discuss the effect of friction during this *ILD*, you won't have time and may confuse student's who are not clear about the acceleration not going to zero at the turnaround point. (If you have more time on another day, show the effect of substantial friction on these graphs. (See the Teacher's Guide for *Energy of a Cart on a Ramp* for the result.)

Figure II-15 shows typical graphs.

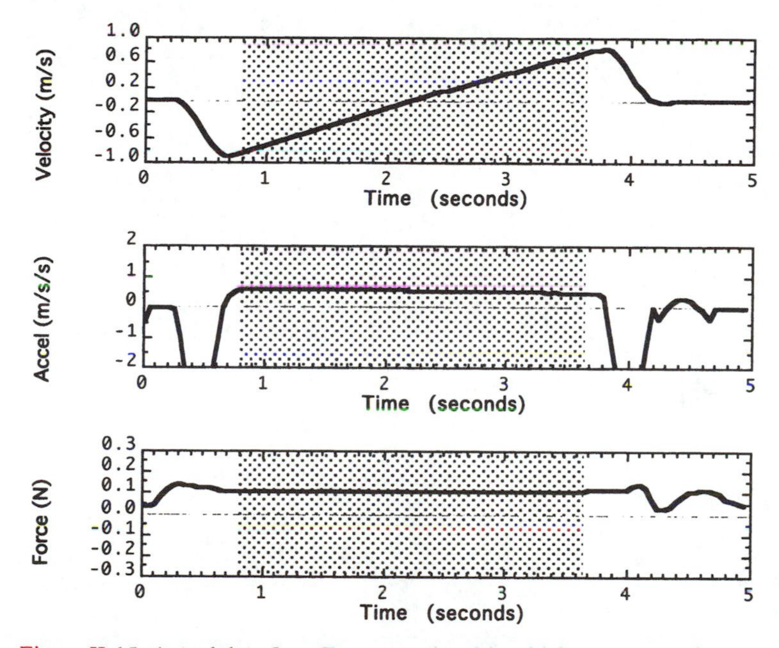

Figure II-15: Actual data from Demonstration 6 in which a cart was given a quick push toward the motion detector with the force acting away. The portion of the motion that the students were asked to predict is selected.

Discussion after the graphs are displayed: Point out that acceleration and force are not zero where the cart reverses direction. Discuss in the context of the coin toss and for motion up and down an inclined ramp. Remind students of previous *ILDs* with the cart on the inclined ramp and the basketball from *Kinematics 2: Motion of Carts.*

NEWTON'S 1ST & 2ND LAWS (N1&2)
TEACHER PRESENTATION NOTES

Classroom introduction to force probe:
Demonstrate the operation of the force probe by graphing as you pull and push on the hook. Note that a pull is positive and a push is negative. Explain that the falling weight exerts a constant force.

Demonstration 1: The cart (with very small friction) is pulled with a constant force so that it moves away from the motion detector, speeding up at a steady rate. Use experiment configuration file **N1&2D1**. Prediction from just after cart leaves hand to just before the cart is stopped.

- Select region you are discussing. Explain applied and net force. Why are they the same here?
- Which is constant for a constant applied force—velocity or acceleration?
- Save the data to display the graphs persistently, and then hide them.

Demonstration 2: The cart with large friction (friction pad in contact with ramp) is pulled so that it still moves away from the motion detector, speeding up at a steady rate. Use the same experiment configuration file as in Demonstration 1. Prediction from just after cart leaves hand to just before the cart is stopped. Show students the friction pad and push the cart to show it slows down.

- Note that *we are measuring only the applied force* while the *net* force determines acceleration.
- Display graphs from Demonstration 1 and compare. Why is the acceleration smaller now?

Demonstration 3: The cart with equal and opposite forces moves away from the motion detector. Use experiment configuration file **N1&2D3**. Prediction from just after cart leaves hand to just before the cart is stopped. Show that cart accelerates in either direction when only one fan unit is on. With both fans balanced, the cart does not move. Now push and release and observe velocity and acceleration.

- Ask what net or resultant force is after release.
- Discuss in context of bicycle and/or car moving down road at constant velocity. Why is it necessary to pedal or step on the accelerator?

Demonstration 4: Cart with very small frictional force is given a pull away from the motion detector and released. Use experiment configuration file **N1&2D4**. Prediction includes pull of hand and ends just before the cart is stopped.

- Is net force same as applied force (with negligible friction)?
- Many students believe force continues after the pull (or push). Newton's First Law!

Demonstration 5: The cart (with very small friction) moves toward the motion detector slowing down at a steady rate. Use experiment configuration file **N1&2D5**. Prediction from just after cart leaves hand to just before the cart is stopped. A force acts in the direction away from the motion detector.

- Note that acceleration and force are positive even though cart is slowing down. "Deceleration" is not necessarily *negative* acceleration. Sign just shows direction.
- When acceleration (force) is in direction opposite to velocity, cart slows down. When all are in the same direction, cart speeds up.
- Applied and net force are the same.

Demonstration 6: The cart (with very small friction) moves toward the motion detector slowing down at a steady rate, comes to rest *momentarily*, and then moves away from the motion detector. Use the same experiment configuration file as in Demonstration 5. Prediction from just after cart leaves hand to just before the cart is stopped.

- **Point out that acceleration and force are not zero where cart reverses direction.**
- **How does the acceleration at the point the cart reverses direction compare to the acceleration just before it reverses direction? How does the force at the point the cart reverses direction compare to the force just before it reverses direction?**
- **Discuss in context of coin toss and for cart motion up and down inclined ramp.**

NEWTON'S 3RD LAW (N3)

Hand in this sheet **Name**__________________________________

INTERACTIVE LECTURE DEMONSTRATIONS
PREDICTION SHEET--**NEWTON'S 3RD LAW**

Directions: This sheet will be collected. Write your name at the top to record your presence and participation in these demonstrations. Follow your instructor's directions. You may write whatever you wish on the attached Results Sheet and take it with you.

Someone pushes a block on a smooth surface. The block experiences a constant frictional force opposite to its motion. Compare the following two forces in *magnitude and direction,* $\vec{F}_{H \to B}$ (the force of the Hand on the Block) and $\vec{F}_{B \to H}$ (the force of the Block on the Hand) during each of the three demonstrations described below.

Demonstration 1: The block is being pushed at a *constant velocity*. How do $\vec{F}_{H \to B}$ and $\vec{F}_{B \to H}$ compare?

How does $\vec{F}_{H \to B}$ compare to the force of friction? What is the net force *on the block*?

Demonstration 2: The block is pushed so that it *speeds up*. How do $\vec{F}_{H \to B}$ and $\vec{F}_{B \to H}$ compare?

How does $\vec{F}_{H \to B}$ compare to the force of friction? What is the net force *on the block?*

Demonstration 3: The block is pushed so that it *slows down*. How do $\vec{F}_{H \to B}$ and $\vec{F}_{B \to H}$ compare?

How does $\vec{F}_{H \to B}$ compare to the force of friction? What is the net force *on the block?*

Demonstration 4: Two people press their hands together. First person A pushes person B's hand in the positive direction. Then person B pushes person A's hand back in the negative direction. How does $\vec{F}_{B \to A}$ (the force of person B's hand on person A) compare to $\vec{F}_{A \to B}$ (the force of person A's hand on person B) for each movement?

Demonstration 5: One cart (called Car A) pushes (not a collision) against (Car B) which is trapped up against a barrier. How does $\vec{F}_{B \to A}$ (the force of Car B on Car A) compare to $\vec{F}_{A \to B}$ (the force of Car A on Car B, as Car B is pushed first easier and then harder against Car A?

Demonstration 6: Two identical carts (called Car A and Car B) are pushed toward each other at the same speed. How does $\vec{F}_{B \to A}$ (the force of Car A on Car B) compare to $\vec{F}_{A \to B}$ (the force of Car B on Car A) during the collision? How do the directions of the forces compare?

Demonstration 7: A massive (heavy) cart (called Truck) is pushed toward a light cart (called Car) that isn't moving. Describe in words how $\vec{F}_{T \to C}$ (the force of Truck on the Car) compares to $\vec{F}_{C \to T}$ (the force of the Car on the Truck) during the collision. Make a rough graph of the forces over time.

Demonstration 8: A light cart (called Car) is pushed toward a massive (heavy) cart (called Truck) that isn't moving. Describe in words how the $\vec{F}_{T \to C}$ (the force of Truck on the Car) compares to $\vec{F}_{C \to T}$ (the force of the Car on the Truck) during the collision.

Can an object at rest exert a force?

If you push the car into the truck, is it possible to push the car fast enough that the force exerted by the car on the truck is greater than the force exerted by the truck on the car? Explain.

Keep this sheet

INTERACTIVE LECTURE DEMONSTRATIONS
PREDICTION SHEET--**NEWTON'S 3RD LAW**

You may write whatever you wish on this sheet and take it with you.

Someone pushes a block on a smooth surface. The block experiences a constant frictional force opposite to its motion. Compare the following two forces in *magnitude and direction,* $\vec{F}_{H\to B}$ (the force of the Hand on the Block) and $\vec{F}_{B\to H}$ (the force of the Block on the Hand) during each of the three demonstrations described below.

Demonstration 1: The block is being pushed at a *constant velocity*. How do $\vec{F}_{H\to B}$ and $\vec{F}_{B\to H}$ compare?

How does $\vec{F}_{H\to B}$ compare to the force of friction? What is the net force *on the block*?

Demonstration 2: The block is pushed so that it *speeds up*. How do $\vec{F}_{H\to B}$ and $\vec{F}_{B\to H}$ compare?

How does $\vec{F}_{H\to B}$ compare to the force of friction? What is the net force *on the block?*

Demonstration 3: The block is pushed so that it *slows down*. How do $\vec{F}_{H\to B}$ and $\vec{F}_{B\to H}$ compare?

How does $\vec{F}_{H\to B}$ compare to the force of friction? What is the net force *on the block?*

Demonstration 4: Two people press their hands together. First person A pushes person B's hand in the positive direction. Then person B pushes person A's hand back in the negative direction. How does $\vec{F}_{B\to A}$ (the force of person B's hand on person A) compare to $\vec{F}_{A\to B}$ (the force of person A's hand on person B) for each movement?

Demonstration 5: One cart (called Car A) pushes (not a collision) against (Car B) which is trapped up against a barrier. How does $\vec{F}_{B\to A}$ (the force of Car B on Car A) compare to $\vec{F}_{A\to B}$ (the force of Car A on Car B, as Car B is pushed first easier and then harder against Car A?

Demonstration 6: Two identical carts (called Car A and Car B) are pushed toward each other at the same speed. How does $\vec{F}_{B \to A}$ (the force of Car A on Car B) compare to $\vec{F}_{A \to B}$ (the force of Car B on Car A) during the collision? How do the directions of the forces compare?

Demonstration 7: A massive (heavy) cart (called Truck) is pushed toward a light cart (called Car) that isn't moving. Describe in words how $\vec{F}_{T \to C}$ (the force of Truck on the Car) compares to $\vec{F}_{C \to T}$ (the force of the Car on the Truck) during the collision. Make a rough graph of the forces over time.

Demonstration 8: A light cart (called Car) is pushed toward a massive (heavy) cart (called Truck) that isn't moving. Describe in words how the $\vec{F}_{T \to C}$ (the force of Truck on the Car) compares to $\vec{F}_{C \to T}$ (the force of the Car on the Truck) during the collision.

Can an object at rest exert a force?

If you push the car into the truck, is it possible to push the car fast enough that the force exerted by the car on the truck is greater than the force exerted by the truck on the car? Explain.

NEWTON'S 3RD LAW (N3)
TEACHER'S GUIDE

Prerequisites:

Kinematics 1: Human Motion, Kinematics 2: *Motion of Carts*, and *Newton's 1st & 2nd Laws ILDs*. If students have done *RealTime Physics Mechanics* Labs 1 and 2 or *Tools for Scientific Thinking Motion and Force* Labs1 and 2, you can skip the *Human Motion ILDs*. We have the *Motion of Carts ILDs* to be a useful review even for students who have done the suggested kinematics labs, but both sequences could be skipped if the students have done the labs.

Note that the following *ILD* sequence is almost as much about the 1st and 2nd laws as the 3rd since the difference between these must be well established in the students' minds.

Equipment:

- computer-based laboratory system
- *ILD* experiment configuration files
- motion detector
- two or three low-friction kinematics carts (See below.)
- track or ramp (See below.)
- at least two but preferably four force probes with collision bumpers or rubber stoppers (See below.)
- block with felt bottom (with mass of at least1 kg) to which a force probe is attached (See below.)
- two 0.5 kg masses (See below.)

General Notes on Preparation and Equipment:

Force probes, carts, track, block and masses: See the *Newton's 1st & 2nd Laws* Teachers' Guide for information on force probes. Use a range of 0-10 N for Demonstrations 1-4, and a 0-50 N range for the other demonstrations (collisions). Remember to always zero the force probes before each demonstration.

The PASCO (www.pasco.com) friction block that comes with the track, felt side down may be used with two PASCO 0.5 mass bars on top. Or a similar block may be fabricated with mass at least 1 kg. Be sure the block is heavy so that you get less "stuttering" when you push with the other force probe. A heavy block will also let you use a calibration that is large enough to do Demonstrations 4 and 5 with the same probes. The force probe is taped to the mass bars, and the mass bars to the friction block. Any dynamics carts could be used, since low-friction is not essential for the collisions. The low-friction carts used in Demonstrations 1-3 are ideal because of the ease of mounting the force probes. The track makes it easier to align the carts for collisions, but the demonstrations could be done without it. It may help to clean the track or table with alcohol to remove any grease that might give you varying friction.

Use some care in collisions between two carts. The force probes should be solidly attached to the carts and aligned with each other. If the carts lift off the track or get forced sideways due to an overzealous collision, the forces may not match due to asymmetric torques on the probes. It is also important that the cart and weights are essentially rigid in later demonstrations or you will get very strange force patterns during the collision. Be sure that the collision is not so hard that one or both probes saturate. As always, be sure to zero the force probes before each collision.

Triggering: The collisions in Demonstrations 6-8 are so fast that it is necessary to trigger the graph display by starting the display when a non-zero force is detected in force probe A. Data just before the trigger will be displayed for probe A and only data after the trigger for probe B. To avoid an obvious gap, the triggering level is set in the experiment configuration files to approximately ±0.5 N. If there is instability for any reason and the probes are triggering before the collision, the value can be set higher.

Demonstrations and Sample Graphs:

(**Note:** Diagrams are shown looking toward *(not as seen by)* the students for the teacher's convenience. It is assumed that the experiment is between you and the students.) With care, you may be able to use the same data set for Demonstrations 1, 2, and 3. This is a good way to save time.

Classroom introduction to the demonstrations:
If not already done as part of the *Newton's 1st & 2nd Laws ILDs*, demonstrate how a force probe reacts to a push or a pull. Define positive and negative directions. Remember to follow all eight steps of the *ILD* procedure (see the Introduction) for each of the short demonstrations.

Demonstration 1, 2, 3: Demonstration 1: Push the block in the positive direction so that it moves with a constant velocity (as much as possible). Demonstrations 2 and 3: Push the block in the positive direction so that it speeds up and then slows down after you get it moving. (Use experiment configuration file **N3D1**.)

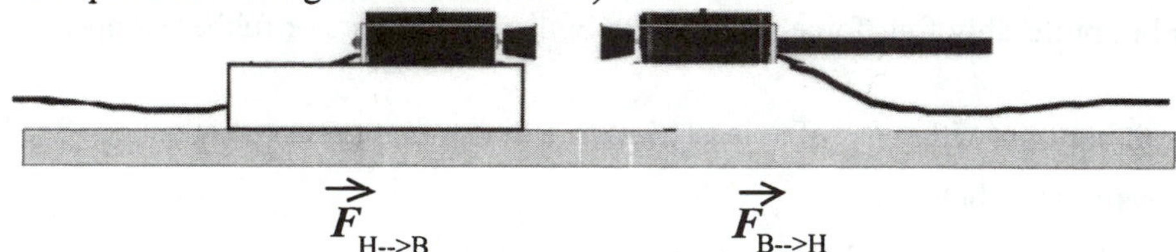

$\vec{F}_{H\text{-->}B}$ $\vec{F}_{B\text{-->}H}$

For Demonstration 1, ask the students to compare the force of the hand on the block and the force of the block on the hand. Make sure the students think the force assignments are reasonable (e.g.. force probe 2 measures force of hand on block).

For Demonstration 2, accelerate the block uniformly from rest and slow down quickly. For Demonstration 3, accelerate quickly and then allow the block to slow down but be sure that the two probes remain in contact. (Of course in all three demonstrations there are times when the block is speeding up and slowing down.) Figure II-16 shows typical graphs for Demonstration 1.

Discussion after the graphs are displayed: For Demonstration 1, ask students to compare the forces of the hand on the block and the block on the hand. What must the *net* force on the block be if it moves with a constant velocity? Which force on the block are we measuring? Also ask how the force of the hand on the block compares to the force of friction. (Note that we can derive the frictional force from the result since if the block moves at constant velocity, net the force is zero and the frictional force must be equal and opposite to the force of the hand on the block. Mention that this is an indirect measurement using Newton's 1st Law. We are not measuring it directly.

For Demonstration 2, ask the same questions. Be sure to establish that the force of the hand on the block is larger than the force of friction in this case since the block is speeding up (Newton's 2nd Law), and yet the force of the hand on the block is again equal and opposite to the force of the block on the hand. For Demonstration 3, ask the same questions. Be sure to establish that the force of the

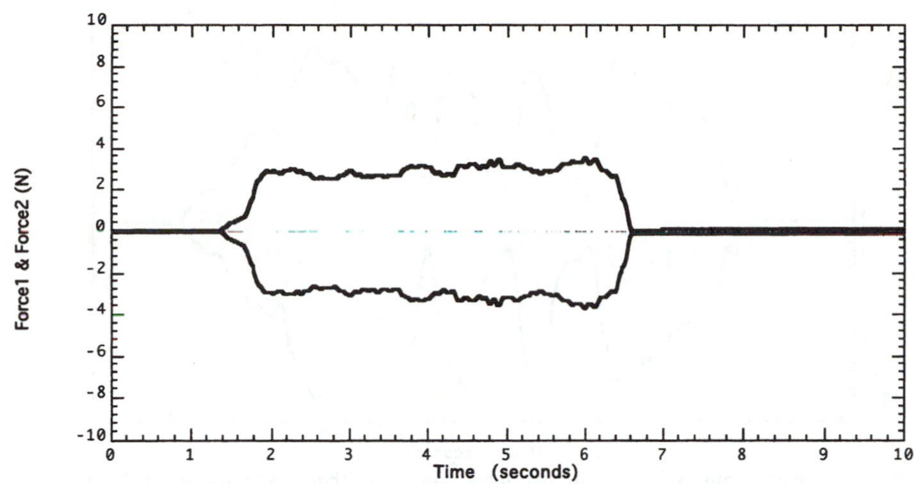

Figure II-16: Typical graphs for pushing a block at a constant velocity with friction, as in Demonstration 1.

hand on the block is smaller than the force of friction in this case since the block is speeding up (Newton's 2nd Law), and yet the force of the hand on the block is again equal and opposite to the force of the block on the hand.

Demonstration 4: First push so assistant's arm moves (stage) left then so s/he moves your arm right. (Use experiment configuration file **N3D4.**) Ask students to predict which force will be larger for each case.

With some care, you can use the same probes used in Demonstrations 1 to 3 without removing the probe from the block. Beware of pushing so hard that you saturate the force probe. Be sure to push straight on. Although it is nice to involve students, our experience is that students push rather wildly. It may be better to get help from a colleague with whom you have practiced or arrange to practice with a student.

Discussion after the graphs are displayed: Why are the forces the same? How can one person's push overcome the others and move his/her hand back? Which forces are not equal?

Demonstration 5: Push cart A (not a collision) into cart B (which is up against a barrier) with varying force. (Use experiment configuration file **N3D5.**)

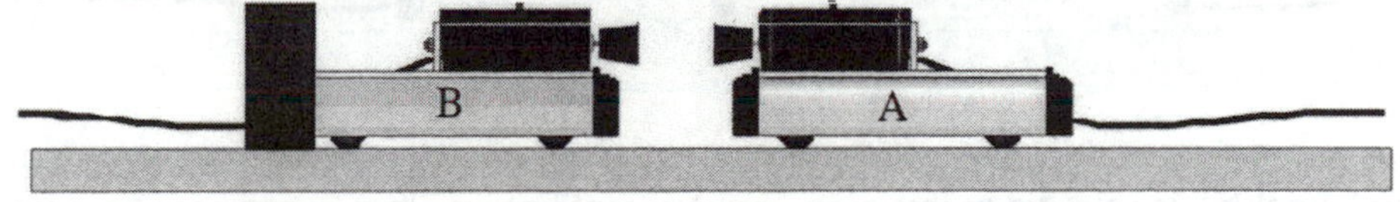

You can conveniently use the carts you will later use for collisions. Be careful not to saturate the force probe reading. Figure II-17 shows typical graphs for Demonstration 5.

Discussion after the graphs are displayed: Notice that Cart A always pushes back with just enough force to equal the force applied to it by Cart B. This is a very quick introduction to passive forces. Discuss other examples of passive forces: force exerted by wall when you lean on it, book on table, etc.

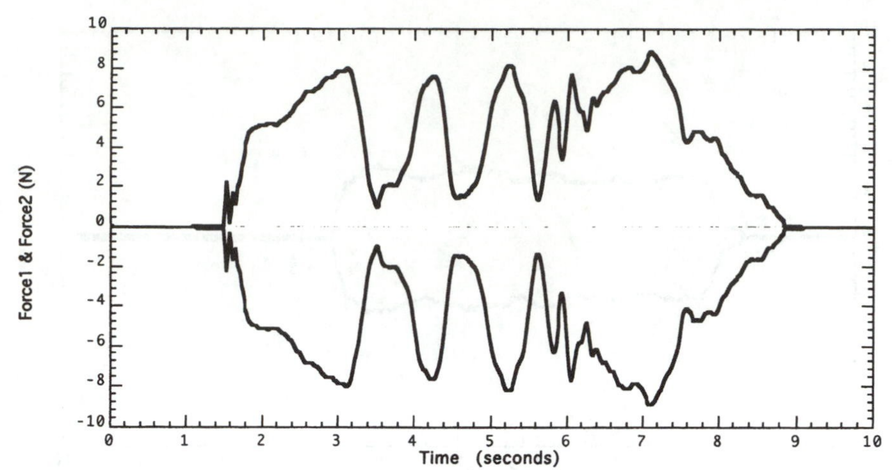

Figure II-17: Typical graphs for pushing two force probes together in a "push of war" as in Demonstration 5.

<u>Demonstration 6:</u> Collide two identical carts (Car A and Car B) at equal speeds. (Use experiment configuration file **N3D6**.)

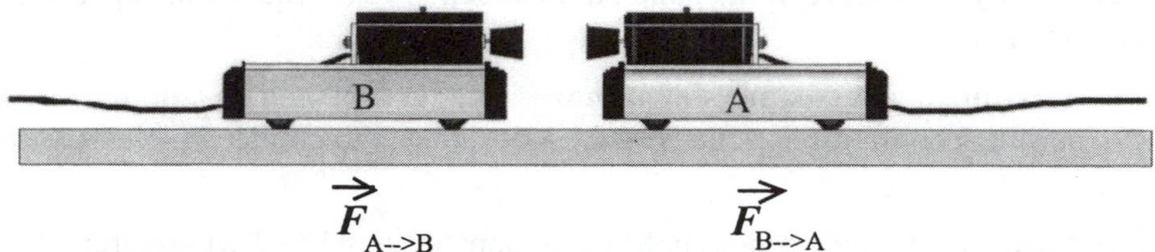

Discussion after the graphs are displayed: This symmetric case is one case when students and physicists almost always agree the forces are equal and opposite.

<u>Demonstration 7:</u> Collide a massive (heavy) cart (labeled Truck) that is moving quickly into a light cart (labeled Compact Car) that is at rest. (Use the same experiment configuration file as in Demonstration 6.)

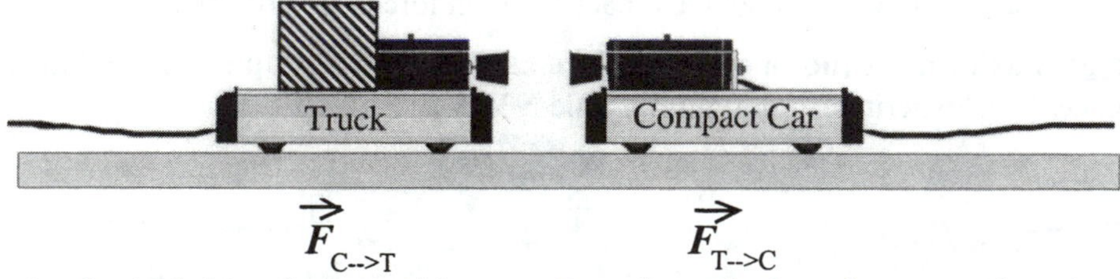

"Truck" moving "quickly" is relative in this case. Experiment before class to see how hard you can push the truck before the probes saturate or the carts jump up or sideways. See notes for Demonstration 6. It is a good idea to label the cart as a Truck after you add the weight. Drop the mass you are adding onto the table to show the students you are adding substantial weight.

Discussion after the graphs are displayed: Ask students to describe the two forces, both magnitudes and directions. Why are the results the same as for the symmetrical case?

<u>Demonstration 8:</u> Collide a light cart (labeled Compact Car) into a massive (heavy) cart (labeled Truck) that is at rest (first slower, than faster). (Use the same experiment configuration file as in Demonstration 6.)

Discussion after the graphs are displayed: Ask students to describe the two forces, both magnitudes and directions. Why are the results the same as for the symmetrical case? Can an object at rest exert a force? Ask if the "car" is pushed fast enough, could the force of the car on the truck be greater than the force of the truck on the car? Do the speed of collision or the fact that one object is at rest make any difference as far as the interaction pair of forces between then two objects?

Finding Impulses, an added feature for calculus-based classes: If you have discussed impulse and the students understand integration or at least area under the curve as a sum, you can show that the impulses of the two interaction forces are equal in all cases by using the integration feature in the software. (You may wish to save this for a later demonstration). Figure II-19 shows the result of integrating under the force graphs in Figure II-18.

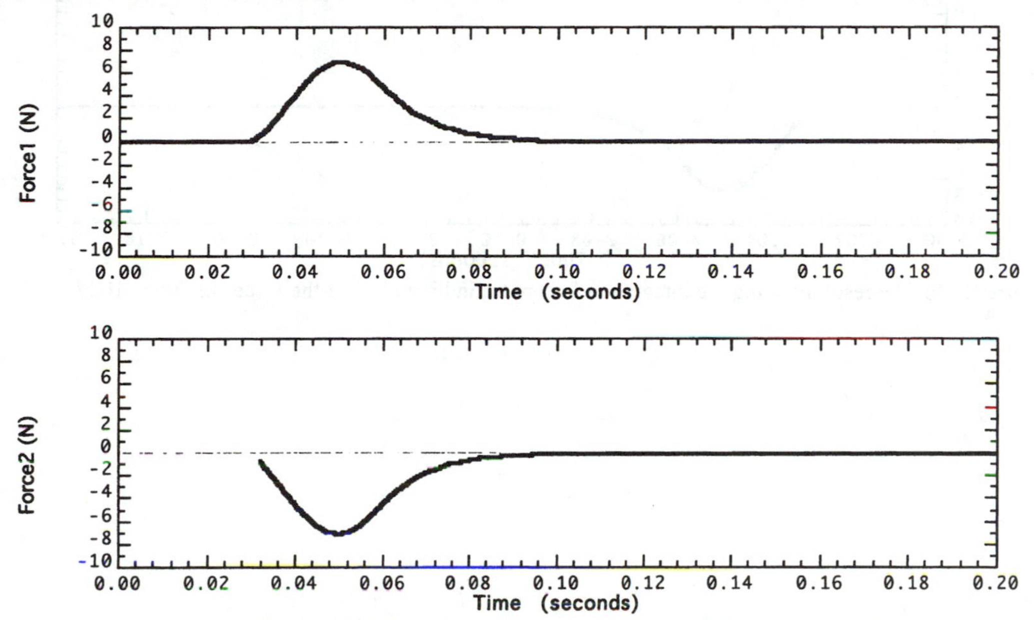

Figure II-18: Typical graphs for the collision in Demonstration 7.

Summary:

Is there any way to make 3rd Law interaction force pairs unequal? No! 3rd law pairs are always equal and act on different objects. Third law pairs do not combine to determine net force and thus acceleration. Only the one of the pair that is acting on the object of interest must be added into net force.

Net force on a single object is what determines motion (acceleration) of the object. If the *net* force is zero, the velocity is constant. If the *net* force is different from zero, then the object accelerates. When we look at all forces on a single object, this is where the 2nd and 1st laws apply. If we are comparing interaction forces between two different objects, then the 3rd law applies.

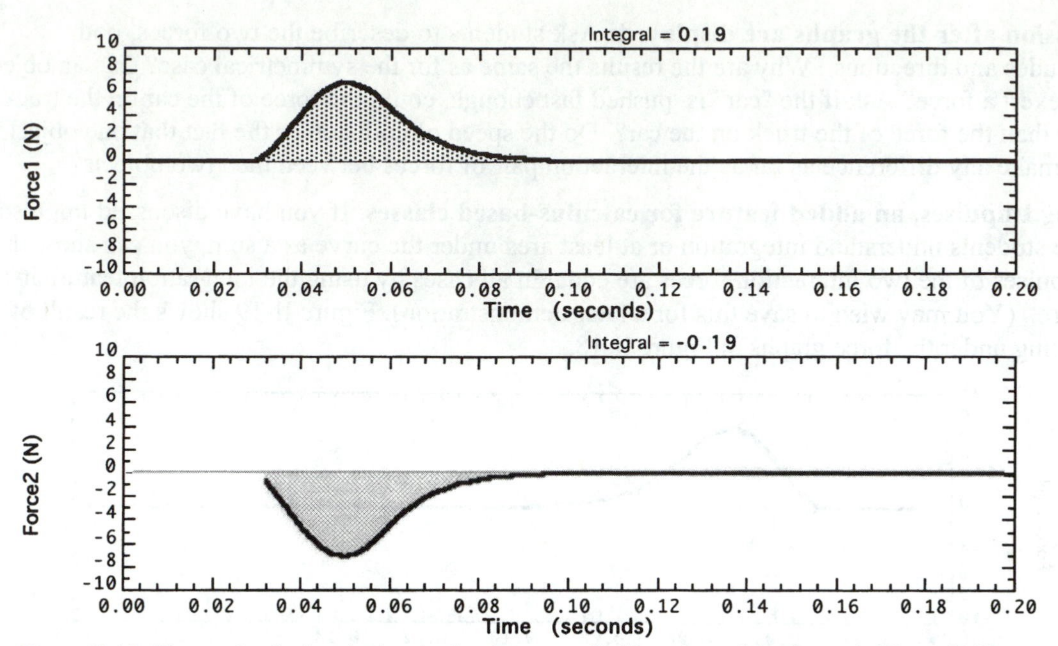

Figure II-19: The result of using the integration feature to find impulses of the forces in Figure II-18.

NEWTON'S 3RD LAW (N3)
TEACHER PRESENTATION NOTES

Classroom introduction to the *Newton's 3rd Law ILDs*:
If not already done as part of the *Newton's 1st & 2nd Law ILDs* and Second law (which should be done before this sequence), demonstrate how a force probe reacts to a push or a pull. Define positive and negative directions. Remember to follow all eight steps of the *ILD* procedure for each demonstration.

Demonstration 1: Push the block in the positive direction so that it moves with a constant velocity (as much as possible). Use experiment configuration file **N3D1**. With care, you may be able to use the same data set for Demonstrations 1, 2 and 3. Remember to zero the force probes before each demonstration.

- Compare the force of the hand on the block and the force of the block on the hand.
- What is the net force *on the block*? Also, which force are we measuring?
- What is the frictional force? Can indirectly measure it using Newton's 1st Law. It is equal to the force of the hand on the block.

Demonstrations 2 & 3: Push the block in the positive direction so that it speeds up and then slows down after you get it moving). Use the same experiment configuration file as in Demonstration 1.)

- Compare the force of the hand on the block and the force of the block on the hand.
- What is the net force *on the block*? Also, which force are we measuring.
- What is the frictional force?
- No matter how cart moves, force of hand on block and the force of block on hand are equal.
- Contrast 3rd law and 2nd law forces and remind students of the relationship to frictional force.

Demonstration 4: First push so assistant's arm moves (stage) left then so s/he moves your arm right. Use experiment configuration file **N3D4**. Predict which force will be larger for each case.

- Why are forces the same?
- Which forces are not equal? How can one person's push overcome the others?

Demonstration 5: Push cart A (not a collision) into cart B (which is up against a barrier) with varying force. Use experiment configuration file **N3D5.**

- Cart A always pushes back with just enough force to equal pushing force from cart B.
- Discuss other examples of passive forces, force when you lean on wall, book on table, etc.

Demonstration 6: Collide two identical carts (Car A and Car B) at equal speeds. Use experiment configuration file **N3D6.**

- Almost all students believe the forces to be equal and opposite in this symmetrical case.

Demonstration 7: Collide a massive (heavy) cart (labeled Truck) that is moving quickly into a light cart (labeled Compact Car) that is at rest. Use the same experiment configuration file as in Demonstration 6.

- Ask students to describe the two forces, both magnitudes and directions.
- Why are the results the same as for the symmetrical case?

<u>Demonstration 8:</u> Collide a light cart (labeled Compact Car) into a massive (heavy) cart (labeled Truck) that is at rest (first slower, than faster). Use the same experiment configuration file as in Demonstration 6.

- Ask students to describe the two forces, both magnitudes and directions. Why are the results the same as for the symmetrical case?
- Can an object at rest exert a force?
- If the "car" is pushed fast enough, could the force of the car on the truck be greater than the force of the truck on the car? Do the speed of collision or the fact that one object is at rest make any difference?

Summarize:

- Is there any way to make Third Law pairs unequal? No. Third law pairs are always equal and act on different objects.
- Third law pairs do not combine to determine net force and thus acceleration. Only the one of pair that is acting on object, must be added into the net force.
- *Net* force on a single object is what determines motion (acceleration) object. If the *net* force is zero, velocity is constant. If the net force is different from zero, then the object accelerates.
- When looking at forces on a single object, the 2nd and 1st laws apply. If we are comparing forces on different objects that are interacting, then the 3rd law applies.

Vectors (VECT)

Hand in this sheet **Name**________________________________

INTERACTIVE LECTURE DEMONSTRATION
PREDICTION SHEET—**VECTORS**

Directions: This sheet will be collected. Write your name at the top to record your presence and participation in these demonstrations. Follow your instructor's directions. You may write whatever you wish on the attached Results Sheet and take it with you.

Demonstration 1:

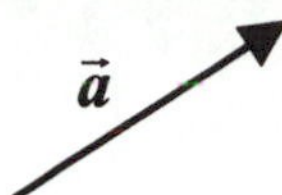

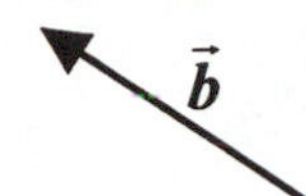

Given the two vectors $\vec{a}$ and $\vec{b}$ shown above, sketch to the right your prediction for their sum $\vec{c} = \vec{a} + \vec{b}$. Be certain to label the vectors.

Demonstration 2:

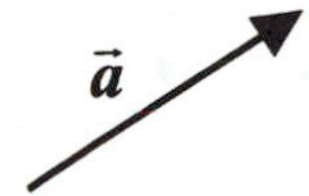

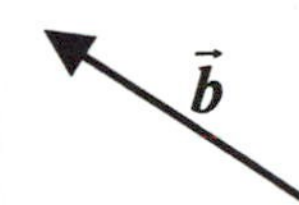

Given the two vectors $\vec{a}$ and $\vec{b}$ shown above, sketch to the right your prediction for their difference $\vec{c} = \vec{a} - \vec{b}$. Be certain to label the vectors.

Demonstration 3:

Given the two vectors $\vec{v}_1$ and $\vec{v}_2$ shown above, sketch to the right your prediction for the vector $\Delta\vec{v}$ that must be added to $\vec{v}_1$ to produce $\vec{v}_2$. Note that $\vec{v}_1 + \Delta\vec{v} = \vec{v}_2$.

Demonstration 4:

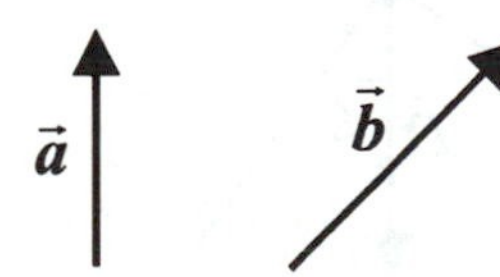

Given the two vectors $\vec{a}$ and $\vec{b}$ shown above, sketch to the right your prediction for the vector $\Delta\vec{b}$ that must be added to $\vec{b}$ to produce $\vec{a}$. Note that $\vec{b} + \Delta\vec{b} = \vec{a}$.

Demonstration 5:

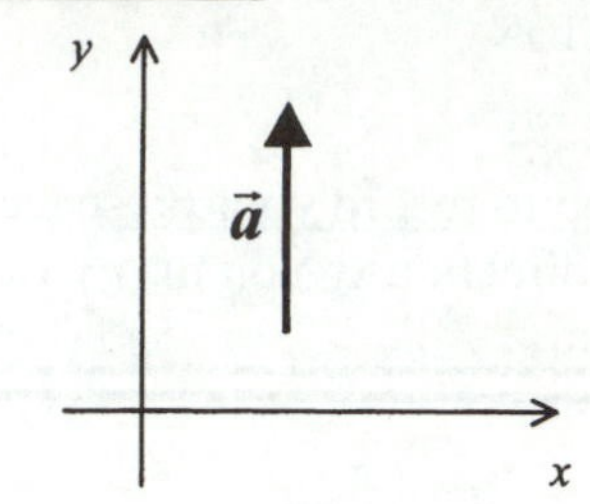

Vector $\vec{a}$ with the *x-y* axes shown only has a y-component. The *x*-component of $\vec{a}$ is zero. Draw another set of *x-y* axes near the vector $\vec{b}$ such that $\vec{b}$ only has an *x*-component in this new coordinate system.

Demonstration 6:

The vector $\vec{C}$ is shown on the right.
Show the *x*-component, C_x on the diagram.
Is the *x*-component positive or negative?

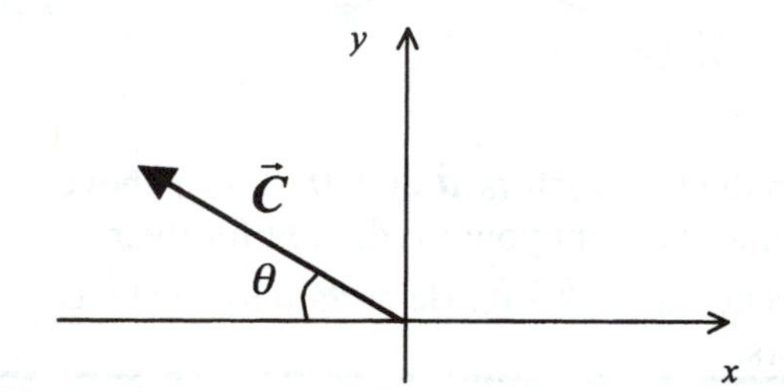

Show the *y*-component, C_y on the diagram.
Is the *y*-component positive or negative?

In terms of C and θ,
Write an expression for C_x:

Write an expression for C_y:

Demonstration 7:

The vector $\vec{D}$ is shown on the right.
Show the *x*-component, D_X on the diagram.
Is the *x*-component positive or negative?

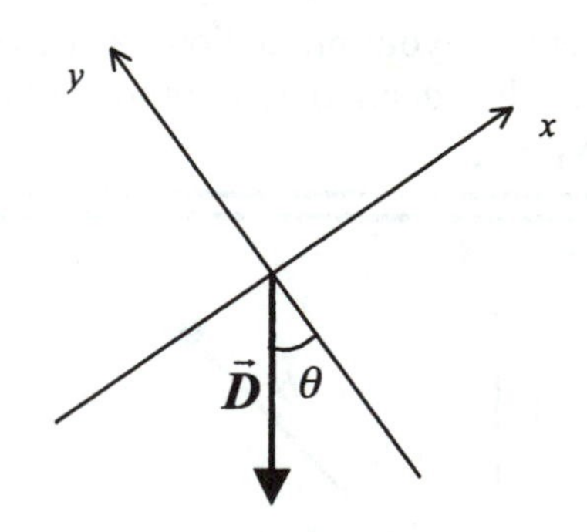

Show the y-component, D_y on the diagram.
Is the y-component positive or negative?

In terms of D and θ,
Write an expression for D_x:

Write an expression for D_y:

Keep this sheet

INTERACTIVE LECTURE DEMONSTRATION
RESULTS SHEET—**VECTORS**

You may write whatever you wish on this sheet and take it with you.

Demonstration 1:

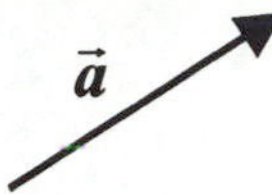
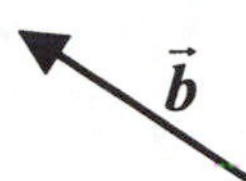

Given the two vectors $\vec{a}$ and $\vec{b}$ shown above, sketch to the right your prediction for their sum $\vec{c} = \vec{a} + \vec{b}$. Be certain to label the vectors.

Demonstration 2:

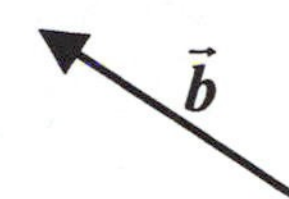

Given the two vectors $\vec{a}$ and $\vec{b}$ shown above, sketch to the right your prediction for their diffference $\vec{c} = \vec{a} - \vec{b}$. Be certain to label the vectors.

Demonstration 3:

Given the two vectors $\vec{v}_1$ and $\vec{v}_2$ shown above, sketch to the right your prediction for the vector $\Delta\vec{v}$ that must be added to $\vec{v}_1$ to produce $\vec{v}_2$. Note that $\vec{v}_1 + \Delta\vec{v} = \vec{v}_2$.

Demonstration 4:

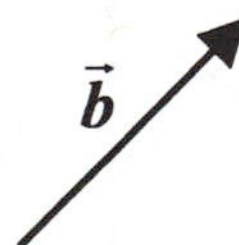

Given the two vectors $\vec{a}$ and $\vec{b}$ shown above, sketch to the right your prediction for the vector $\Delta\vec{b}$ that must be added to $\vec{b}$ to produce $\vec{a}$. Note that $\vec{b} + \Delta\vec{b} = \vec{a}$.

Demonstration 5:

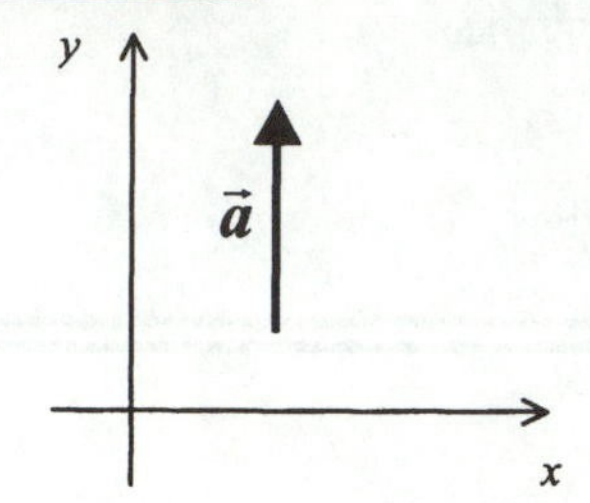

Vector $\vec{a}$ with the *x*-*y* axes shown only has a *y*-component. The *x*-component of $\vec{a}$ is zero. Draw another set of *x*-*y* axes near the vector $\vec{b}$ such that $\vec{b}$ only has an *x*-component in this new coordinate system.

Demonstration 6:

The vector $\vec{C}$ is shown on the right.
Show the *x*-component, C_x on the diagram.
Is the *x*-component positive or negative?

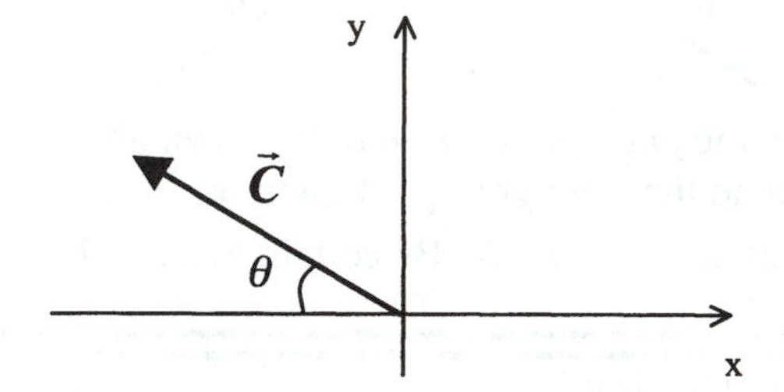

Show the *y*-component, C_y on the diagram.
Is the *y*-component positive or negative?

In terms of C and θ,
Write an expression for C_x:

Write an expression for C_y:

Demonstration 7:

The vector $\vec{D}$ is shown on the right.
Show the *x*-component, D_x on the diagram.
Is the *x*-component positive or negative?

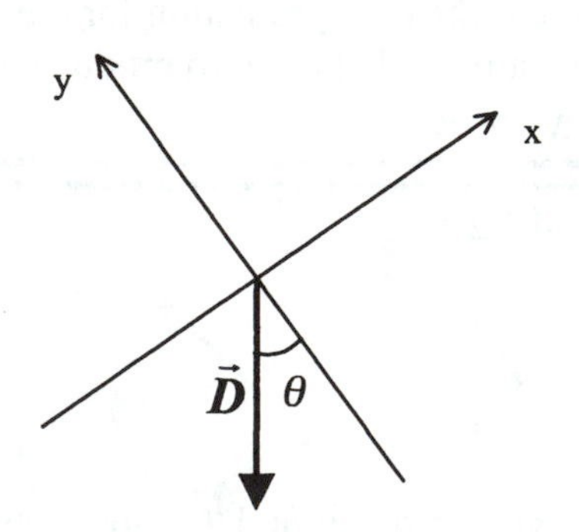

Show the *y*-component, D_y on the diagram.
Is the *y*-component positive or negative?

In terms of D and θ,
Write an expression for D_x:

Write an expression for D_y:

VECTORS(VECT)
TEACHER'S GUIDE

Prerequisites:

There are no prerequisites to these *ILDs*. It will be helpful if students have been introduced to vectors and vector algebra.

Equipment:

Visualizer software (See below.)

General Notes on Preparation and Equipment:

***Visualizer* software:**
Visualizer is a software package that provides a "vector playground" where vectors can be displayed and manipulated. It is ideal for the *Vectors ILDs*. *Visualizer* is available free, for Macintosh and Windows computers from the Center for Science and Mathematics Teaching at Tufts University (http://ase.tufts.edu/csmt/). Also available is a *Vector Tutorial*. The *Tutorial* may be used as a homework assignment to supplement the work in this *ILD* sequence.

You can also do the *Vectors ILDs* without computer support by following the eight-step procedure, and drawing the results on a transparency or on the board.

Demonstrations and Sample Results:

Demonstration 1: Vector addition. Use *Visualizer*, and draw the two vectors in the vector playground. Ask students to predict the sum of the two vectors.

Figure II-20 shows the two vectors and their sum in a window from *Visualizer*. The triangle rule for the addition of the two vectors is shown using "ghost vectors."

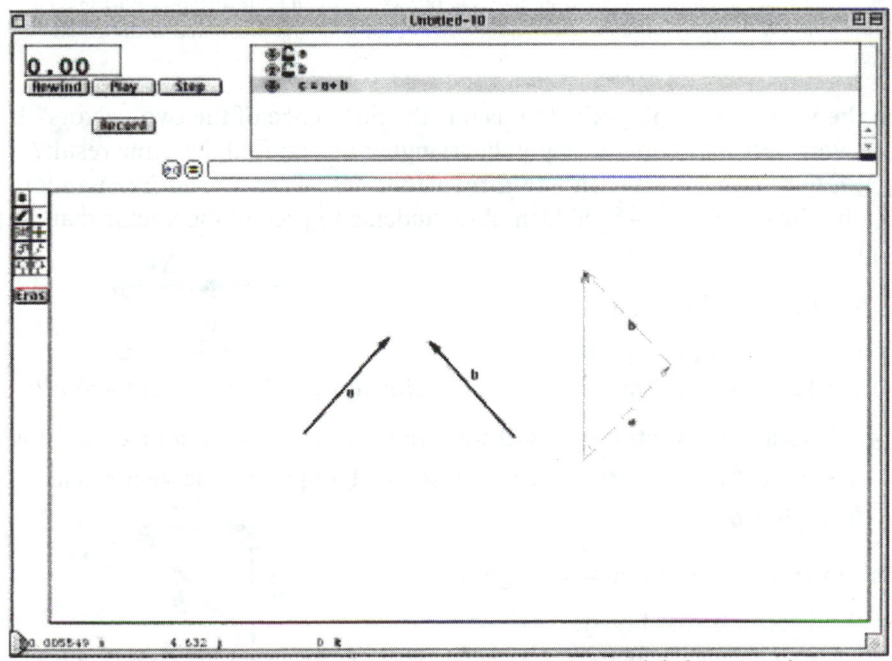

Figure II-20: Window from *Visualizer* showing two vectors and their sum, as in Demonstration 1. The triangle rule for vector addition is illustrated with "ghost vectors."

Discussion after the vector is displayed: How is this the sum of the two vectors? How does the triangle rule work? How would you apply the parallelogram rule and find the same result?

Demonstration 2: Vector subtraction. Use the same *Visualizer* window. Delete the vector sum. Ask students to predict the difference of the two vectors.

Figure II-21 shows the two vectors and their difference in a window from *Visualizer*. The parallelogram rule for the subtraction of the two vectors is shown using "ghost vectors."

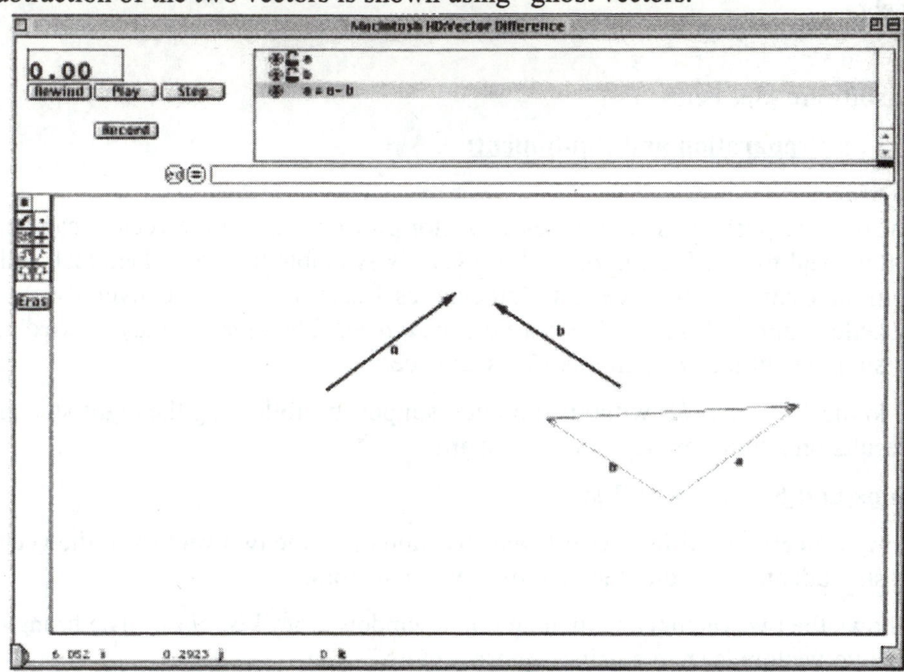

Figure II-21: Window from *Visualizer* showing two vectors and their difference, as in Demonstration 2. The parallelogram rule for vector subtraction is illustrated with "ghost vectors."

Discussion after the vector is displayed: How is this the difference of the two vectors? How does the parallelogram rule work? How would you apply the triangle rule and find the same result?

Demonstration 3: Change in a vector in one-dimension. Use a new *Visualizer* window. Display $\vec{v}_1$ and $\vec{v}_2$, but have $\Delta\vec{v} = \vec{v}_2 - \vec{v}_1$ hidden. Ask students to predict the vector that must be added to $\vec{v}_1$ to give $\vec{v}_2$.

$\vec{v}_1$ $\Delta\vec{v}$ $\vec{v}_2$

The vector $\Delta\vec{v}$ is illustrated on the right.

Discussion after the vector is displayed:
Ask students to describe why this is the correct vector. How do you find the vector $\Delta\vec{v}$?

Demonstration 4: Change in a vector in two-dimensions. Use a new *Visualizer* window. Display $\vec{a}$ and $\vec{b}$, but have $\Delta\vec{b} = \vec{a} - \vec{b}$ hidden. Ask students to predict the vector that must be added to $\vec{b}$ to give $\vec{a}$.

$\Delta\vec{b}$ $\vec{a}$ $\vec{b}$

The vector $\Delta\vec{b}$ is shown in the diagram on the right.

Discussion after the vector is displayed:
Ask students to describe why this is the correct vector. How do you find the vector $\Delta\vec{b}$?

Demonstration 5: Rotation of coordinate axes. Use a new *Visualizer* window. Display $\vec{b}$ with normal x and y axes. Ask students to draw the new axes that will make $\vec{b}$ have only an x-component. The correct rotation of the axes is shown on the right.

Discussion after the axes are displayed:
Students have difficulty with components. This discussion will help with the next two demonstrations. How are components defined? Why is there no y-component with this orientation of the axes? Is the x-component positive or negative?

<u>**Demonstration 6:**</u> **Vector components.** Use a new *Visualizer* window. Display $\vec{C}$ with the x and y axes. Ask students to show the x-component and y-component on the diagram, and state whether each is positive or negative. Figure II-22 shows how the components of $\vec{C}$ are displayed in *Visualizer*.

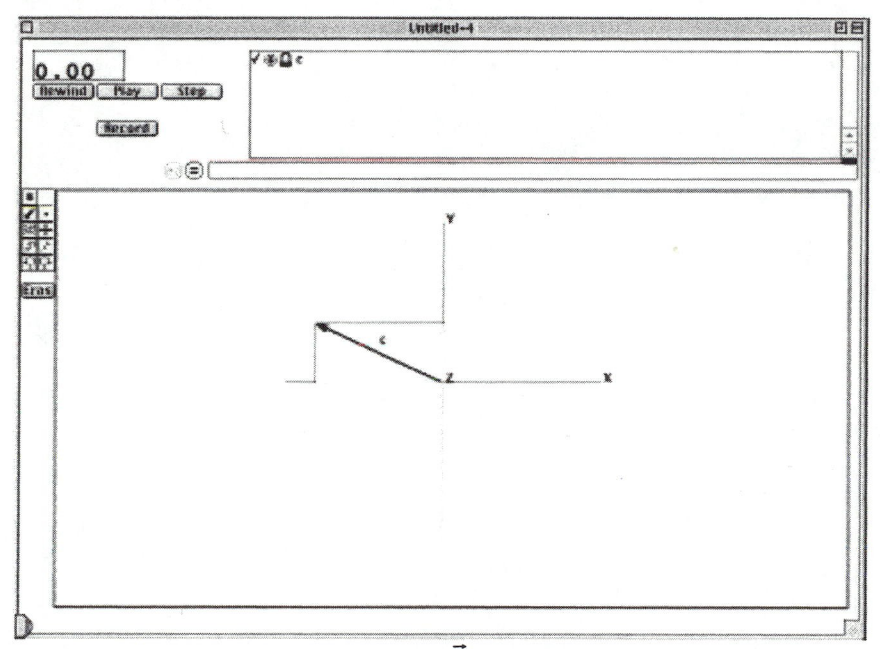

Figure II-22: Vector components of $\vec{C}$ displayed in *Visualizer*.

The components are $C_x = C\cos\theta$ and $C_y = C\sin\theta$. C_x is negative and C_y is positive.

Discussion after the components are displayed:
How are components defined? Where do the trigonometric functions come from? Why is C_x negative?

<u>**Demonstration 7:**</u> **Vector components.** Use a new *Visualizer* window. Display $\vec{D}$ with the x and y axes rotated as shown. Ask students to show the x-component and y-component on the diagram, and state whether each is positive or negative.

Discussion after the components are displayed: How are components defined? Where do the trigonometric functions come from?
The components are $D_x = D\cos\theta$ and $D_y = D\sin\theta$. Why are both D_x and D_y negative?

VECTORS (VECT)
TEACHER PRESENTATION NOTES

Demonstration 1: Vector addition. Draw the two vectors in the *Visualizer* vector playground.

- After showing the sum and the "ghost vectors," ask how the sum of two vectors is defined.
- How does the triangle rule work?
- How would you apply the parallelogram rule to find the same result?

Demonstration 2: Vector subtraction. Use a new *Visualizer* window, with the same two vectors.

- After showing the sum and the "ghost vectors." ask how the difference of two vectors is defined.
- How does the parallelogram rule work?
- How would you apply the triangle rule to find the same result?

Demonstration 3: Change in a vector in one dimension. Use a new *Visualizer* window, with the two vectors, $\vec{v}_1$ and $\vec{v}_2$ displayed, but their difference, $\Delta\vec{v}$, hidden.

- After showing the difference, ask students how you find $\Delta\vec{v}$.
- How do you know this is the right vector?

Demonstration 4: Change in a vector in two dimensions. Use a new *Visualizer* window, with the two vectors, $\vec{a}$ and $\vec{b}$ displayed, but their difference, $\Delta\vec{b}$, hidden.

- After showing the difference, ask students how you find $\Delta\vec{b}$.
- How do you know this is the right vector?

Demonstration 5: Rotation of coordinate axes. Use a new *Visualizer* window, with $\vec{b}$, displayed, and with normal axes (y vertical and x horizontal).

- After showing the rotated axes, ask students why there is no y-component?
- Is the x-component positive or negative?

Demonstration 6: Vector components. Use a new *Visualizer* window. Display $\vec{C}$ with the axes.

- How are components defined? Where do the trigonometric functions come from? The components are $C_x = C\cos\theta$ and $C_y = C\sin\theta$. C_x is negative and C_y is positive.
- Why is C_x negative?

Demonstration 7: Vector components. Use a new *Visualizer* window. Display $\vec{D}$ with the x and y axes rotated as shown.

- How are components defined? The components are $D_x = D\cos\theta$ and $D_y = D\sin\theta$.
- Why are both D_x and D_y negative?

PROJECTILE MOTION (PROJ)

Hand in this sheet **Name**______________________________

INTERACTIVE LECTURE DEMONSTRATION
PREDICTION SHEET—PROJECTILE MOTION

Directions: This sheet will be collected. Write your name at the top to record your presence and participation in these demonstrations. Follow your instructor's directions. You may write whatever you wish on the attached Results Sheet and take it with you.

Today's physical demonstration involves a ball thrown in the air with an initial velocity upward and to the right.

The *trajectory* of the ball looks approximately like the sketch to the right. Note that the origin of the coordinate system has been chosen to be the initial position of the ball.

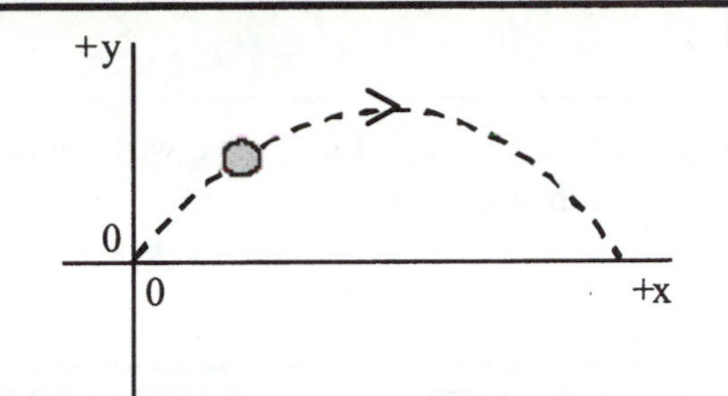

Demonstration 1: Sketch on the axes on the right your predictions for the *x* coordinate of the ball as a function of time and the y coordinate of the ball as a function of time.

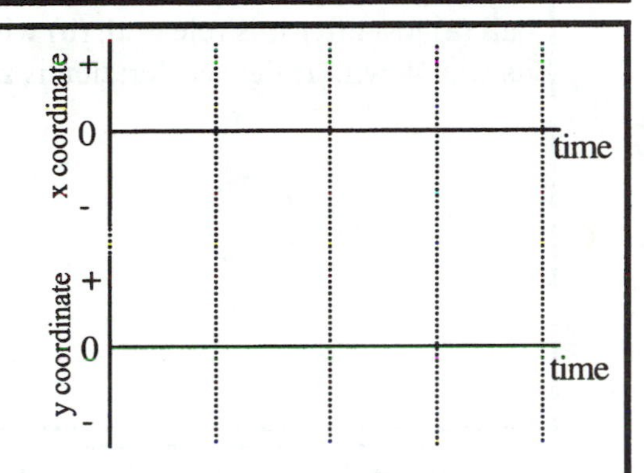

Based on your graph of *x* vs. *t*, write a kinematic equation for x as a function of time.

$x =$ ____________________

Based on your graph for *y* vs. *t*, write a kinematic equation for *y* as a function of time.

$y =$ ____________________

Question 1: When is the speed of the ball a maximum? A minimum?

Question 2: At the highest point in its motion, is the speed of the ball zero?

Question 3: When is the *x*-component of the velocity a maximum? A minimum?

Question 4: When is the *y*-component of the velocity a maximum? A minimum?

Demonstration 2: On the axes to the right, sketch your predictions for the *x*-component of the velocity as a function of time and the y-component of the velocity as a function of time.

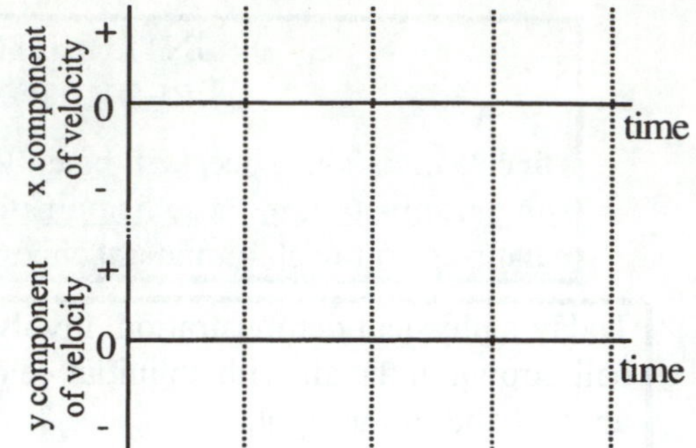

Based on your graph for v_x vs. t, write an equation for v_x as a function of time.

v_x = ______________________

Based on your graph for v_y vs. t, write an equation for v_y as a function of time.

v_y = ______________________

Question 5: In the space below, draw an arrow that represents the direction of the acceleration of the ball (a) just after it is released, (b) when it reaches the highest point in its trajectory, (c) while it is on its way down. If the acceleration is zero, write ZERO above the ball.

(a) 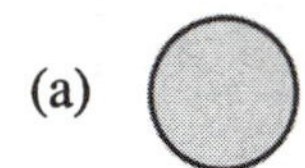 (b) 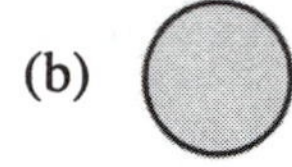 (c) 

Question 6: In the space below, draw the free-body (force) diagram for the ball (a) just after it is released, (b) when it reaches the highest point in its trajectory, (c) while it is on its way down. If there are no forces acting on the ball, write NONE above the ball. If the net force on the ball is zero, write ZERO above the ball.

(a) 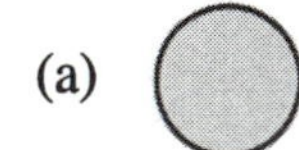 (b) 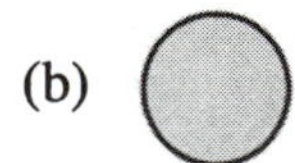 (c) 

Keep this sheet

INTERACTIVE LECTURE DEMONSTRATION
RESULTS SHEET—**PROJECTILE MOTION**

You may write whatever you wish on this sheet and take it with you.

Today's physical demonstration involves a ball thrown in the air with an initial velocity upward and to the right.

The *trajectory* of the ball looks approximately like the sketch to the right. Note that the origin of the coordinate system has been chosen to be the initial position of the ball.

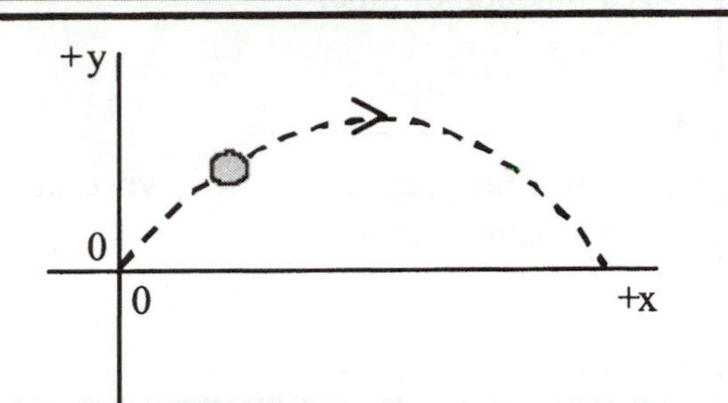

Demonstration 1: Sketch on the axes on the right your predictions for the x coordinate of the ball as a function of time and the y coordinate of the ball as a function of time.

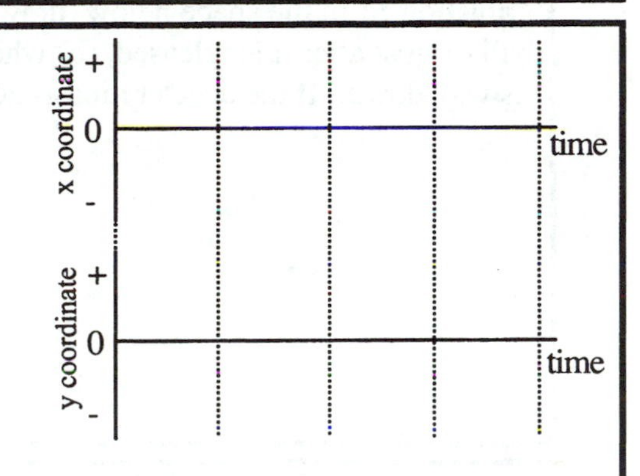

Based on your graph of x vs. t, write a kinematic equation for x as a function of time.

$x =$ ________________________

Based on your graph for y vs. t, write a kinematic equation for y as a function of time.

$y =$ ________________________

Question 1: When is the speed of the ball a maximum? A minimum?

Question 2: At the highest point in its motion, is the speed of the ball zero?

Question 3: When is the x-component of the velocity a maximum? A minimum?

Question 4: When is the y-component of the velocity a maximum? A minimum?

Demonstration 2: On the axes to the right, sketch your predictions for the *x*-component of the velocity as a function of time and the y-component of the velocity as a function of time.

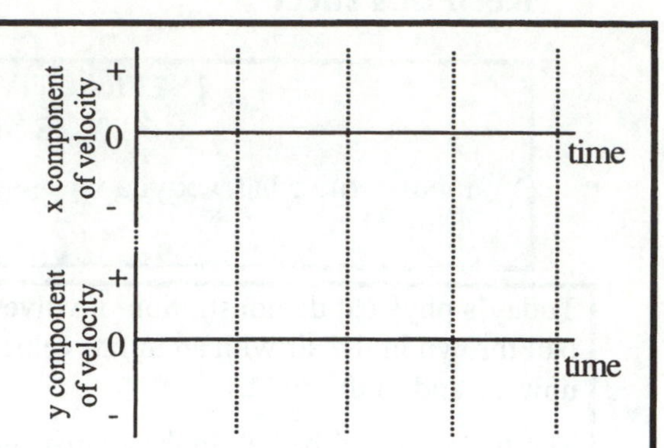

Based on your graph for v_x vs. t, write an equation for v_x as a function of time.

v_x = 

Based on your graph for v_y vs. t, write an equation for v_y as a function of time.

v_y = ______________________

Question 5: In the space below, draw an arrow that represents the direction of the acceleration of the ball (a) just after it is released, (b) when it reaches the highest point in its trajectory, (c) while it is on its way down. If the acceleration is zero, write ZERO above the ball.

(a) (b) (c)

Question 6: In the space below, draw the free-body (force) diagram for the ball (a) just after it is released, (b) when it reaches the highest point in its trajectory, (c) while it is on its way down. If there are no forces acting on the ball, write NONE above the ball. If the net force on the ball is zero, write ZERO above the ball.

(a) (b) (c)

PROJECTILE MOTION(PROJ)
TEACHER'S GUIDE

Prerequisites:

ILD sequences *Kinematics 1—Human Motion*, *Kinematics 2—Motion of Carts* and *Vectors* are prerequisite to these *ILDs*.

Equipment:

digital video camera

video collection software

tennis ball

dark background

in place of the above, you can use the *ILD* experiment configuration file PROJD1 (See below.)

video analysis software (See below.)

General Notes on Preparation and Equipment:

The projectile motion video:
If you have a digital video camera and video collection software, you can easily make a video of the motion of a thrown tennis ball and have it loaded directly into your computer for analysis. The details of video collection will not be discussed here. You can find lots of information at the *Workshop Physics* web-site at http://physics.dickinson.edu/~wp_web/WP_homepage.html.

If you don't have the required equipment, or simply don't want to collect a live video, you can use the *QuickTime* video in experiment configuration file PROJD1. We suggest that you do the live demonstration throwing the tennis ball in front of a dark background, and then show the video.

Video analysis software:
There are a number of video analysis software packages available. You will need one with basic features that allow you to collect position and time data of an object by clicking on its location in each frame of a *QuickTime* movie. A package that has the capability to calculate motion quantities such as velocity and acceleration, and graph them vs. time will be most useful. Logger Pro 3 from Vernier Software and Technology (www.vernier.com) includes such a video analysis package. *VideoPoint*, another available package has many additional advanced features for more advanced video analysis. You can find more information on *VideoPoint*, and on their video capture program, *VideoPoint Capture* from *Lenox Softworks* (www.lsw.com/videopoint). You can download a free demonstration copy of *VideoPoint*. (While the demonstration copy will not allow you to analyze the video in the experiment configuration file for this demonstration, there is a projectile motion movie included, Prjctiledemo.mov, that could be used instead.)

Demonstrations and Sample Graphs:

Demonstration 1: Graphs of the *x*-coordinate and *y*-coordinate of the motion. (Use experiment configuration file **PROJD1**.) Throw the ball and ask students to sketch their predictions on the axes.

Figure II-23 shows a typical frame from the movie with the trajectory displayed. Figure II-24 shows typical *x*-coordinate and *y*-coordinate vs. time graphs. The kinematic equations should be something like $x = x_o + v_{ox}t$ and $y = y_o + v_{ox}t - \frac{1}{2}gt^2$.

Discussion after the graphs are displayed: Show the students the graphs. Ask them to describe the motion represented by the x-coordinate-time graph. Does it represent constant velocity or constant acceleration? What is the net force on the ball in the horizontal (x) direction? What does this tell you about the acceleration component in the horizontal direction? Repeat these questions for the vertical (y) direction.

Ask students to describe the kinematic equations. Write them on the board. Ask students to define the quantities x_o, y_o, v_{ox}, v_{oy}, and g.

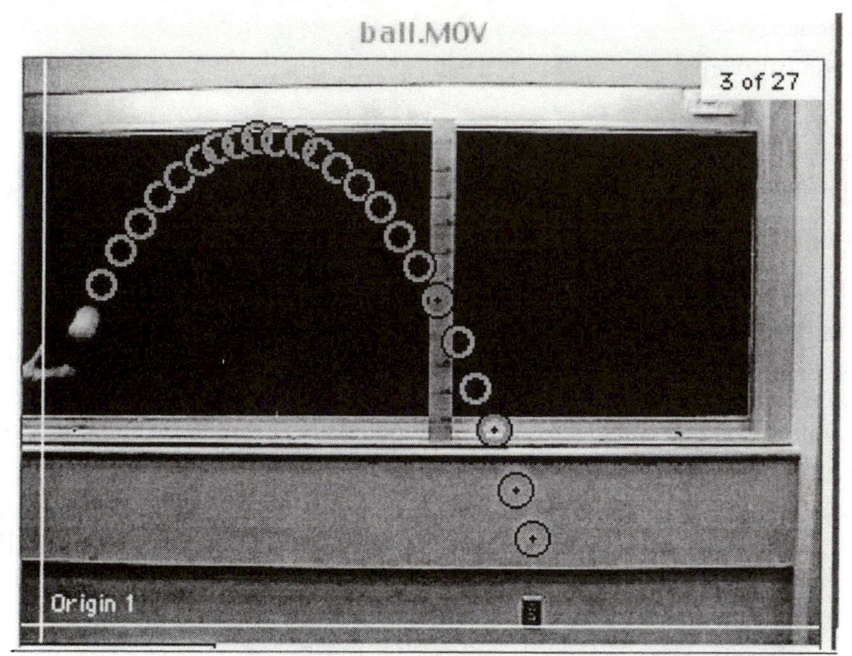

Figure II-23: Trajectory of a thrown ball superimposed on a frame from a *QuickTime* video of its motion, for the motion in Demonstration 1. Recorded using VideoPoint.

Question 1: The speed of the ball is maximum at the beginning position and ending position in the video. It is minimum at the highest point. How do you know this from the graphs?

Question 2: The speed at the highest point is not zero. The ball has zero vertical component of velocity there, but it still has the same horizontal component that it started with. How do you know this from the graphs?

Question 3: The x-component of the velocity always has the same value—it is constant since the x-component of the acceleration is zero. How do you know this from the graphs?

Question 4: The y-component of the velocity has its maximum magnitude at the beginning position (where the y-velocity component is positive) and ending position (where the y-velocity component is negative). How do you know this from the graphs?

Demonstration 2: Graphs of the x-component of velocity and y-component velocity. (Use the same experiment configuration file as in Demonstration 1.) Throw the ball again and ask students to sketch their predictions on the axes.

Figure II-24 shows a typical graphs for the x- and y-components of the velocity vs time. The kinematic equations should be something like $v_x = v_{ox} = \text{constant}$ and $v_y = v_{oy} - gt$.

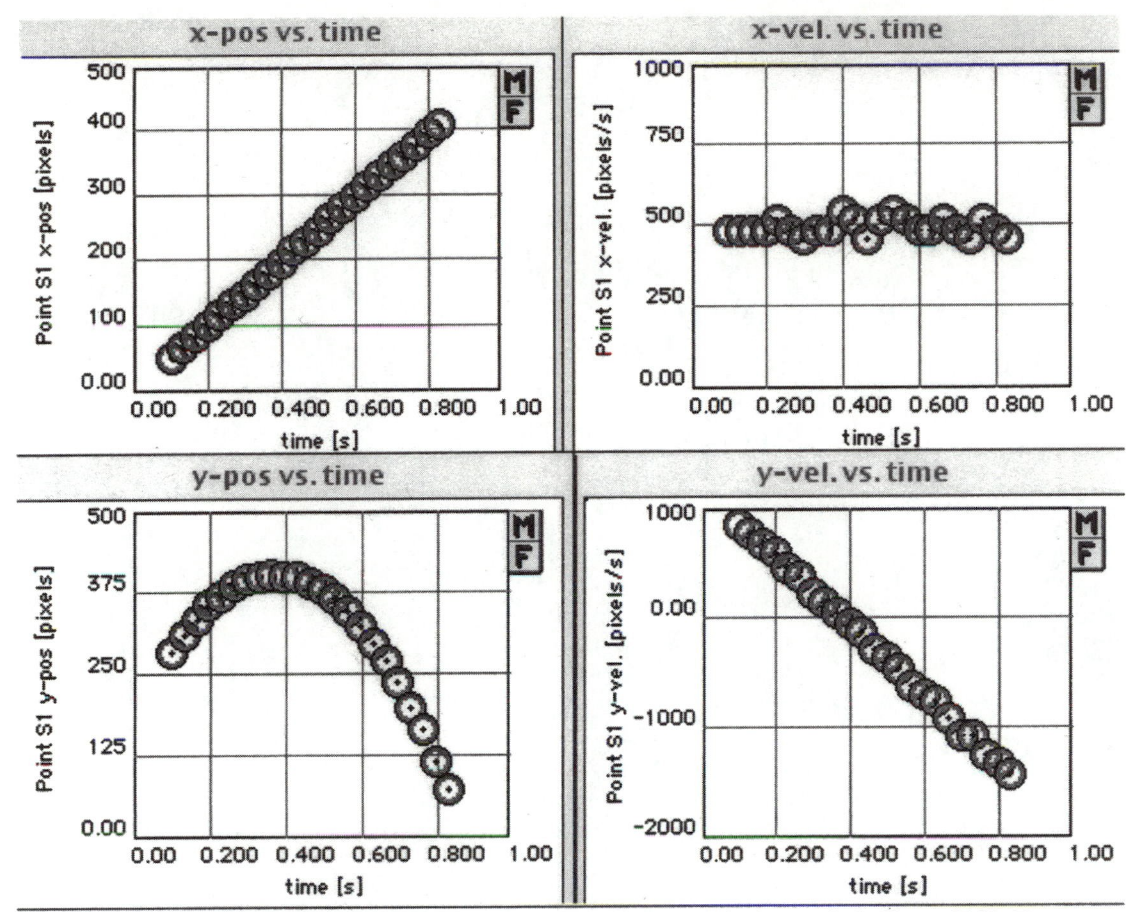

Figure II-24: Graphs of position and velocity vs. time for the ball in the movie in Figure II-23. Top:x components, Bottom: y components.

Discussion after the graphs are displayed: Show the students the graphs. Ask them to describe the motion represented by the *x*-velocity component-time graph. Does it represent constant velocity or constant acceleration? How can you tell from this graph that the *x*-component of the acceleration is zero? Repeat these questions for the vertical (*y*) component of the velocity. How could you find the *y*-component of the acceleration from this graph?

Ask students to describe the kinematic equations. Write them on the board. Ask students to explain how they represent the graphs and the motion.

Question 5: The acceleration vector points down and has the same magnitude in (a), (b) and (c). How can you tell this from the graphs? How can you tell this from the net force on the ball? Remind them of the earlier kinematics demonstrations.

Question 6: The force vector points down and has the same magnitude in (a), (b) and (c). How can you tell this from the answers to Question 5? Remind them of the earlier Newton's Law demonstrations.

Summary:

Because of the vector nature of position, velocity, acceleration and force, it is possible to examine the *x*-components and *y*-components independently. The graphs in Figure II-24 show these separate components and how they behave under the influence of the gravitational force on the ball.

PROJECTILE MOTION (PROJ)
Teacher Presentation Notes

Classroom introduction to *Projectile Motion ILDs*:
You will throw the ball, and camera will record the motion. The software will record and analyze the position of the ball in successive instants in time, and break this down into *x*- and *y*-components.

Demonstration 1: Graphs of the *x*-coordinate and *y*-coordinate of the motion. Use experiment configuration file **PROJD1**. Throw the ball and ask students to sketch their predictions on the axes.

- After showing students the graphs, ask them to describe motion represented by the *x*-coordinate-time graph. Does it represent constant velocity or constant acceleration? What is the net force on the ball in the horizontal (*x*) direction? What does this tell you about the acceleration component in the horizontal direction?
- Repeat these questions for the vertical (*y*) direction.
- Ask students to describe the kinematic equations. ($x = x_o + v_{ox}t$, $y = y_o + v_{ox}t - \frac{1}{2}gt^2$.) Write them on the board, and ask students to define x_o, y_o, v_{ox}, v_{oy}, and g.
- **Question 1:** Speed is maximum at the beginning and ending positions in the video. It is minimum at the highest point. How do you know this from the graphs?
- **Question 2:** Sspeed at the highest point is not zero. How do you know this from the graphs?
- **Question 3:** The *x*-component always has the same value—it is constant. How do you know this from the graphs?
- **Question 4:** The *y*-component of the velocity has its maximum magnitude at the beginning and ending positions. How do you know this from the graphs?

Demonstration 2: Graphs of the *x*-component of velocity and *y*-component velocity. (Use the same experiment configuration file as in Demonstration 1.) Throw the ball again and ask students to sketch their predictions on the axes.

- After showing students the graphs, ask them to describe motion represented by the *x*-velocity component-time graph. Does it represent constant velocity or constant acceleration? How can you tell from this graph that the *x*-component of the acceleration is zero?
- Repeat these questions for the vertical (*y*) component of the velocity. How could you find the *y*-component of the acceleration from this graph?
- Ask students to describe the kinematic equations. ($v_x = v_{ox} = \text{constant}$, $v_y = v_{oy} - gt$.) Write them on the board, and ask how they represent the graphs and the motion.
- **Question 5:** The acceleration vector points down and has the same magnitude in (a), (b) and (c). How can you tell from the graphs? How can you tell from the net force on the ball? Remind them of the earlier kinematics demonstrations.
- **Question 6:** The force vector points down and has the same magnitude in (a), (b) and (c). How can you tell this from Question 5? Remind them of earlier Newton's Law demonstrations.

Summary:
Position, velocity, acceleration and force are vectors. The graphs show the independent *x*- and *y*-components of these quantities, and how they behave under the influence of the gravitational force.

ENERGY OF A CART ON A RAMP (ENER)

Hand in this sheet **Name**_____________________________________

INTERACTIVE LECTURE DEMONSTRATION
PREDICTION SHEET—ENERGY OF A CART ON A RAMP

Directions: This sheet will be collected. Write your name at the top to record your presence and participation in these demonstrations. Follow your instructor's directions. You may write whatever you wish on the attached Results Sheet and take it with you.

Demonstration 1: Consider a cart (with almost no friction) given a quick push up an inclined ramp away from the motion detector which is chosen to be the origin. Sketch on the right your prediction of the position-time, velocity-time and acceleration-time graphs for the cart. It moves up, slows down, then rolls back down the ramp where it is stopped at the same place it was pushed. Include the push and the catch on your graphs.

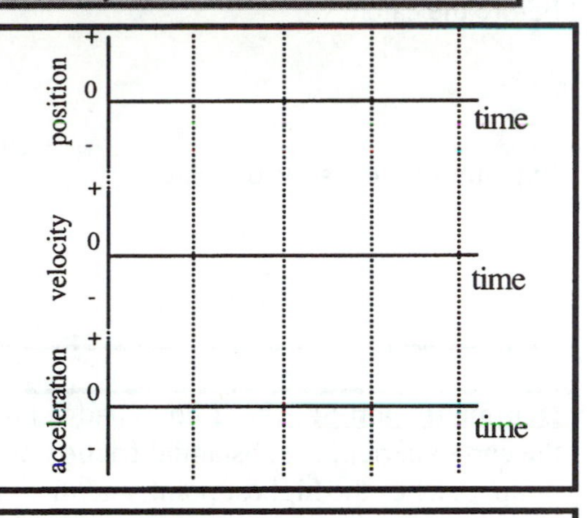

Demonstration 2: Sketch on the right your prediction of the kinetic energy of the cart (the energy due to motion) over time as it moves as described above. Keep in mind that the kinetic energy $K = \frac{1}{2}mv^2$, where m is the mass of the cart and v is its velocity.

Where is the kinetic energy zero?

Where is it a maximum?

Is kinetic energy conserved?

Demonstration 3: Sketch on the right your prediction of the potential energy of the cart (the energy due to raising a mass in the gravitational field of the earth) over time as it moves as described above. Define the potential energy to be zero at the height where the cart is first pushed. Then $U^{grav}=mgh$ where h is the height above the starting point, g is the acceleration due to gravity, and m is the cart's mass. h can be calculated from x--the cart's distance from the motion detector using $h=(x-x_o)\sin ø$ where x_o is the initial position and $ø$ is the angle the track makes with the horizontal.

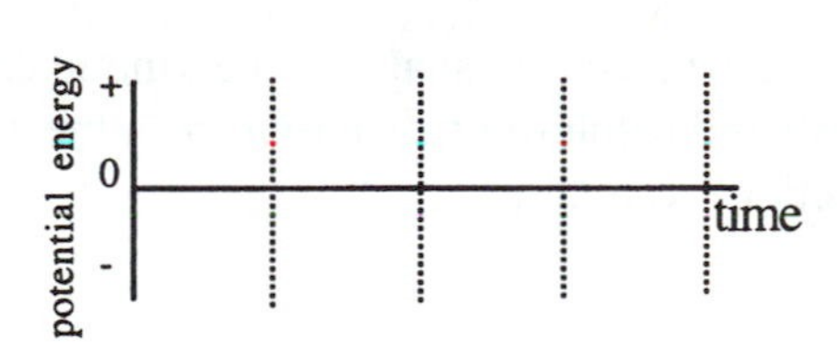

Where is the potential energy zero?

Where is it a maximum?

Is the potential energy conserved?

Demonstration 4: Sketch on the right your prediction of the mechanical energy (the sum of the kinetic and potential energies) of the cart over time as it moves as described above.

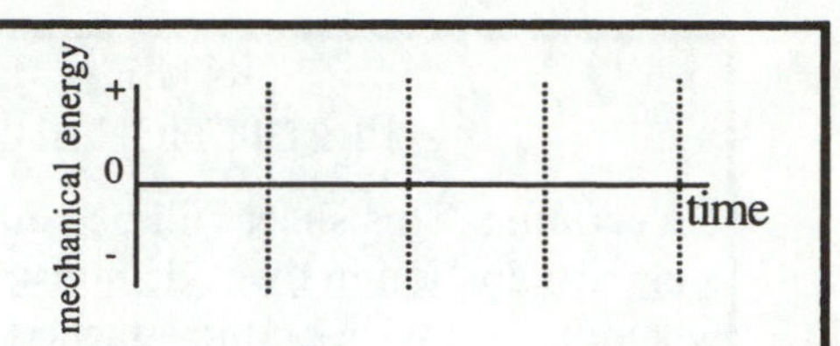

Describe the mechanical energy of the cart after the push and before the catch.

Is the mechanical energy conserved?

Explain what conserved means.

Where does the cart get its initial energy?

Demonstration 5: The friction pad is now lowered so that the cart experiences substantial friction as it moves up the ramp, reaches the highest point, and then moves back down. Sketch your predictions on the right for the velocity-time and acceleration-time graphs of the cart with friction. Also predict the shapes of the kinetic energy, potential energy, and mechanical energy (sum of U and K) for the motion. Include the push and the catch for all quantities. *Remember that friction can no longer be ignored.*

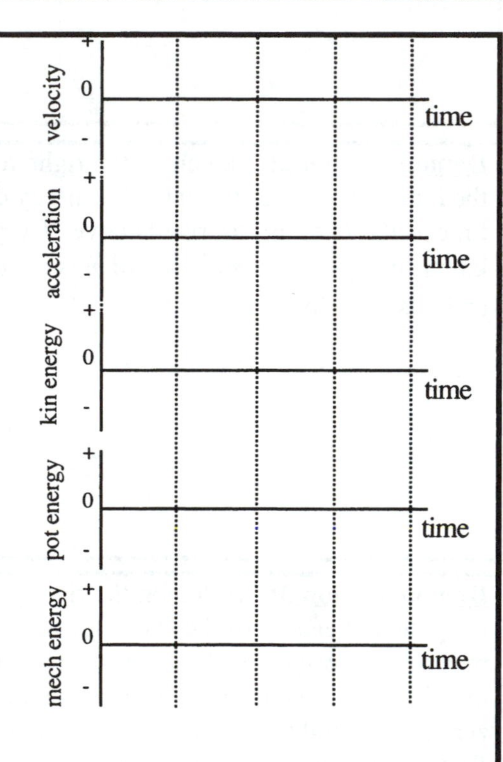

Is the mechanical energy conserved? Explain.

Do you expect the shape of the kinetic energy graph for the push and catch to be different from the case with no friction? Explain.

Keep this sheet

INTERACTIVE LECTURE DEMONSTRATION
RESULTS SHEET—**ENERGY OF A CART ON A RAMP**

You may write whatever you wish on this sheet and take it with you.

Demonstration 1: Consider a cart (with almost no friction) given a quick push up an inclined ramp away from the motion detector which is chosen to be the origin. Sketch on the right your prediction of the position-time, velocity-time and acceleration-time graphs for the cart. It moves up, slows down, then rolls back down the ramp where it is stopped at the same place it was pushed. Include the push and the catch on your graphs.

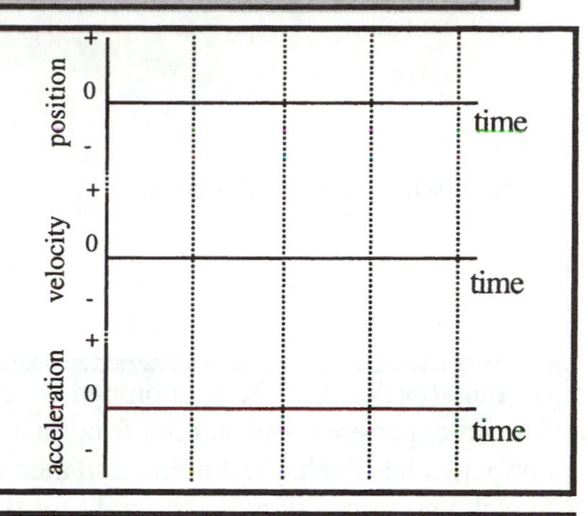

Demonstration 2: Sketch on the right your prediction of the kinetic energy of the cart (the energy due to motion) over time as it moves as described above. Keep in mind that the kinetic energy $K = \frac{1}{2}mv^2$, where m is the mass of the cart and v is its velocity.

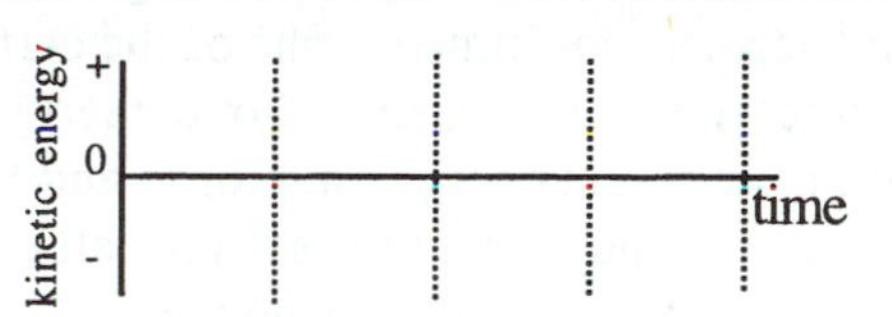

Where is the kinetic energy zero?

Where is it a maximum?

Is kinetic energy conserved?

Demonstration 3: Sketch on the right your prediction of the potential energy of the cart (the energy due to raising a mass in the gravitational field of the earth) over time as it moves as described above. Define the potential energy to be zero at the height where the cart is first pushed. Then $U^{grav}=mgh$ where h is the height above the starting point, g is the acceleration due to gravity, and m is the cart's mass. h can be calculated from x--the cart's distance from the motion detector using $h=(x-x_o)\sin\phi$ where x_o is the initial position and ϕ is the angle the track makes with the horizontal.

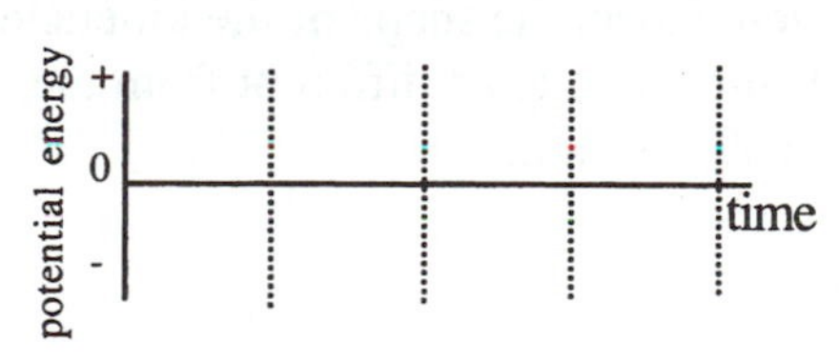

Where is the potential energy zero?

Where is it a maximum?

Is the potential energy conserved?

Demonstration 4: Sketch on the right your prediction of the mechanical energy (the sum of the kinetic and potential energies) of the cart over time as it moves as described above.

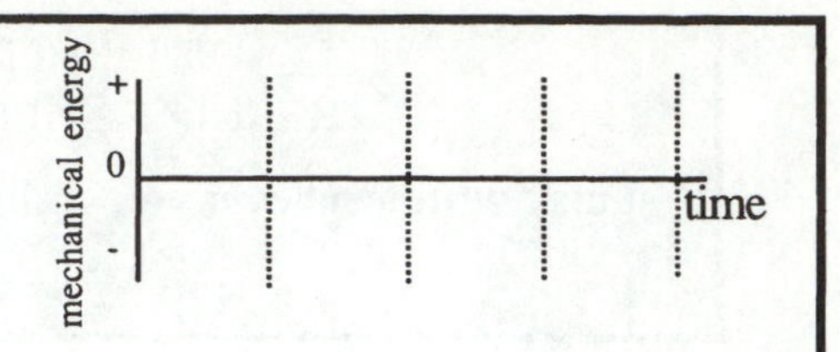

Describe the mechanical energy of the cart after the push and before the catch.

Is the mechanical energy conserved?

Explain what conserved means.

Where does the cart get its initial energy?

Demonstration 5: The friction pad is now lowered so that the cart experiences substantial friction as it moves up the ramp, reaches the highest point, and then moves back down. Sketch your predictions on the right for the velocity-time and acceleration-time graphs of the cart with friction. Also predict the shapes of the kinetic energy, potential energy, and mechanical energy (sum of U and K) for the motion. Include the push and the catch for all quantities. *Remember that friction can no longer be ignored.*

Is the mechanical energy conserved? Explain.

Do you expect the shape of the kinetic energy graph for the push and catch to be different from the case with no friction? Explain.

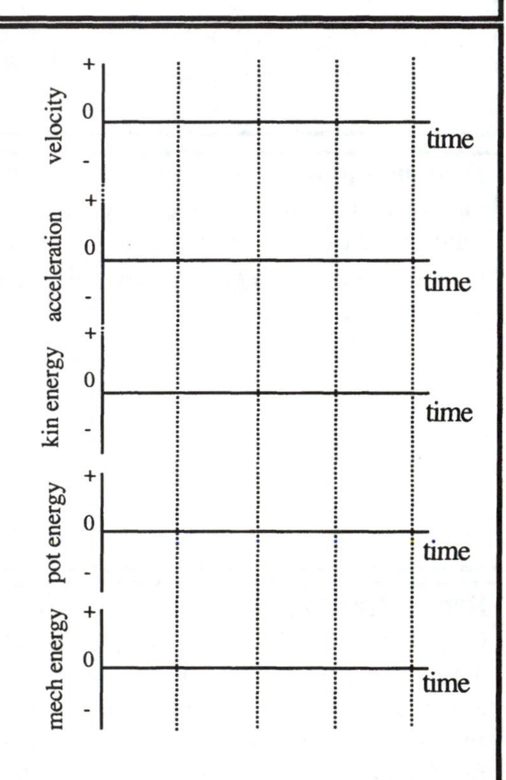

ENERGY OF A CART ON A RAMP (ENER)
TEACHER'S GUIDE

Prerequisites: The four *Kinematics* and *Newton's Laws ILD* sequences are prerequisites. If students have done *RealTime Physics Mechanics* Labs1-6 or *Tools for Scientific Thinking Force and Motion* Labs 1-4, they should also be prepared for this sequence.

Equipment:

computer-based laboratory system

ILD experiment configuration files

motion detector

two low-friction kinematics carts are better (If you only have one, it should have an adjustable friction pad mounted on it.) (See below.)

mass bar or other mass to make cart mass 1 kg

2.2 meter aluminum track, long door threshold or a very smooth, level table or ramp (See below.)

General Notes on Preparation and Equipment:

Carts, mass bar and track:
See the Teacher's Guide for the *Motion of Carts ILDs* for detailed information about ramps and carts including the friction pad. You will need an adjustable friction pad on one cart. You can get by with only one cart with a friction pad, but it is highly preferable to adjust the pad appropriately before class rather than trying to do so in class. Any cart with very low friction will work but truly low friction carts are difficult to find. Likewise, any smooth, level ramp will work.

Experimental setup:
We elevate a PASCO 2.2 m track on a rolling table so it can be easily seen and check to see that the track is level when the table is in the demonstration position. Then we slope the track approximately 5.5 cm (thickness of a convenient block) over 122 cm (the distance between the feet of the track in this case). You may vary the slope somewhat but don't make the slope too steep or the motion happens too quickly.

Take some of the geometrical measurements ahead of time for convenience. Mark the starting point and the height from the starting point at a known distance along the track for use in potential energy calculation. We use cards taped to the appropriate points. Be sure to enter these quantities for calculating potential energy into the setup files.

As you practice the following *ILDs*, remember that for many motion detectors, the cart must never be closer than 0.5 m from the motion detector.

Demonstrations and Sample graphs:

Demonstration 1: Cart with almost no friction is given a quick push up the inclined track. It moves up, reaches its highest point, rolls down and is caught at the same point it was released. Predictions include the push and the catch. (Use experiment configuration file **ENERD1**.) After the data are collected, save them so that the graphs are persistently displayed for Demonstrations 2-5. Figure II-25 shows typical graphs.

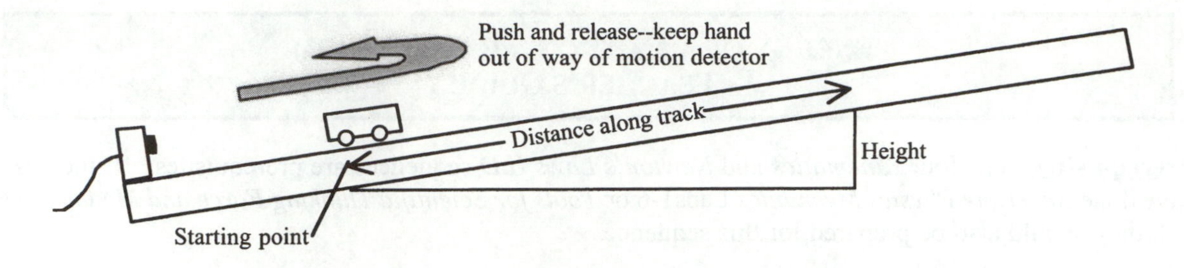

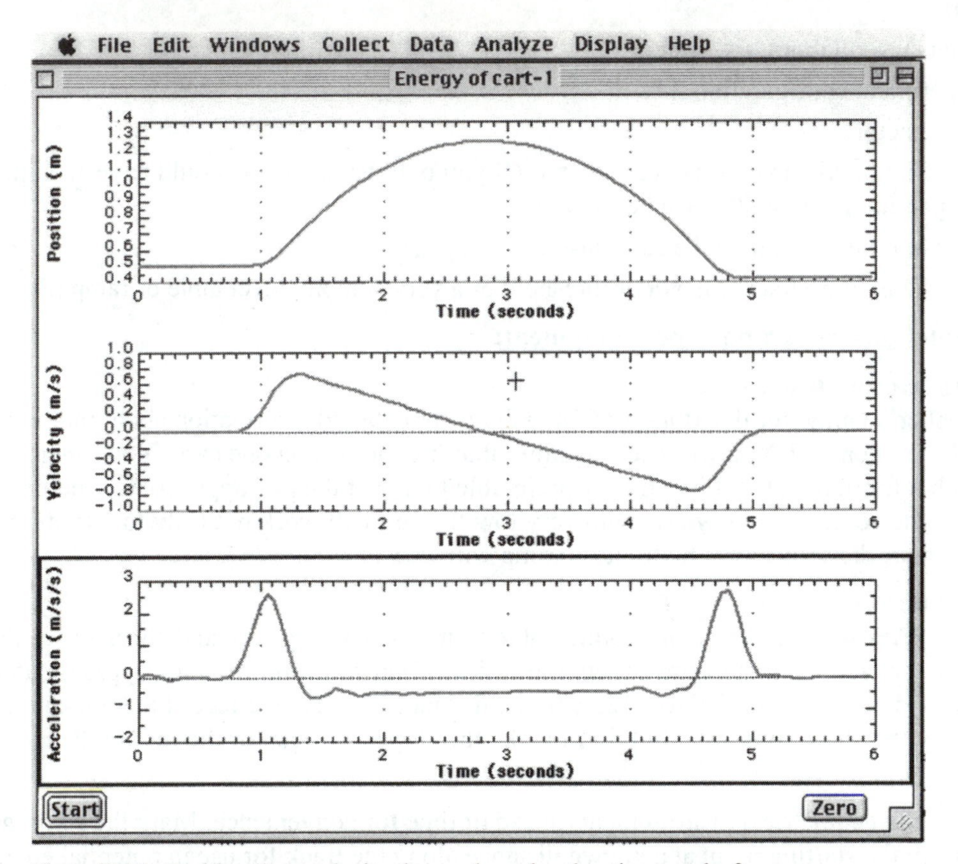

Figure II-25: Position-time, velocity-time and acceleration-time graphs for a low-friction cart moving up an inclined ramp, away from the motion detector, and then back down, as in Demonstration 1.

Discussion after the graphs are displayed: What do position-time, velocity-time, and acceleration-time graphs look like including the push and the catch? Remind the students that they observed this same motion in the *Kinematics—Motion of Carts ILDs.* Direct attention to the point where the cart reverses direction. Acceleration is not zero at the top. Speed is about equal right after the push and before the catch. (This is review except for the push and the catch.)

Note: You do not need to keep taking new data. You can select the appropriate window from experiment configuration file ENERD1 for each demonstration and use one set of data. You should however always send the cart up and down (without taking data) before the student's predictions.

<u>Demonstration 2:</u> Predict kinetic energy for the same motion. (Use the same experiment configuration file as in Demonstration 1.) Select the second window with the velocity showing and the <u>kinetic energy hidden</u>. Remind students that $K = \frac{1}{2}mv^2$, where m is the mass of the cart (in this case 1 kg) and v is the velocity. Roll cart up and down inclined track without taking data before students make their predictions. Figure II-26 shows typical graphs.

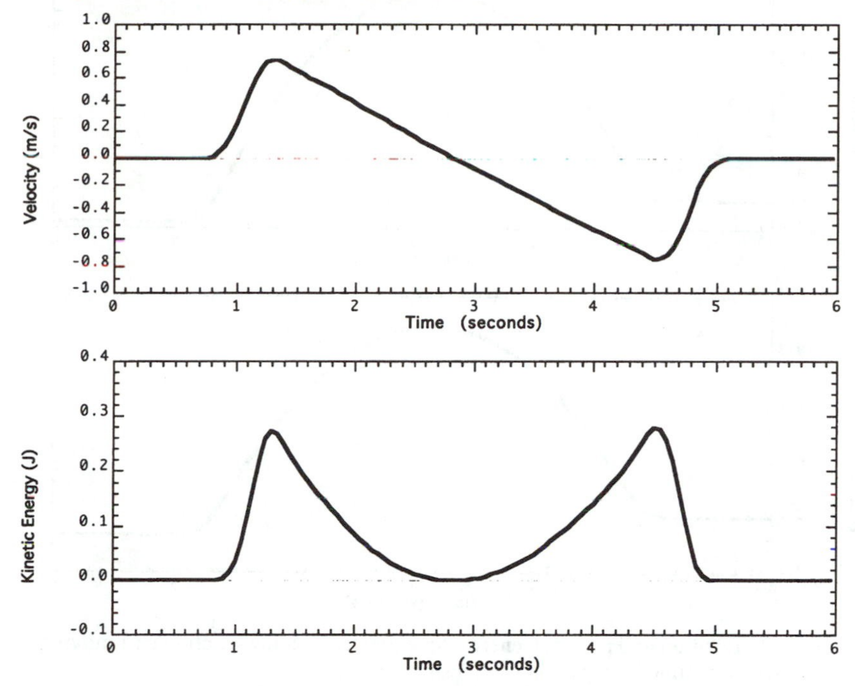

Figure II-26: Velocity-time and kinetic energy-time graphs for a low-friction cart moving up and down an inclined ramp as in Demonstration 1.

Discussion after the kinetic energy graph is displayed: Show students the equation in the data column used to calculate kinetic energy, e.g. (0.5*1*"vel"^2). Many students will think kinetic energy is conserved even though the value changes. Two reasons they think so: 1) they think all energy is conserved, 2) and kinetic comes back to the same value. (It's not an unreasonable idea to think that this is "conservation"). Make sure to point out that it isn't conserved even after the push and before the catch, since conservation means remaining constant, and the kinetic energy isn't a constant.

You will find that some number of students don't realize that kinetic energy, which is a function of velocity squared, must always be positive, and they show kinetic energy as negative on their graphs when velocity is. This is a good time to point out that the square of a negative number is positive and that kinetic energy is always positive.

<u>Demonstration 3:</u> Predict potential energy for same motion. (Use the same experiment configuration file as in Demonstration 1.)

Select the third window with the position showing and the <u>potential energy hidden</u>. Remind students that $U^{grav} = mgh$, where m is the mass of the cart (in this case 1 kg) and h is the height above its starting position. (The starting position is defined to have zero potential energy.) Show result of releasing the cart from different heights. Then roll the cart up and down the inclined track without taking data before students make their predictions. Figure II-27 shows typical graphs.

Discussion after the potential energy graph is displayed: Show students the equation in the data column used to calculate potential energy, e.g. (1*9.8*("Dist"-Dist0)*sin(ø)). Note that the sine is the rise divided by the distance along the track between any two points. Using the positions of the support legs may help students understand. (Take measurements ahead of time and use large labels.)

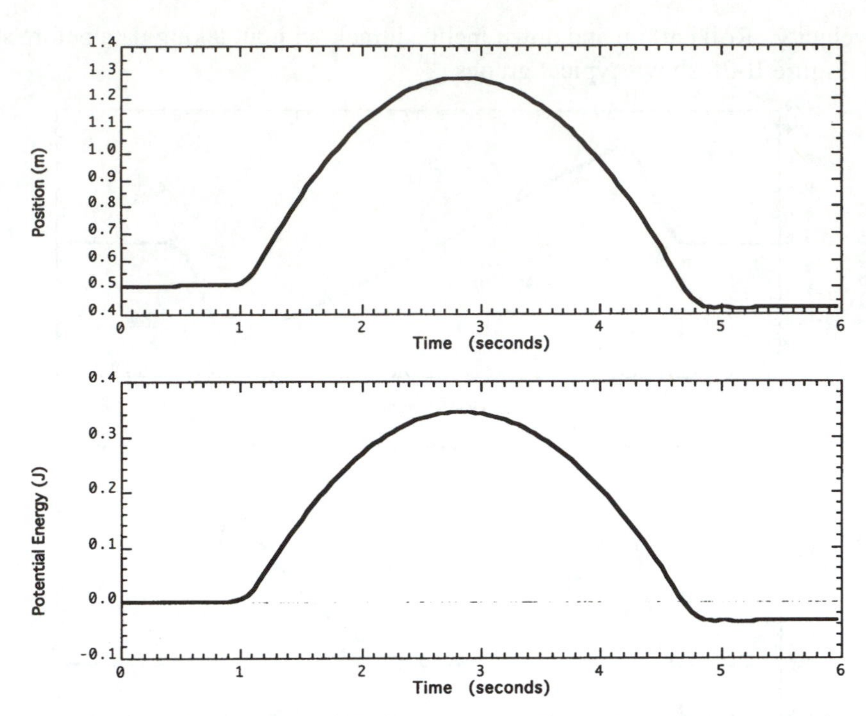

Figure II-27: Position and potential energy-time graphs for a low-friction cart moving up and down an inclined ramp as in Demonstration 1.

Even though students have seen the position vs. time graph, they may have trouble predicting the shape of the potential energy graph. The fact that this is difficult to predict does not seem to harm the understanding. The values at the endpoints (zero) and the middle (maximum) are most important.

When you demonstrate the cart rolling down from different heights, just have them notice that it moves increasingly faster at the end as you release from further up the track. This may motivate the idea of "potential energy."

Notice in Figure II-27 the potential energy goes negative since the cart was caught slightly below the defined zero point. If you have time and you catch the cart this way, you can explain that a potential energy below zero means that you must do work on the cart to return it to the zero point. If you let it go it will only go more negative.

Again, students may think that potential energy is conserved because it begins and ends at the same value. Remind them of the definition of conservation, and point out that the potential energy is <u>not constant</u> during the motion.

Note: Potential energy vs. time may be unfamiliar to students since it is rarely shown in textbooks. However, looking at the three energies as functions of time and talking about "when" various energies are conserved or not seems to make a lot of sense to the students. Giving them the formula for potential energy as a function of position and showing position vs. time during the prediction period is definitely helpful. If you have more than 40 to 50 minutes to spend on this *ILD* sequence, you could explore additional ideas. You can easily plot potential energy as a function of position by replacing time on the horizontal axis with position. The result is shown in Figure II-28. As expected the potential energy is a linear function of position. You could also look at kinetic energy and, later, total mechanical energy as functions of position as well, but you will need to help the students interpret the graphs.

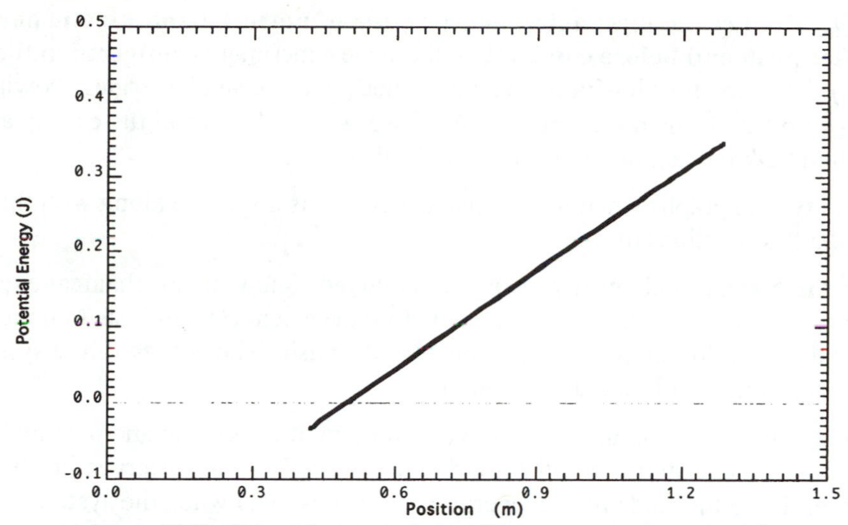

Figure II-28: Potential energy of a low-friction cart moving up and down an inclined track as in Demonstration 1, graphed as a function of position (instead of time) from the defined zero (starting point). These are the same data as shown in Figures II-27.

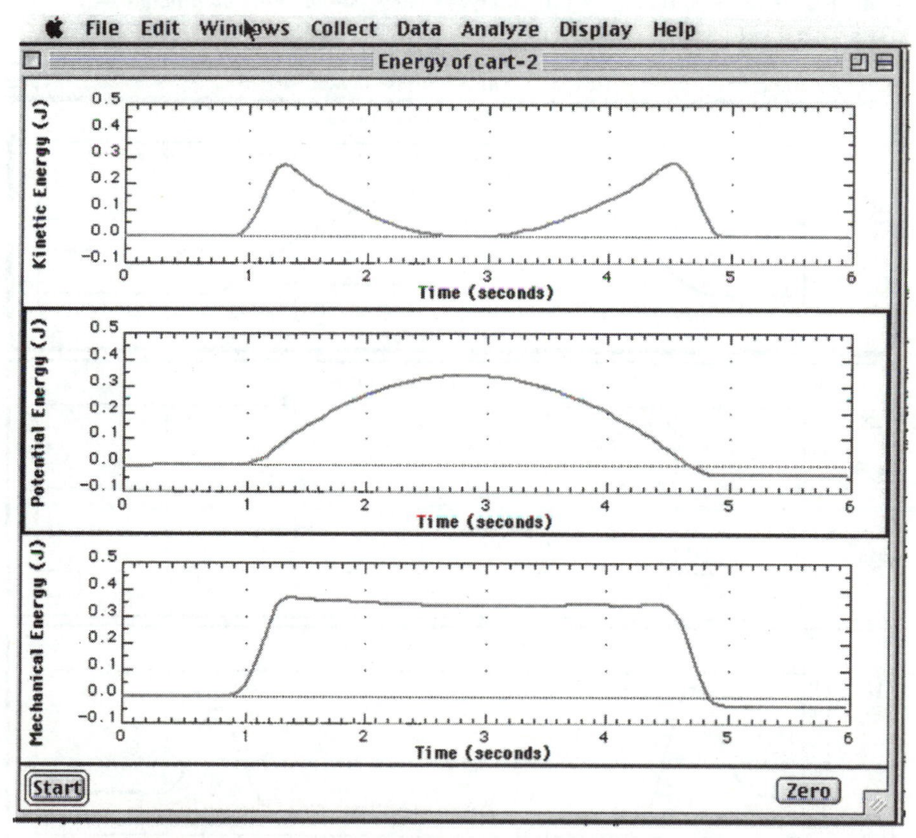

Figure II-29: Kinetic, potential, and total mechanical energy vs. time for a low-friction cart moving up and down an inclined ramp as in Demonstration 1.

<u>Demonstration 4:</u> Predict the mechanical energy (sum of potential and kinetic energy) for the same motion (after push and before catch). (Use the same experiment configuration file as in Demonstration 1.) Select the fourth window with the kinetic and potential energies showing, and the <u>mechanical energy hidden</u>. Remind students that $E^{mech} = K + U^{grav}$. Then roll the cart up and down the inclined track without taking data before students make their predictions.

Figure II-29 shows typical graphs when the mechanical energy is displayed along with the kinetic and potential energies, all as functions of time.

Discussion after the mechanical energy graph is displayed: Show the mechanical energy. Total Mechanical Energy is conserved after the push and before the catch. Get students to notice that energy is not conserved in the system if you look before and after the push. The energy of the system comes from your push (e.g,, from your breakfast or lunch).

We have finally come to something that is conserved--the sum of the kinetic and potential energies. As mentioned in the notes, many students will think that K or even U^{grav} is conserved just because they are energies. Note carefully to the students that energy is only conserved when the system is essentially isolated, that is, when you are not touching the cart. (It is slightly difficult to consider a system that includes the "earth" as isolated.) The fact the mechanical energy starts at zero when you initially hold the cart and ends at zero when you catch it, is not why energy is considered conserved.

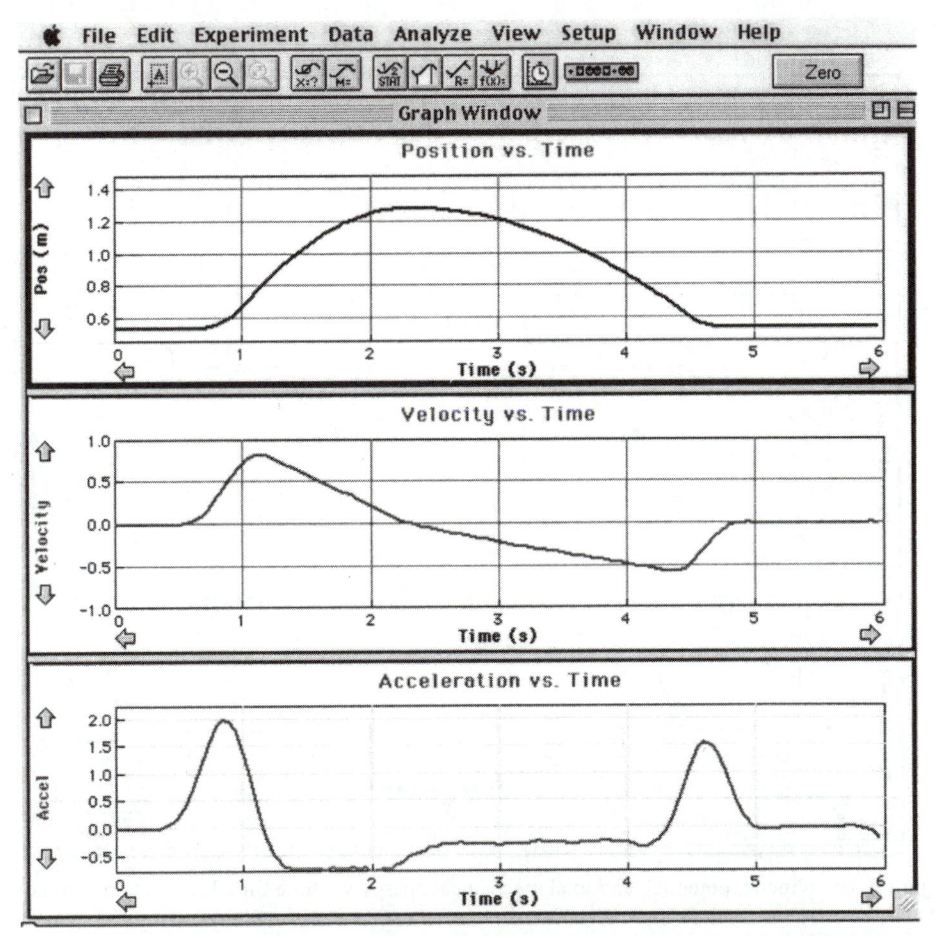

Figure II-30: Position-time, velocity-time and acceleration-time graphs for a high-friction cart moving up and down an inclined track as in Demonstration 5.

Students may have detected a slight decrease in mechanical energy due to friction, and you can address the effect of friction in the next demonstration.

<u>Demonstration 5:</u> We now lower the friction pad so that the cart experiences substantial friction as it moves up the ramp, reaches the highest point and then moves back down the ramp. ((Use the same experiment configuration file as in Demonstration 1, and keep Window 4 open on the screen.) Students should be told to include the push and the catch in their predictions for all quantities, and to remember that friction can no longer be ignored.

Figure II-30 shows typical position-time, velocity-time and acceleration-time graphs, while Figure II-31 shows graphs of kinetic energy, potential energy and mechanical energy vs. time.

Discussion after the graphs are displayed: First display just the position-time, velocity-time and acceleration-time graphs. Remind students that the friction force always acts in a direction opposed to the motion. On the way up, the friction force and the component of the gravitational force along the ramp act in the same direction (down the ramp or negative) producing a large acceleration as shown in Figure II-30. After the cart starts back down the ramp, friction is in the positive direction (up the ramp) while the gravitational force component is still down the ramp. Thus the net force is small and the acceleration decreases almost to zero. (Since it is so small, slight variations in friction due to grease or dirt on the track show up and the acceleration can be a little rough.)

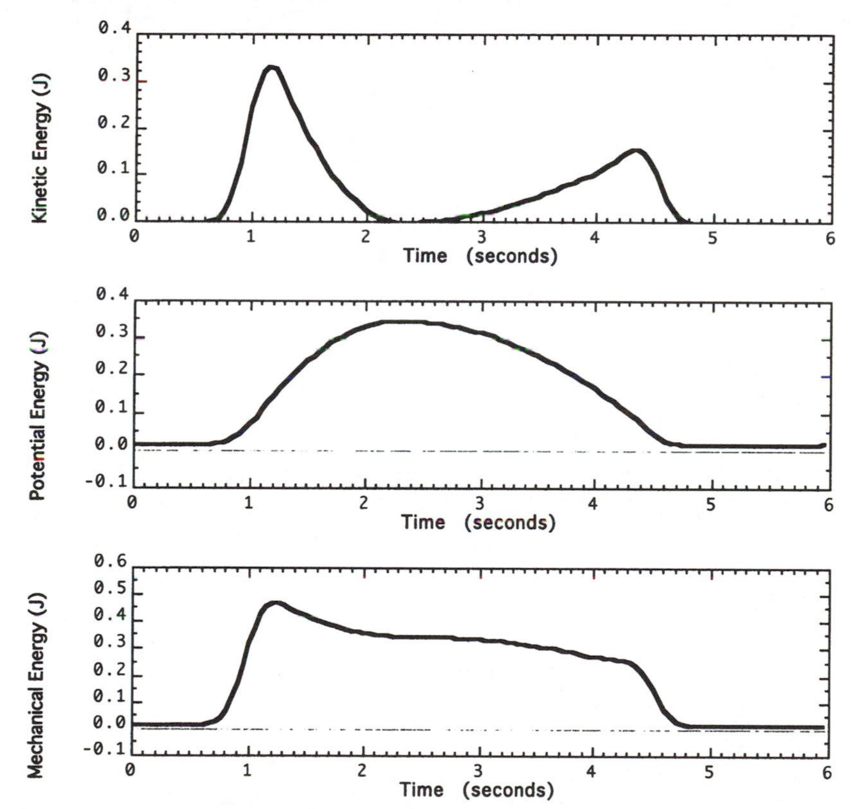

Figure II-31: Kinetic, potential, and total mechanical energy vs. time for a high-friction cart moving up and down an inclined track as in Demonstration 5.

Ask students if mechanical energy is conserved even after the push and before the catch. As you can see in Figure II-31, mechanical energy is not conserved when there is significant friction. What happened to the energy of the push?

If you have time, you can discuss that the rate of change of mechanical energy is proportional to the velocity. Little energy is lost per unit time near the top. Most is lost at beginning. Display as function of position to show that mechanical energy is nearly linear with position.

Show that the final speed is lower than the initial speed and acceleration to stop the cart is less than that to start it up the ramp. The time to come down is much greater than time to go up. There are two separate parabolas since the net force going up is so different from that coming down.

Energy of a Cart on a Ramp (ENER)
Teacher Presentation Notes

Classroom introduction to the *Energy of a cart on a Ramp ILDs*:
Explain the setup (students should be familiar with it by now). Remind them that the positive direction is away from the motion detector. Explain that you will use a low friction cart in Demonstrations 1-4, and the effects of friction can be ignored for these demonstrations. You may wish to talk about what energy conservation means (or postpone it until Demonstration 2).

Demonstration 1: Cart with almost no friction is given a quick push up the inclined track. It moves up, reaches its highest point, rolls down and is caught at the same point it was released. Use the first window in experiment configuration file **ENERD1**. You may need to bring it to the front. Predictions include the push and the catch.

- Remind students that they observed this motion before in the *Motion of Carts ILDs*.
- Direct student attention to the time when cart reverses direction. Acceleration not zero there.
- Starting and ending points are the same, so speed is equal right after push and before catch.
- Save data so that the graphs are persistently displayed on the screen for Demonstrations 2-5.

Note: While you do not need to keep taking new data, you should always send the cart up and down (without taking data) before the individual predictions.

Demonstration 2: Predict kinetic energy for same motion. Use the second window in the same experiment configuration file as in Demonstration 1 with velocity shown and kinetic energy hidden. Roll cart up and down inclined track before prediction without taking data.

- Show kinetic energy-time and velocity-time graphs. Show the equation used, (0.5*1*"vel"^2).
- Many students think kinetic energy is conserved because 1) they think all energy is conserved or 2) kinetic energy returns to same value. Conserved means remaining constant.

Demonstration 3: Predict potential energy for same motion. Use the third window in the same experiment configuration file as in Demonstration 1 with position shown and potential energy hidden. Define starting position to have zero potential energy. Show result of releasing cart from different heights. Roll cart up and down inclined track before prediction without taking data.

- Show potential energy-time graph along with position-time graph.
- Show the equation in the data column used to calculate potential energy, e.g. (1*9.8*("Dist2"-Dist1)*sin(ø)). Note that the sine is the rise over the track length between any two points.
- Students may have trouble predicting the shape of the potential energy-time graph.
- Many students will think potential energy is conserved because it comes back to the same value. Remind them what "conserved" means.

Demonstration 4: Predict mechanical energy (sum of potential and kinetic energy) for same motion (after push and before catch). Use the fourth window in the same experiment configuration file as in Demonstration 1 with kinetic and potential energies shown and mechanical energy hidden. Roll cart up and down inclined track without taking data before predictions.

- Show the mechanical energy graph. Show the equation used, ($E = K + U$).
- Mechanical Energy is conserved after the push and before the catch.
- Point out that energy is not conserved in the system if you look before and after the push. Energy for system comes from your push. (from your breakfast or lunch)

<u>Demonstration 5:</u> We now lower the friction pad so that the cart experiences substantial friction as it moves up the ramp, reaches the highest point and then moves back down the ramp. Leave window 4 from the same experiment file displayed. Tell the students to include the push and catch in their predictions, and to remember friction can no longer be ignored.

- Ask students if mechanical energy is conserved even after the push and before the catch.
- What happened to the energy of the push?

MOMENTUM (MOM)

Hand in this sheet **Name**________________________________

INTERACTIVE LECTURE DEMONSTRATION
PREDICTION SHEET—**MOMENTUM**

Directions: This sheet will be collected. Write your name at the top to record your presence and participation in these demonstrations. Follow your instructor's directions. You may write whatever you wish on the attached Results Sheet and take it with you.

Demonstration 1: A ball is dropped to the ground. On the diagram to the right, draw the force(s) acting on the ball as it is falling. Is there a net force acting on the ball as it falls?

(The ball is in the air, falling)

Is the momentum of the ball conserved (constant) as it falls?

If you think momentum is conserved, state why. If not, describe a system, that includes the ball, in which the total momentum is conserved.

Demonstration 2 A ball is dropped to the ground. The ball hits the ground and bounces up. It has the same speed just before and just after it hits the ground. On the diagram to the right, draw the force(s) acting on the ball *while it is in contact with the ground.* Is there a net force acting on the ball when it is in contact with the ground?

(The ball is in contact with the ground, bouncing)

Is the momentum of the ball conserved (constant) from just before it hits the ground until just after it leaves the ground?

If you think momentum is conserved, state why. If not, describe a system, that includes the ball, in which the total momentum is conserved.

Demonstration 3: Two objects A and B, of equal mass, move towards each other along a frictionless track. They have the same initial speed ($v_A = v_B$), when they collide and bounce apart. In the lower half of the box on the right, draw the force(s) acting on each object *as they collide* (while they are in contact with each other).

Before the collision

$\vec{v}_A$ $\vec{v}_B$

A B

Is the momentum of either object *just before* the collision equal to its momentum *just after* the collision? Why or why not?

Draw a FBD for each object *during the collision*

A B

Is the *combined* momentum of the two objects before the collision equal to their combined momentum after the collision? Why or why not?

Demonstration 4: Two objects A and B, of equal mass, collide on a frictionless track and *stick together*. One of the objects is initially traveling to the right, and the other object is initially at rest. In the lower half of the box on the right, draw the force(s) acting on each object *as they collide* (while they are in contact with each other).

Before the collision

$\vec{v}_A$ $\vec{v}_B = 0$

A B

Is the momentum of either object *just before* the collision equal to its momentum *just after* the collision? Why or why not?

Draw a FBD for each object
during the collision

Is the *combined* momentum of the two objects before the collision equal to their combined momentum after the collision? Why or why not?

A B

Demonstration 5: Two objects A and B, of *unequal* mass, initially at rest, suddenly spring apart from each other. In the lower half of box on the right, draw the forces acting on each object while they are still touching just *as they begin to fly apart*.

Initial situation:

$\vec{v}_A = \vec{v}_B = 0$

A B

Is the momentum of either object *just before* the collision equal to its momentum *just after* the collision? Why or why not?

Draw below a FBD for each object
while they are still in contact as they begin to spring apart

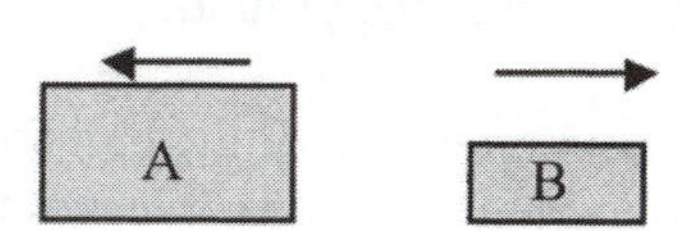

Is the *combined* momentum of the two objects before they spring apart equal to their combined momentum after they spring apart? Why or why not?

Demonstration 6: A cart slides along a frictionless track, strikes a wall, and bounces back with speed equal to its initial speed. In the space to the right, draw the forces acting on both the cart and the wall *while the cart is in contact with the wall.*

wall

cart

Is the momentum of the cart conserved? Why or why not?

Is the momentum of the wall conserved? Why or why not?

Draw a FBD for each object
during the collision:

Is the *combined* momentum of the cart and the wall conserved? Why or why not?

Keep this sheet

INTERACTIVE LECTURE DEMONSTRATION
RESULTS SHEET—**MOMENTUM**

You may write whatever you wish on this sheet and take it with you.

Demonstration 1: A ball is dropped to the ground. On the diagram to the right, draw the force(s) acting on the ball as it is falling. Is there a net force acting on the ball as it falls?

(The ball is in the air, falling)

Is the momentum of the ball conserved (constant) as it falls?

If you think momentum is conserved, state why. If not, describe a system, that includes the ball, in which the total momentum is conserved.

Demonstration 2 A ball is dropped to the ground. The ball hits the ground and bounces up. It has the same speed just before and just after it hits the ground. On the diagram to the right, draw the force(s) acting on the ball *while it is in contact with the ground*. Is there a net force acting on the ball when it is in contact with the ground?

(The ball is in contact with the ground, bouncing)

Is the momentum of the ball conserved (constant) from just before it hits the ground until just after it leaves the ground?

If you think momentum is conserved, state why. If not, describe a system, that includes the ball, in which the total momentum is conserved.

Demonstration 3: Two objects A and B, of equal mass, move towards each other along a frictionless track. They have the same initial speed ($v_A = v_B$), when they collide and bounce apart. In the lower half of the box on the right, draw the force(s) acting on each object *as they collide* (while they are in contact with each other).

Before the collision

$\vec{v}_A$ $\vec{v}_B$

A B

Is the momentum of either object *just before* the collision equal to its momentum *just after* the collision? Why or why not?

Draw a FBD for each object *during the collision*

A B

Is the *combined* momentum of the two objects before the collision equal to their combined momentum after the collision? Why or why not?

Demonstration 4: Two objects A and B, of equal mass, collide on a frictionless track and *stick together*. One of the objects is initially traveling to the right, and the other object is initially at rest. In the lower half of the box on the right, draw the force(s) acting on each object *as they collide* (while they are in contact with each other).

Before the collision

$\vec{v}_A$ $\vec{v}_B = 0$

A B

Is the momentum of either object *just before* the collision equal to its momentum *just after* the collision? Why or why not?

Draw a FBD for each object *during the collision*

Is the *combined* momentum of the two objects before the collision equal to their combined momentum after the collision? Why or why not?

A B

Demonstration 5: Two objects A and B, of *unequal* mass, initially at rest, suddenly spring apart from each other. In the lower half of box on the right, draw the forces acting on each object while they are still touching just *as they begin to fly apart*.

Initial situation:

$\vec{v}_A = \vec{v}_B = 0$

A B

Is the momentum of either object *just before* the collision equal to its momentum *just after* the collision? Why or why not?

Draw below a FBD for each object *while they are still in contact as they begin to spring apart*

Is the *combined* momentum of the two objects before they spring apart equal to their combined momentum after they spring apart? Why or why not?

A B

Demonstration 6: A cart slides along a frictionless track, strikes a wall, and bounces back with speed equal to its initial speed. In the space to the right, draw the forces acting on both the cart and the wall *while the cart is in contact with the wall.*

wall

cart

Is the momentum of the cart conserved? Why or why not?

Draw a FBD for each object *during the collision*:

Is the momentum of the wall conserved? Why or why not?

Is the *combined* momentum of the cart and the wall conserved? Why or why not?

MOMENTUM (MOM)
TEACHER'S GUIDE

Prerequisites: The four *Kinematics* and *Newton's Laws ILD* sequences are prerequisites. If students have done *RealTime Physics Mechanics* Labs1-6 or *Tools for Scientific Thinking Force and Motion* Labs 1-4, they should also be prepared for this sequence. This would be a good review for students who have completed through *RealTime Physics Mechanics* Lab 8.

Equipment:

any bouncy rubber ball

a hard (not carpeted) floor

two low-friction kinematics cart , at least one with a spring bumper (See below.)

mass bar or other mass that you can fasten on a cart securely to approximately double the cart mass

2.2 meter aluminum track, long door threshold or a very smooth, level table or ramp (See below.)

a massive brick or wall or heavy concrete block that the end of the track can be placed up against

General Notes on Preparation and Equipment:

Carts, mass bar and track:
See the Teacher's Guide for the *Motion of Carts ILDs* for detailed information about ramps and carts. Note that only the M-9430 and M-6950 PASCO (www.pasco.com) carts have spring bumpers. Any cart with very low friction and a spring bumper will work but truly low friction carts are difficult to find. Likewise, any smooth, level surface or ramp will work.

Note on computer-supported measurements:
The results of these demonstrations could be displayed quantitatively using a computer-based laboratory system. In some cases you would need two motion detectors. We have run these demonstrations without quantitative measurements, and believe that they are effective in getting the concepts across when used in this manner. Students are able to observe the velocities of the objects before and after the collisions, and can make reasonable comparisons visually.

Demonstrations and Results:

Demonstration 1: Ball is dropped to the ground. Predictions begin after the ball is released and end before it hits the floor.

F^{grav}

Discussion: Get students to discuss the net gravitational force on the ball as it falls, and the consequence that the momentum of the ball increases as it falls. Ask what conservation means. Is momentum conserved? Start to relate the occurrence of a net force on the ball with a non-conservation of its momentum. For what system is the gravitational force of the earth on the ball an *internal* force? (The system made up of the ball and the earth. There is no net *external* force on that system, and its momentum is conserved.)

Demonstration 2: Ball bouncing off the floor. Show the demonstration again. Predictions begin when the ball first contacts the floor and end when it leaves the floor moving upward.

F^{N}

F^{grav}

Discussion: The force or free-body diagram for the ball when it is in contact with the floor is shown on the right. Ask students to consider what the ball is doing, in order to answer whether there is a net force on the ball. When the ball first hits the floor, it is slowing down. What must be the direction of the net force? After the ball momentarily comes to rest, it is speeding up moving upward. What must be the direction of the net force?

There is a net upward force on the ball. Can its momentum be conserved? Many students will think that momentum is conserved because the ball has the same *speed* after it leaves the floor that it had on the way down before it hit the floor. Is momentum a vector? Is momentum conserved if the ball's momentum was negative (downward) before it hit the floor and positive (upward) after it left the floor?

In what system is there no net force on the ball? (The system must include the ball, floor and the earth in order for there to be no net *external* force on the ball. The momentum of *that* system *is* conserved.)

<u>Demonstration 3:</u> Symmetric, elastic collision between two equal mass carts. Collide the two equal mass carts with the springy bumper between them. Try to give them both the same speeds. Free body diagrams are during the time when the objects are in contact.

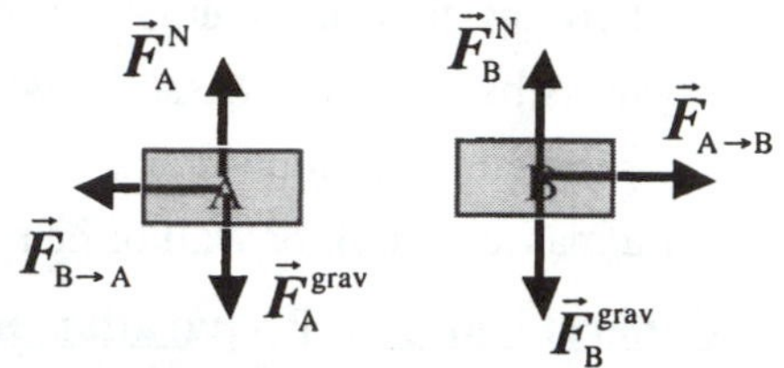

Discussion: The free-body diagrams should look something like those on the right. Ask students if the momentum of either cart is the same before and after the collision. Students may still think that the momentum of one cart is conserved because each has the same *speed* before and after the collision. Is momentum a vector? Is the momentum of Cart A the same before and after the collision? Cart B? Ask about the system made up of the two carts. Are there any net external forces? Is the momentum of this system conserved? What can be said about $\vec{F}_{A\to B}$ and $\vec{F}_{B\to A}$? How do you know that the forces are equal and opposite?

<u>Demonstration 4:</u> Asymmetric, in-elastic collision between two equal mass carts. Collide the two equal mass carts with the Velcro between them so that they stick together after the collision. Cart B should be initially at rest. Free body diagrams are during the time when the objects are in contact.

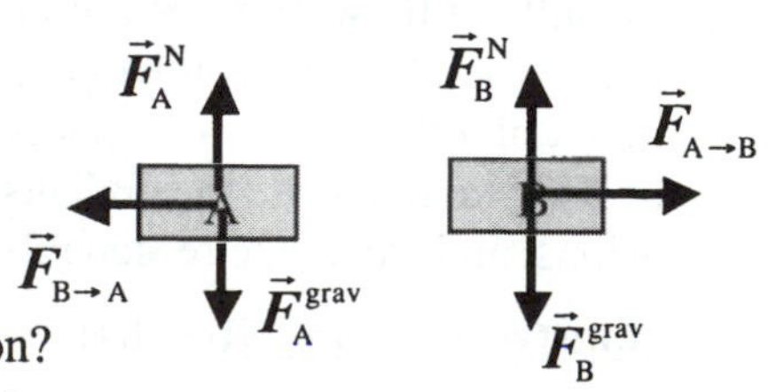

Discussion: The free-body diagrams should look something like those on the right. Ask students if the momentum of either cart is the same before and after the collision. Is the momentum of the system including the two carts conserved? Many students will not believe that it is because the collision is so asymmetrical. Ask how much mass is moving before the collision? After the collision? Also ask them to compare the velocity of A before the collision to the velocity of A plus B after. Could the momentum of the system including the two carts be conserved? Are there any net external forces? Is the momentum of this system conserved? What can be said about $\vec{F}_{A\to B}$ and $\vec{F}_{B\to A}$? How do you know that the forces are equal and opposite?

<u>Demonstration 5:</u> Unequal mass "explosion." Use the two carts with the springy bumper loaded between them. Put the mass bar or extra mass on Cart A. Both carts should be initially at rest, and should spring apart when the springy bumper is released. Free body diagram is during time when the two carts are in contact.

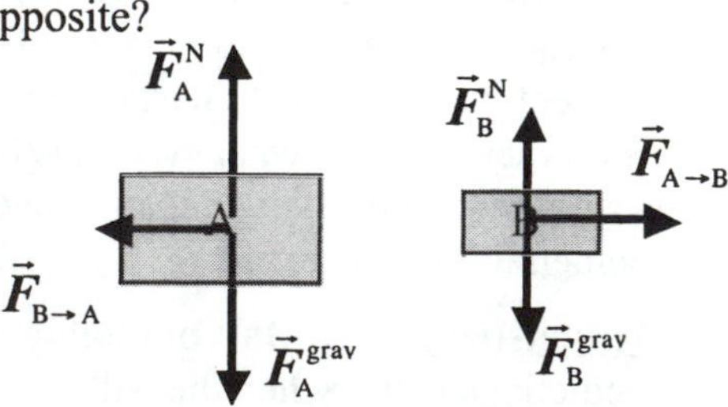

Discussion: The free-body diagrams should look something like those on the right. Ask students if the momentum of either cart is the same before and after the explosion. (Remind them that each had zero momentum before the explosion.) Is the momentum of the system including the two carts conserved? Many students will not believe that it is because the explosion is so asymmetrical. Ask them to compare the velocities of A and B after the explosion and the masses of A and B. Could the momentum of the system including the two carts be conserved? Are there any net external forces? Is the momentum of this system conserved? What can be said about $\vec{F}_{A\to B}$ and $\vec{F}_{B\to A}$?

<u>Demonstration 6:</u> Collision with a wall. Use a cart with the springy bumper out and facing toward the wall (or heavy concrete block). Collide the cart with the wall. Free body diagram is during time when the cart is in contact with the wall.

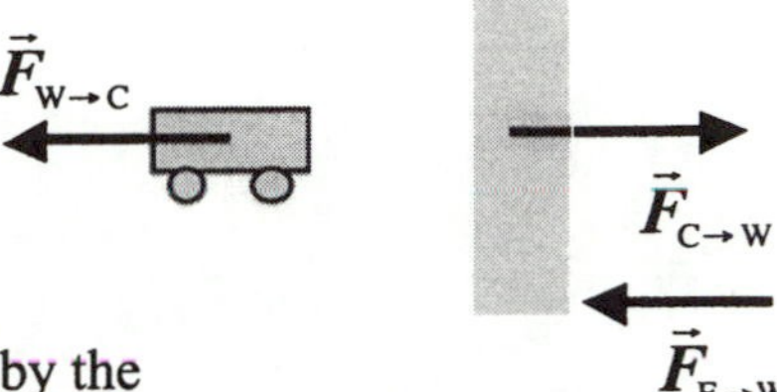

Discussion: This time we have only included the forces in the direction of the motion. Note that there are two such forces on the wall (concrete block), one exerted by the cart and one exerted by the floor or table. Some students may still think that the momentum of the cart is conserved. Again remind them that momentum is a vector! Does the wall move? Is its momentum conserved? Is the combined momentum of the cart and wall conserved? Are there any net *external* forces acting on this system? (The system's momentum is not conserved, since there is an *external* force acting on the wall.) What can be said about $\vec{F}_{W\to C}$ and $\vec{F}_{C\to W}$?

MOMENTUM (MOM)
TEACHER PRESENTATION NOTES

Classroom introduction to the *Momentum ILDs:*
Explain that the students will be observing what happens to the momentum of objects when they collide with each other or with rigid objects like walls.

Demonstration 1: Ball is dropped to the ground. Predictions from after the ball is released to before it hits the floor.

- Discuss net gravitational force on ball as it falls, and the momentum increases as the ball falls.
- Ask what conservation means. Is momentum conserved? Relate a net force on the ball with non-conservation of its momentum.
- For what system is the gravitational force of the earth on the ball an *internal* force? Then there is no net *external* force on that system, and its momentum is conserved.

Demonstration 2: Ball bouncing off the floor. Show the demonstration again. Predictions from when the ball first contacts the floor to when it leaves the floor moving upward. Draw free-body diagram for the ball when it is in contact with the floor.

- Ask students to consider what the ball is doing. When the ball is in contact with the floor it slows down, reverses direction and moves up. Net force must be upward.
- There is a net upward force on the ball. Is momentum a vector? Is momentum conserved if it was negative (downward) before it hit the floor and positive (upward) after it left the floor?
- In what system is there no net *external* force on the ball? (The ball, floor and the earth.)

Demonstration 3: Symmetric, elastic collision between two equal mass carts. Collide the two equal mass carts with the springy bumper between them. Try to give them both the same speeds.

- Is the momentum of Cart A the same before and after the collision? Cart B? Ask about the system made up of the two carts. Is momentum a vector?
- Are there any net *external* forces? Is the momentum of this system conserved?
- What can be said about $\vec{F}_{A \to B}$ and $\vec{F}_{B \to A}$? How do you know?

Demonstration 4: Asymmetric, in-elastic collision between two equal mass carts. Collide the two equal mass carts with the velcro between them so that they stick together after the collision. Cart B should be initially at rest.

- Ask if the momentum of either cart is the same before and after the collision.
- Is the momentum of the system including the two carts conserved?
- How much mass is moving before the collision? After? Compare the velocity of A before the collision to the velocity of A plus B after. Could the total momentum conserved?
- Are there any net *external* forces? Is the momentum of this system conserved? What can be said about $\vec{F}_{A \to B}$ and $\vec{F}_{B \to A}$?

Demonstration 5: Unequal mass "explosion." Use the two carts with the springy bumper loaded between them, and extra mass on Cart A. Both carts initially at rest. Should spring apart.

- Ask if momentum of either cart is the same before and after the explosion. (Remind them that each had zero momentum before the explosion.)

- Is the momentum of the system including the two carts conserved? Ask them to compare the velocities of A and B after the explosion and the masses of A and B.
- Are there any net *external* forces? What can be said about $\vec{F}_{A \to B}$ and $\vec{F}_{B \to A}$?

Demonstration 6: Collision with a wall. Use a cart with the springy bumper out and facing toward the wall (or heavy concrete block). Collide the cart with the wall. Draw free body diagram during time when the cart is in contact with the wall.

- There are three forces in the direction of motion: one exerted by the wall on the cart, one exerted by the cart on the wall and one exerted by the floor (or table) on the wall.
- Again remind them that momentum is a vector! Does the wall move? Is its momentum conserved? Is the combined momentum of the cart and wall conserved? Are there any net *external* forces acting on this system?
- What can be said about $\vec{F}_{W \to C}$ and $\vec{F}_{C \to W}$?

Rotational Motion (ROT)

Hand in this sheet **Name**______________________________

INTERACTIVE LECTURE DEMONSTRATION PREDICTION SHEET—ROTATIONAL MOTION

Directions: This sheet will be collected. Write your name at the top to record your presence and participation in these demonstrations. Follow your instructor's directions. You may write whatever you wish on the attached Results Sheet and take it with you.

Demonstration 1: A turntable is rotating counter-clockwise at *constant angular speed*. An object near the edge of the turntable remains there without slipping or flying off. The figure to the right shows the object at one instant (position 1) and then an instant later (position 2). On the picture, draw the object's velocity vector $\vec{v}_1$ at position 1 and $\vec{v}_2$ at position 2. Then, in the space to the right, draw the vector that represents the difference of the two velocity vectors, $\Delta\vec{v} = \vec{v}_2 - \vec{v}_1$.

Draw $\vec{v}_1$ and $\vec{v}_2$ here

1

2

Draw $\Delta\vec{v}$ here

Demonstration 2: The turntable's rotational speed is *increasing*. An object near the edge of the turntable remains there without slipping or flying off. The figure to the right shows the object at one instant (position 1) and then an instant later (position 2). On the picture, draw the object's velocity vector $\vec{v}_1$ at position 1 and $\vec{v}_2$ at position 2. Then, in the space to the right, draw the vector that represents the difference of the two velocity vectors, $\Delta\vec{v} = \vec{v}_2 - \vec{v}_1$.

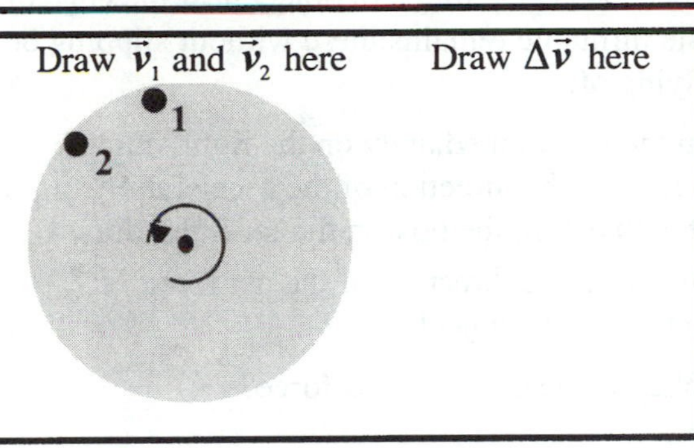

Demonstration 3: The turntable's rotational speed is *decreasing*. An object near the edge of the turntable remains there without slipping or flying off. The figure to the right depicts the object at one instant (position 1) and then an instant later (position 2). On the picture, draw the object's velocity vector $\vec{v}_1$ at position 1 and $\vec{v}_2$ at position 2. Then, in the space to the right, draw the vector that represents the difference of the two velocity vectors, $\Delta\vec{v} = \vec{v}_2 - \vec{v}_1$.

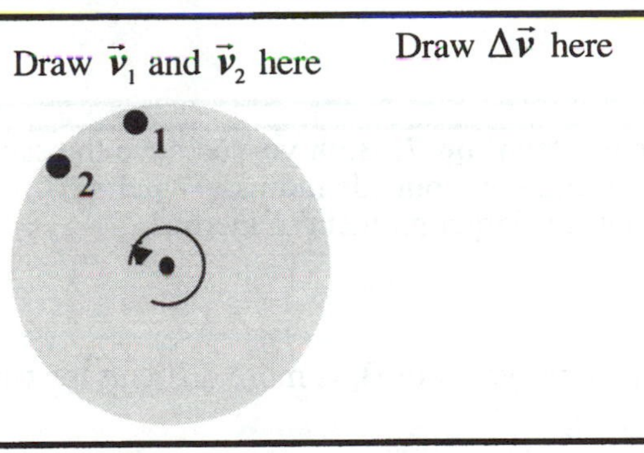

Demonstration 4: The turntable is rotating at *constant speed*. An object near the edge of the turntable remains there without slipping or flying off. In the box immediately on the right, draw a vector in the direction of the acceleration, $\vec{a}$, of the object. In the box on the far right, draw a vector in the direction of the net force, $\vec{F}^{net}$, acting on the object.

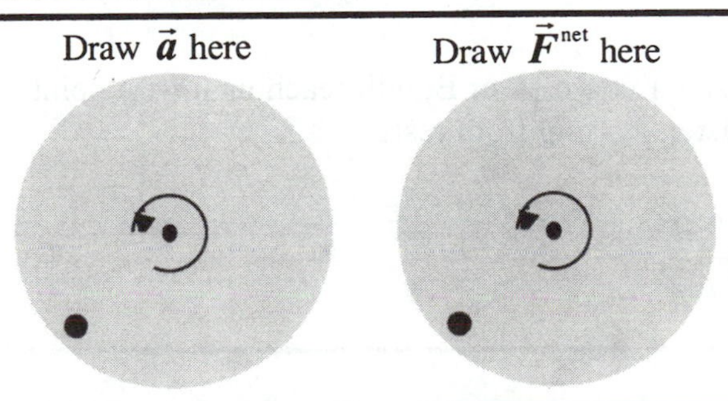

Demonstration 5: The turntable's angular speed is *increasing*. An object near the edge of the turntable remains there without slipping or flying off.

In the box immediately on the right, draw a vector in the direction of the acceleration, $\vec{a}$, of the object. In the box on the far right, draw a vector in the direction of the net force, $\vec{F}^{net}$, acting on the object.

What is the cause of this force?

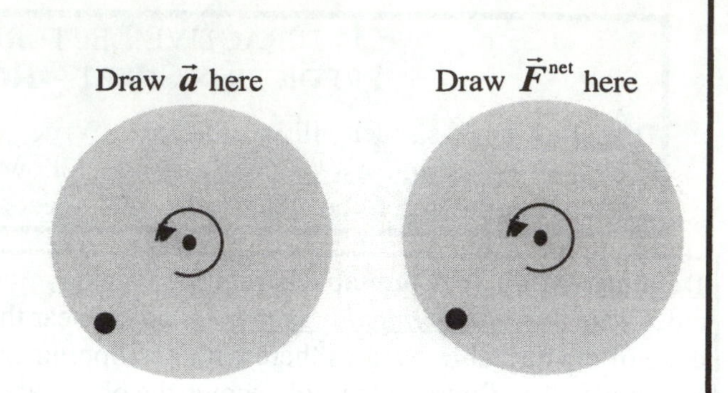

Demonstration 6: The turntable's angular speed is *decreasing*. An object near the edge of the turntable remains there without slipping or flying off.

In the box immediately on the right, draw a vector in the direction of the acceleration, $\vec{a}$, of the object. In the box on the far right, draw a vector in the direction of the net force, $\vec{F}^{net}$, acting on the object.

What is the cause of this force?

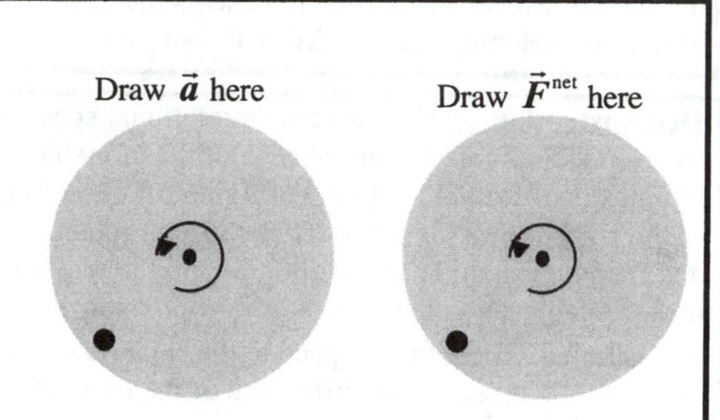

Demonstration 7: Both yo-yos have the same mass and same outside radius. Which yo-yo, A or B, has the larger moment of inertia?

Which yo-yo, A or B, is more difficult to start rotating?

Which yo-yo, A or B, will reach its lowest point faster, starting from rest?

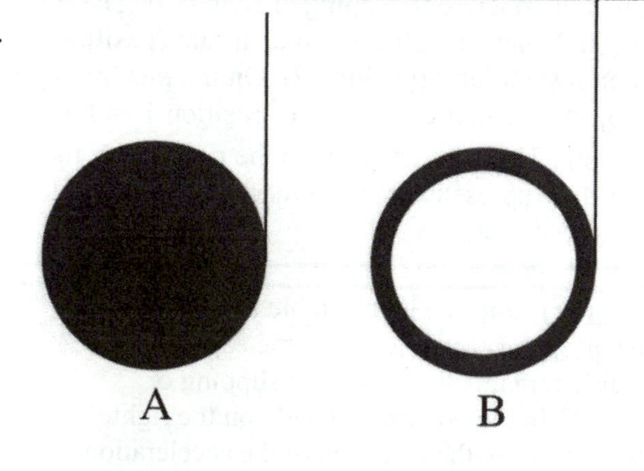

Keep this sheet

INTERACTIVE LECTURE DEMONSTRATION
RESULTS SHEET—**ROTATIONAL MOTION**

You may write whatever you wish on this sheet and take it with you.

Demonstration 1: A turntable is rotating counter-clockwise at *constant angular speed*. An object near the edge of the turntable remains there without slipping or flying off. The figure to the right shows the object at one instant (position 1) and then an instant later (position 2). On the picture, draw the object's velocity vector $\vec{v}_1$ at position 1 and $\vec{v}_2$ at position 2. Then, in the space to the right, draw the vector that represents the difference of the two velocity vectors, $\Delta\vec{v} = \vec{v}_2 - \vec{v}_1$.

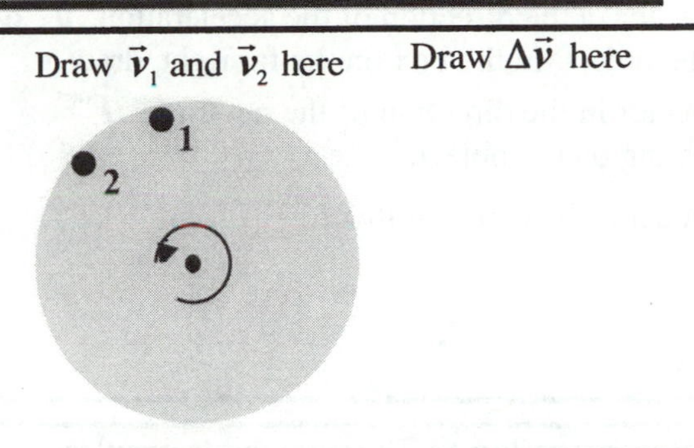

Demonstration 2: The turntable's rotational speed is *increasing*. An object near the edge of the turntable remains there without slipping or flying off. The figure to the right shows the object at one instant (position 1) and then an instant later (position 2). On the picture, draw the object's velocity vector $\vec{v}_1$ at position 1 and $\vec{v}_2$ at position 2. Then, in the space to the right, draw the vector that represents the difference of the two velocity vectors, $\Delta\vec{v} = \vec{v}_2 - \vec{v}_1$.

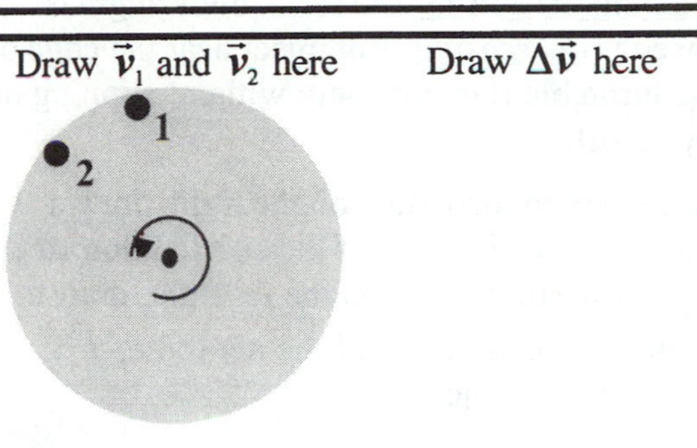

Demonstration 3: The turntable's rotational speed is *decreasing*. An object near the edge of the turntable remains there without slipping or flying off. The figure to the right depicts the object at one instant (position 1) and then an instant later (position 2). On the picture, draw the object's velocity vector $\vec{v}_1$ at position 1 and $\vec{v}_2$ at position 2. Then, in the space to the right, draw the vector that represents the difference of the two velocity vectors, $\Delta\vec{v} = \vec{v}_2 - \vec{v}_1$.

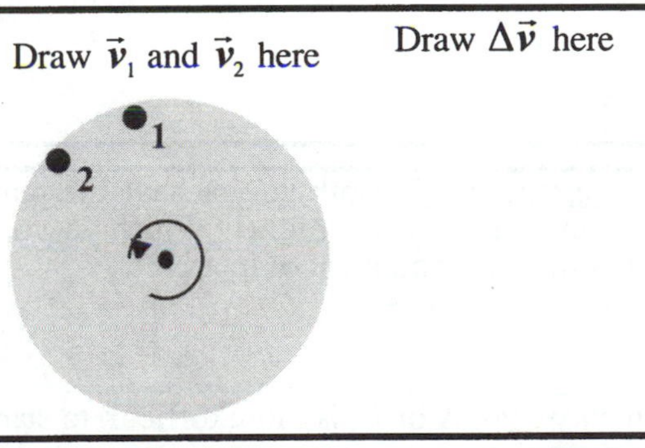

Demonstration 4: The turntable is rotating at *constant speed*. An object near the edge of the turntable remains there without slipping or flying off. In the box immediately on the right, draw a vector in the direction of the acceleration, $\vec{a}$, of the object. In the box on the far right, draw a vector in the direction of the net force, $\vec{F}^{net}$, acting on the object.

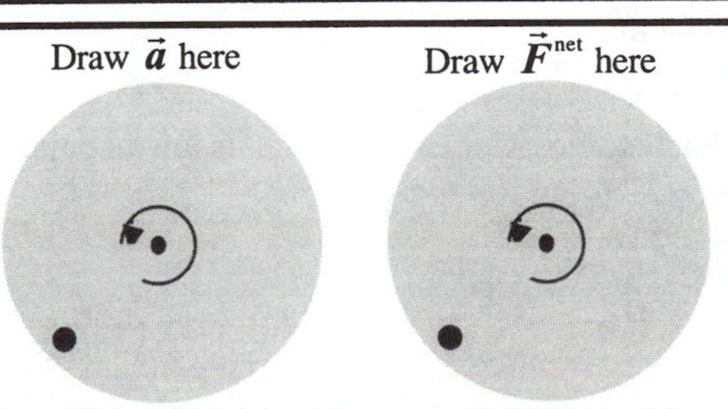

Demonstration 5: The turntable's angular speed is *increasing*. An object near the edge of the turntable remains there without slipping or flying off.

In the box immediately on the right, draw a vector in the direction of the acceleration, $\vec{a}$, of the object. In the box on the far right, draw a vector in the direction of the net force, $\vec{F}^{net}$, acting on the object.

What is the cause of this force?

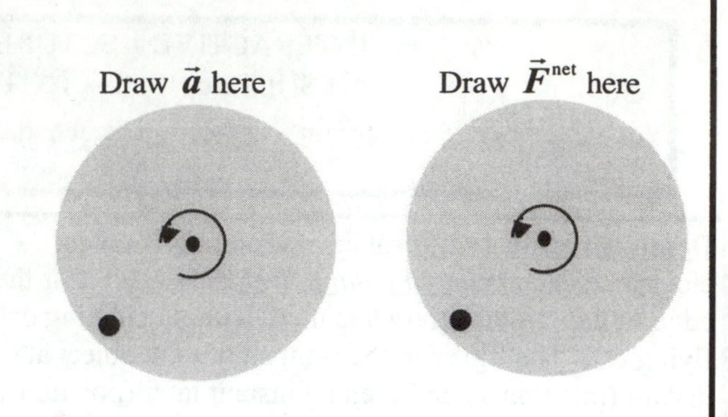

Demonstration 6: The turntable's angular speed is *decreasing*. An object near the edge of the turntable remains there without slipping or flying off.

In the box immediately on the right, draw a vector in the direction of the acceleration, $\vec{a}$, of the object. In the box on the far right, draw a vector in the direction of the net force, $\vec{F}^{net}$, acting on the object.

What is the cause of this force?

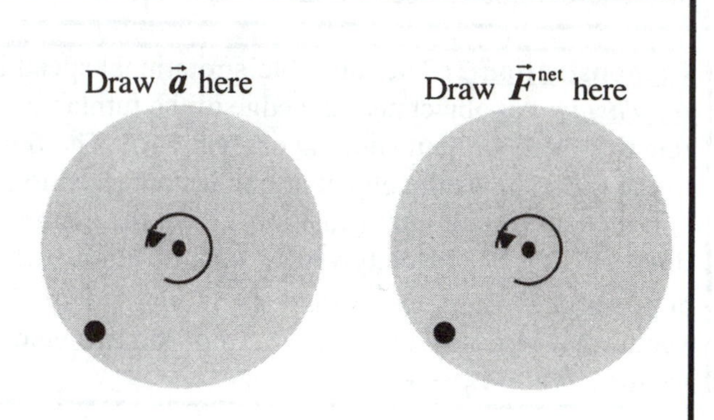

Demonstration 7: Both yo-yos have the same mass and same outside radius. Which yo-yo, A or B, has the larger moment of inertia?

Which yo-yo, A or B, is more difficult to start rotating?

Which yo-yo, A or B, will reach its lowest point faster, starting from rest?

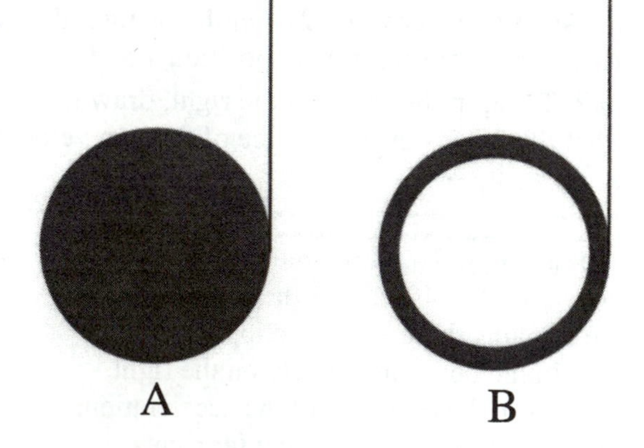

ROTATIONAL MOTION (ROT)
TEACHER'S GUIDE

Prerequisites: The four *Kinematics* and *Newton's Laws ILD* sequences, and the *Vectors ILDs* are prerequisites. If students have done *RealTime Physics Mechanics* Labs1-6 or *Tools for Scientific Thinking Force and Motion* Labs 1-4, they should also be prepared for this sequence.

Equipment:

large, low friction rotating turntable (See below.)

0.5 or 1 kg mass

disk and hoop of equal mass and radius (See below.)

string

General Notes on Preparation and Equipment:

Platform: The apparatus is just to show the demonstrations. No quantitative measurements will be taken. The platform needed is the type that is usually found in a physics department for rotational motion or conservation of angular momentum demonstrations. It should be large enough so that students can see what is going on (at least 25 cm in diameter), and it should have low enough friction that it can rotate for awhile without coming to rest. Examples are the Sargent Welch Rotating Platform (WL0973) or Rotating Support Kit (CP24484-00) (www.sargentwelch.com), the Frey Scientific Rotating Platform (15590262) (www.freyscientific.com), or the PASCO Rotational System (ME-8950A) or Introductory Rotational Apparatus (ME-9341) (www.pasco.com). An old phonograph turntable would also work just fine.

Disk and hoop: These too are usually found in a physics department demonstration room. They should both be at least 10 cm in diameter. One should be a solid disk, and the other should be a ring, with all of the mass concentrated on the outside edge. The larger disk and ring in the PASCO Rotational Inertia Set (ME-9774) should work, or these could easily be fabricated in a machine shop. To make a "yo-yo" just tape the end of a piece of string to the disk or hoop, and wind it around the circumference.

Demonstrations and Results:

Demonstration 1: Mass rotating at constant angular speed. Show students the demonstration, and ask them to focus on the direction in which the mass is moving at any instant.

Discussion: The diagrams are shown on the right. Students should understand that the velocity vectors are of the same length (same magnitude velocity since the turntable is rotating at a constant angular speed), and that they are drawn tangential to the edge of the turntable.

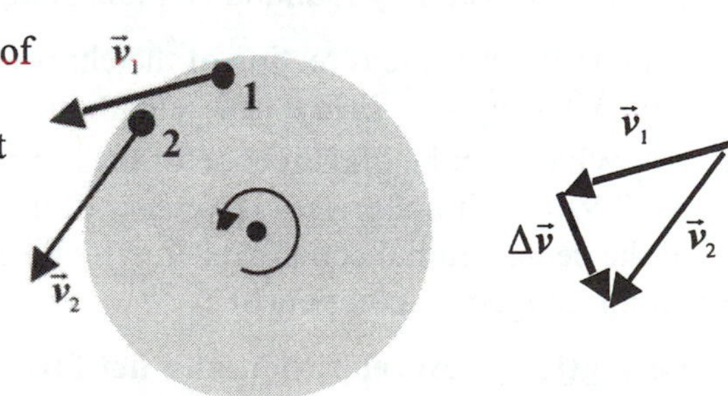

Ask them in which direction the difference (or change) in velocity seems to point relative to the turntable. Is it toward the center? Why? What is the direction of the acceleration of the mass?

Demonstration 2: Mass rotating with increasing angular speed. Show students the demonstration, and ask them to focus on the direction in which the mass is moving at any instant, and how fast it is moving.

Discussion: Now the vector representing $\vec{v}_2$ is longer than the one representing $\vec{v}_1$. This reflects the fact that the turntable is speeding up, and the mass is moving faster all the time. Ask in which direction the difference (or change) in velocity seems to point this time. Is it toward the center? Why not? What is the direction of the acceleration of the mass?

Demonstration 3: Mass rotating with decreasing angular speed. Show students the demonstration, and ask them to focus on the direction in which the mass is moving at any instant, and how fast it is moving.

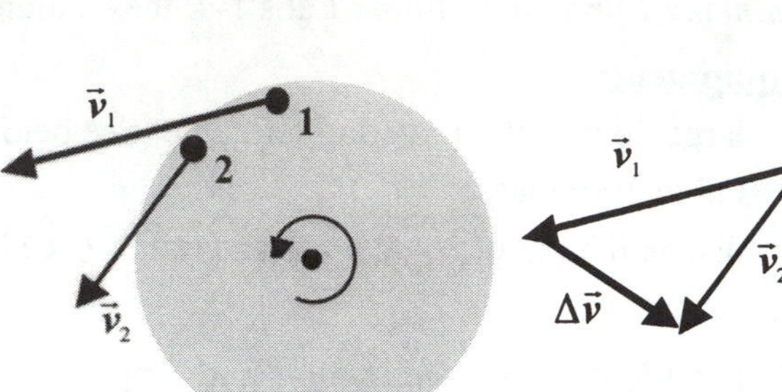

Discussion: Now the vector representing $\vec{v}_1$ is longer than the one representing $\vec{v}_2$. This reflects the fact that the turntable is slowing down, and the mass is moving slower all the time. Ask in which direction the difference (or change) in velocity seems to point this time. Is it toward the center? Why not? What is the direction of the acceleration of the mass?

Demonstration #4: Acceleration and net force for mass rotating with constant angular speed. Show students the demonstration, and ask them to focus on the change in velocity found in Demonstration 1.

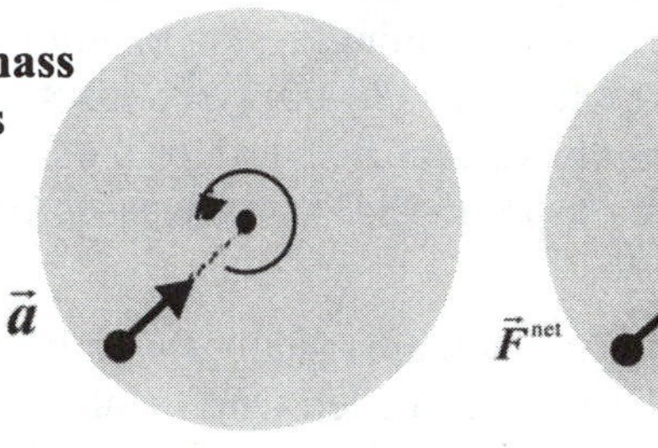

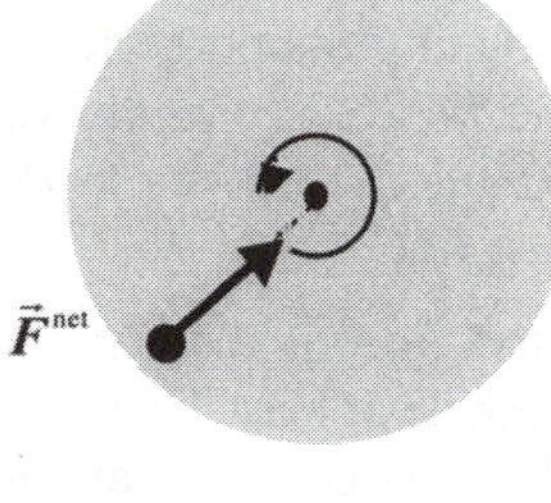

Discussion: What is the direction of the change in velocity, $\Delta\vec{v} = \vec{v}_2 - \vec{v}_1$? How is acceleration defined? What must be the direction of the acceleration of the mass? Why is there an acceleration toward the center if the mass is moving at a constant speed?

Demonstration 5: Acceleration and net force for mass rotating with increasing angular speed. Show students the demonstration, and ask them to focus on the change in velocity found in Demonstration 2.

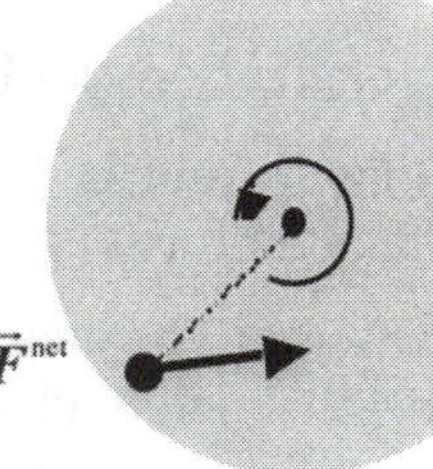

Discussion: What is the direction of the change in velocity, $\Delta\vec{v} = \vec{v}_2 - \vec{v}_1$? How is acceleration defined? What must be the direction of the acceleration of the mass? Why is it not toward the center of the turntable? Why does the acceleration have a component toward the center (radial component)? Why does the acceleration have a component in the direction of the motion (tangential component)?

Demonstration 6: Acceleration and net force for mass rotating with decreasing angular speed. Show students the demonstration, and ask them to focus on the change in velocity found in Demonstration 3.

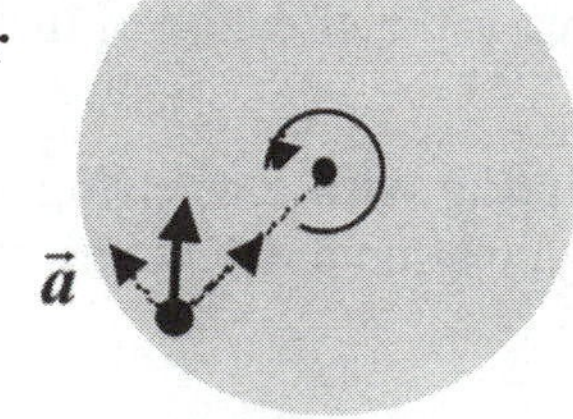

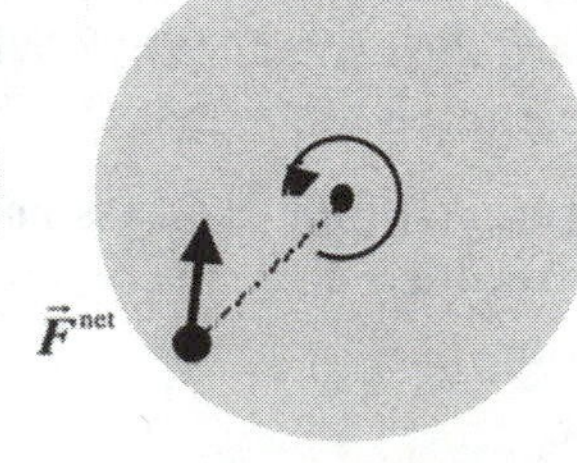

Discussion: What is the direction of the change in velocity, $\Delta\vec{v} = \vec{v}_2 - \vec{v}_1$? How is acceleration defined? What must be the direction of the acceleration of the mass? Why is it not toward the center of the turntable? Why does the acceleration have a component toward the center (radial component)? Why does the acceleration have a component in the direction opposite to the motion (tangential component)?

Demonstration 7: Which yo-yo falls faster? This is a classic demonstration written in a more interactive way. Show the students the disk and the hoop, and tell them they both have the same diameter and mass. Ask them to make all three predictions. Only show the demonstration with the yo-yos falling after all three predictions.

Discussion: How is moment of inertia defined? How is the mass distributed in each yo-yo. Which yo-yo has more mass further from the center?

What does moment of inertia tell you about how difficult it is to start an object rotating?

How is torque defined? Compare the torques on the two yo-yos when they are hanging from the string. Is one larger, or are they equal.

With the same torque applied to each, which yo-yo will rotate the fastest after a given time? Which yo-yo will reach its lowest point first?

ROTATIONAL MOTION (ROT)
TEACHER PRESENTATION NOTES

Classroom introduction to the *Rotation ILDs:*
Explain that the students will be observing an object moving in a circle, and analyzing the way that its velocity changes, its acceleration and the net force acting on it.

Demonstration 1: Mass rotating at constant angular speed. Ask students to observe rotation of mass and focus on the direction it moves at any moment.

- Show the velocity vectors. Ask students to describe them. Why are they the same length? Why are they drawn tangential to the edge of the turntable?
- How do you find the vector difference? What is the direction of the change in the velocity?
- What is the direction of the acceleration of the mass?

Demonstration 2: Mass rotating with increasing angular speed. Ask students to observe rotation of mass and focus on the direction it moves and its speed at any moment.

- Show the velocity vectors. Ask students to describe them. Why is $\vec{v}_2$ longer than $\vec{v}_1$?
- How do you find the vector difference? What is the direction of the change in velocity? Is it toward the center this time? Why not? What is the direction of the acceleration of the mass?

Demonstration 3: Mass rotating with decreasing angular speed. Ask students to observe rotation of mass and focus on the direction it moves and its speed at any moment.

- Show the velocity vectors. Ask students to describe them. Why is $\vec{v}_1$ longer than $\vec{v}_2$?
- How do you find the vector difference? What is the direction of the change in velocity? Is it toward the center this time? Why not? What is the direction of the acceleration of the mass?

Demonstration 4: Acceleration and net force for mass rotating with constant angular speed. Ask students to focus on the change in velocity found in Demonstration 1.

- How is acceleration defined? What must be the direction of the acceleration of the mass?
- Why is there an acceleration toward the center if the mass is moving at a constant speed?

Demonstration 5: Acceleration and net force for mass rotating with increasing angular speed. Ask students to focus on the change in velocity found in Demonstration 2.

- What must be the direction of the acceleration? Why is it not toward the center?
- Why does the acceleration have a component toward the center (radial component)? A component in the direction of the motion (tangential component)?

Demonstration 6: Acceleration and net force for mass rotating with decreasing angular speed. Ask the students to focus on the change in velocity found in Demonstration 3.

- What must the direction of the acceleration of the mass? Why is it not toward the center?
- Why does the acceleration have a component toward the center (radial component)? A component in the direction opposite to the motion (tangential component)?

Demonstration 7: Which yo-yo falls faster? Show the disk and the hoop. Tell students they have the same diameter and mass. Only show yo-yos falling after all three predictions.

- How is moment of inertia defined? How is the mass distributed in each yo-yo. Which yo-yo has more mass further from the center?

- What does moment of inertia tell you about how difficult it is to start an object rotating?
- How is torque defined? Compare the torques on the two yo-yos when they are hanging from the string. Is one larger, or are they equal.
- With the same torque applied to each, which yo-yo will rotate the fastest after a given time? Which yo-yo will reach its lowest point first?

STATICS (STAT)

Hand in this sheet **Name**_________________________

INTERACTIVE LECTURE DEMONSTRATIONS
PREDICTION SHEET—**STATICS**

Directions: This sheet will be collected. Write your name at the top to record your presence and participation in these demonstrations. Follow your instructor's directions. You may write whatever you wish on the attached Results Sheet and take it with you.

Demonstration 1: A block is on a frictionless surface. Two forces act through the center of the block as shown. Under what condition(s) will the block have no tendency to begin rotating?

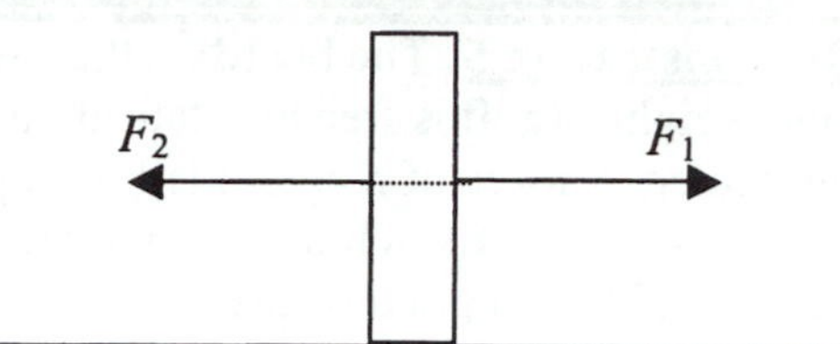

Under what condition(s) will the block have no tendency to begin moving to the right or left parallel to the surface?

Demonstration 2: Now the two forces act on the block as shown. Under what condition(s) will the block have no tendency to begin rotating?

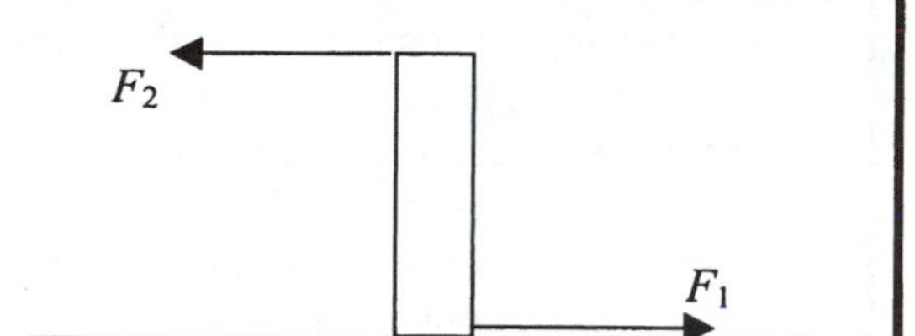

Under what condition(s) will the block have no tendency to begin moving to the right or left parallel to the surface?

Demonstration 3: Now the two forces act on the block as shown. Under what condition(s) will the block have no tendency to begin rotating?

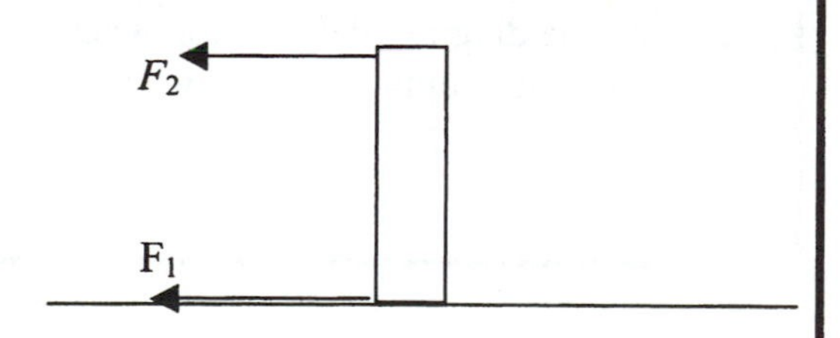

Under what condition(s) will the block have no tendency to begin moving to the right or left parallel to the surface?

Demonstration 4: The block on the right is free to rotate around an axis through its center. The force F is applied to the top of the block as shown. A second force of magnitude $2F$ is also applied to the block. Sketch on the diagram where the second force could be applied so that the block would have no tendency to begin rotating. (Note: there may be more than one possibility.)

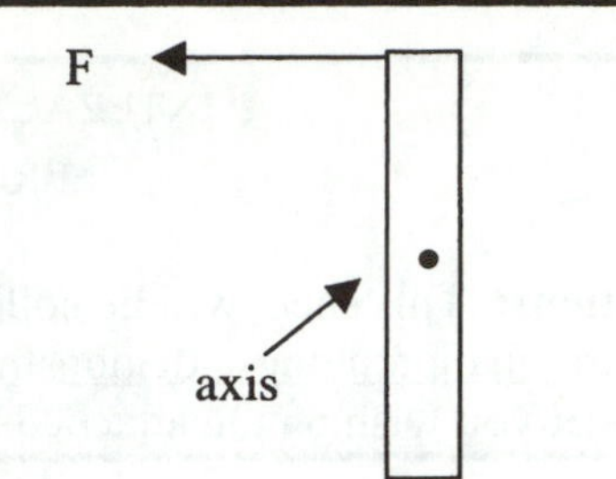

Demonstration 5: The board on the right has mass M and weight Mg. It is free to rotate about an axis through its center. A force of magnitude F is applied to the block as shown. Sketch on the diagram where a second force could be applied so that the block would have no tendency to begin rotating, and give the magnitude of the force below. (Note: there may be more than one possibility.)

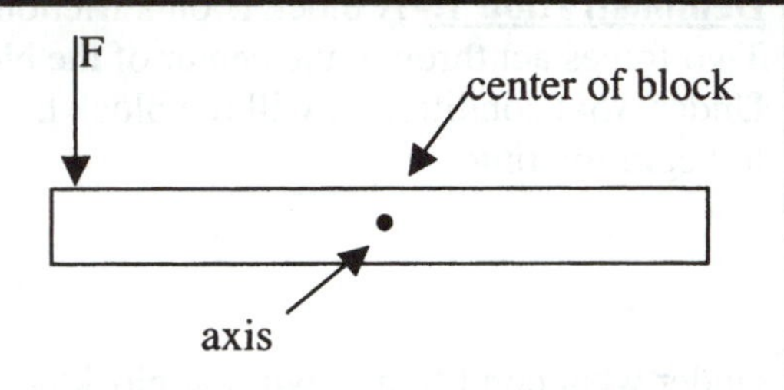

Demonstration 6: The same board as in Demonstration 6 is now free to rotate about an axis midway between its center and end. A force of magnitude F is applied to the block. Sketch on the diagram where a second force could be applied so that the block would have no tendency to begin rotating, and give the magnitude of the force below. (Note: there may be more than one possibility.)

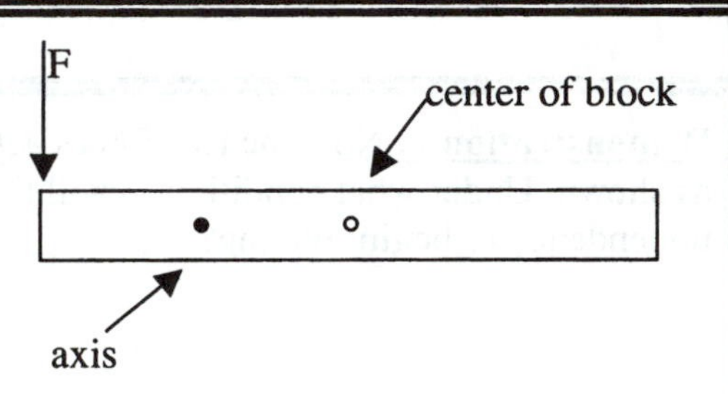

Demonstration 7: A block is standing on a tabletop. It is pushed at the top so that it tips a little bit. When it is released, it returns to its original position. However, if it is tilted enough, it falls down when it is released. Sketch on the diagram the position where it will just not return to its original position when tilted.

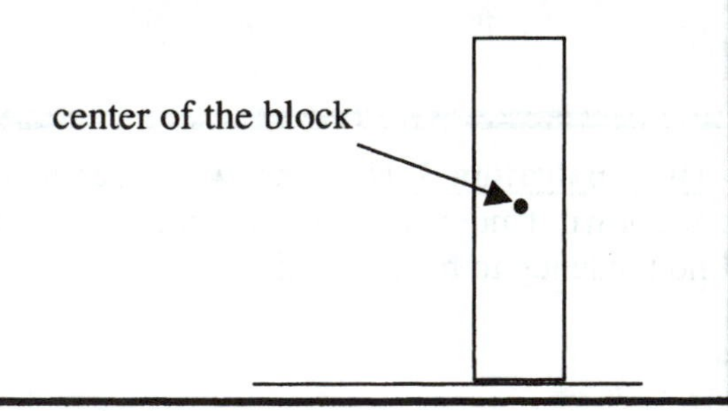

Keep this sheet

INTERACTIVE LECTURE DEMONSTRATIONS
RESULTS SHEET—STATICS

You may write whatever you wish on this sheet and take it with you.

Demonstration 1: A block is on a frictionless surface. Two forces act through the center of the block as shown. Under what condition(s) will the block have no tendency to begin rotating?

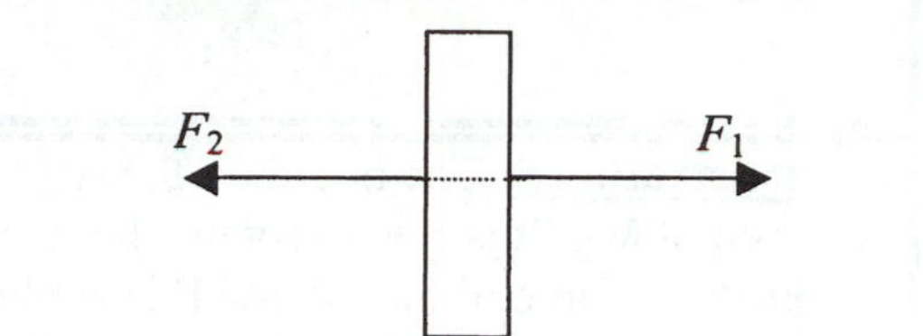

Under what condition(s) will the block have no tendency to begin moving to the right or left parallel to the surface?

Demonstration 2: Now the two forces act on the block as shown. Under what condition(s) will the block have no tendency to begin rotating?

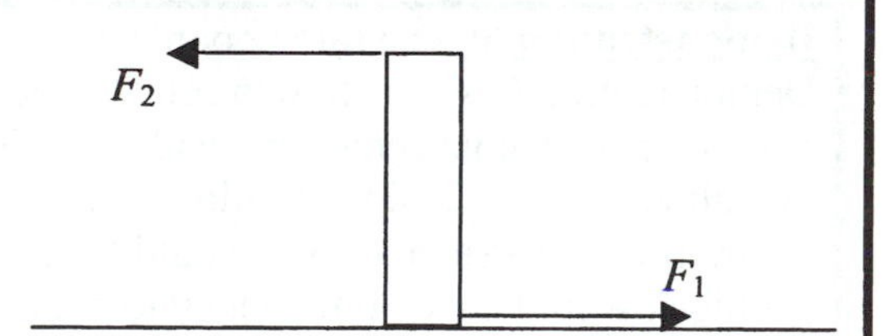

Under what condition(s) will the block have no tendency to begin moving to the right or left parallel to the surface?

Demonstration 3: Now the two forces act on the block as shown. Under what condition(s) will the block have no tendency to begin rotating?

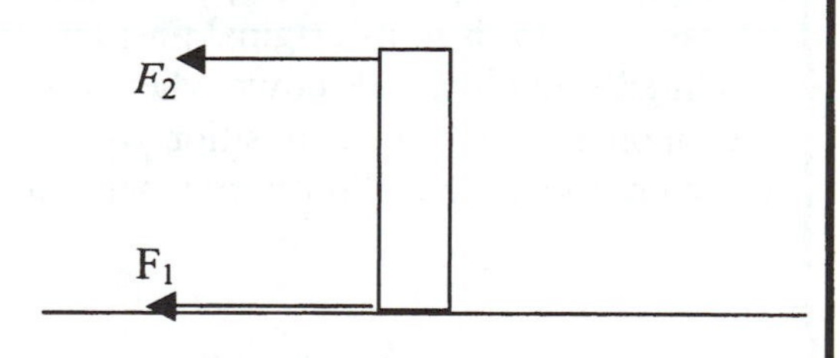

Under what condition(s) will the block have no tendency to begin moving to the right or left parallel to the surface?

Demonstration 4: The block on the right is free to rotate around an axis through its center. The force F is applied to the top of the block as shown. A second force of magnitude $2F$ is also applied to the block. Sketch on the diagram where the second force could be applied so that the block would have no tendency to begin rotating. (Note: there may be more than one possibility.)

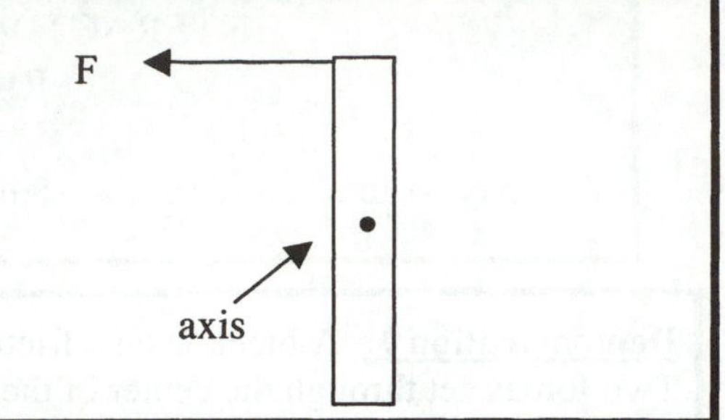

Demonstration 5: The board on the right has mass M and weight Mg. It is free to rotate about an axis through its center. A force of magnitude F is applied to the block as shown. Sketch on the diagram where a second force could be applied so that the block would have no tendency to begin rotating, and give the magnitude of the force below. (Note: there may be more than one possibility.)

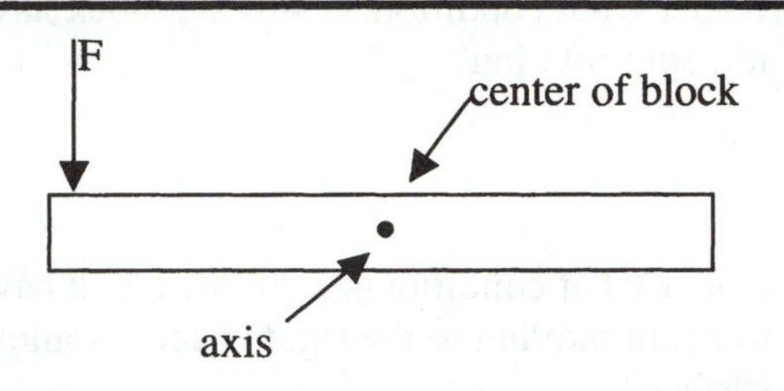

Demonstration 6: The same board as in Demonstration 6 is now free to rotate about an axis midway between its center and end. A force of magnitude F is applied to the block. Sketch on the diagram where a second force could be applied so that the block would have no tendency to begin rotating, and give the magnitude of the force below. (Note: there may be more than one possibility.)

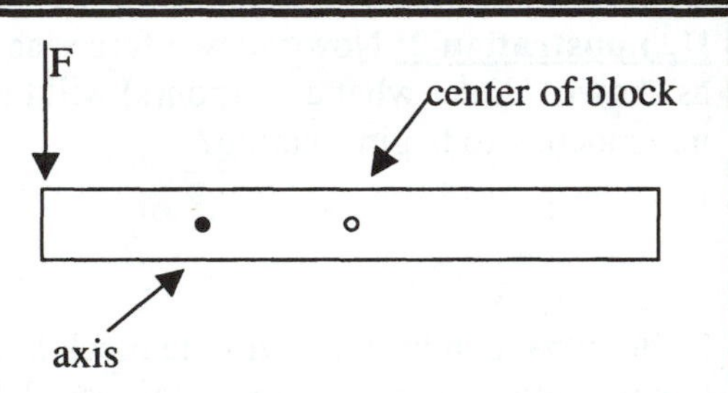

Demonstration 7: A block is standing on a tabletop. It is pushed at the top so that it tips a little bit. When it is released, it returns to its original position. However, if it is tilted enough, it falls down when it is released. Sketch on the diagram the position where it will just not return to its original position when tilted.

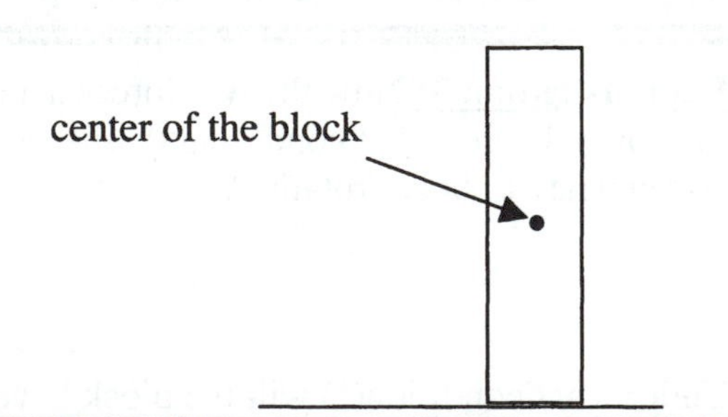

STATICS (STAT)
TEACHER'S GUIDE

Prerequisites:

This ILD sequence can be used as an introduction to statics and equilibrium, or as a review of this topic after lectures and/or text readings. Students should have been introduced to how torques are calculated, and the sign convention for torques.

Equipment:

2" x 4" block about 18" long with screw eyes to attach spring scales (See below.)

two spring scales with 0-10 N ranges with large displays, or two force probes (See below.)

2" x 4" board 2' long with holes through center and midway between center and one end, and with screw eyes to attach spring scales (See below.)

support stand and cross rod slightly smaller than the holes in the board

low friction cart (optional) (See below.)

Block and board:
The screw eyes should be located at the application points of the forces. See the diagrams on the prediction sheet. The spring scales will be hooked onto these screw eyes. The second board should rotate fairly freely on the support rod.

Force probes:
If you choose to use force probes instead of spring scales, see the *Teacher's Guide* for the *Newton's 1st & 2nd Laws ILDs* for recommendations. Set up the software to display large digital displays of both force probes.

Low friction cart:
It is impossible to have zero friction between the block and the table. The friction produces a torque at the bottom of the block that may tend to make it rotate. If the block is standing on top of a low friction cart like the PASCO ME-9430, ME-9454, ME-6950 or ME-6951, the frictional force at the base is nearly eliminated.

Demonstrations and Sample Results:

Demonstration 1: Opposite forces along the same line through center. After the prediction and discussion steps are completed, attach the two spring scales to the screw eyes, and pull with both scales. At first pull with equal forces, then with a stronger force on the right and finally with a stronger force on the left.

Discussion after observing the results: Ask students to describe how the block tends to begin to move. With the equal and opposite forces does it translate? Does it rotate? With the stronger force on the right does it translate? Rotate? With the stronger force on the left does it translate? Rotate? Is there a net force in any of these cases? Is there a net torque about the center of mass? What are the two conditions for the block to not begin translating or rotating?

Demonstration 2: Two forces acting at the ends in opposite directions. After the prediction and discussion steps are completed, attach the two spring scales to the screw eyes, and pull with both scales. At first pull with equal forces, then with a stronger force on the top, and then with a stronger force on the bottom.

Discussion after observing the results: Ask students to describe how the block tends to begin to move. With the equal forces does it translate? Does it rotate? With the stronger force on the top does it translate? Rotate? With the stronger force on the bottom does it translate? Rotate? Is there a net force in any of these cases? Is there a net torque about the center of mass?

__Demonstration 3:__ **Two forces acting at the ends in the same direction.** After the prediction and discussion steps are completed, attach the two spring scales to the screw eyes, and pull with both scales. At first pull with equal forces, then with a stronger force on the top, and then with a stronger force on the bottom.

Discussion after observing the results: Ask students to describe how the block tends to begin to move. With the equal and opposite forces does it translate? Does it rotate? With the stronger force on the top does it translate? Rotate? With the stronger force on the bottom does it translate? Rotate? Is there a net force in any of these cases? Is there a net torque about the center of mass?

__Demonstration 4:__ **Vertical board with axis through center**. Mount the board on the support rod so that it is free to rotate. After the prediction and discussion steps are completed, test the various predictions using the spring scales. The force 2F should be applied so that 1) its moment arm is half that of the force F (1/4 the length of the board). It could act on the top half of the board toward the right or on the bottom half toward the left, in both cases midway between the end of the board and the axis.

Discussion after observing the results: How are torques calculated? Why is the net torques zero in these cases? Why is there no torque caused by the gravitational force on the board?

__Demonstration 5:__ **Horizontal board with axis through center.** Tell the students the actual weight of the board before they make their predictions. After the prediction and discussion steps are completed, apply a downward force on the left end, and find a second force that keeps the board from rotating. There are a number of possibilities, e.g., force F downward on the right end, force 2F downward midway between the center and the right end, 2F upward midway between the center and the left end, F upward at the left end.

Discussion after observing the results: What is the net torque in each of these cases? What are the details of calculating the torques? Why is there no torque caused by the gravitational force of magnitude Mg acting on the board? Is the board in balance when F does not act on it?

__Demonstration 6:__ **Horizontal board with axis off-center.** Tell the students the actual weight of the board before they make their predictions. After the prediction and discussion steps are completed, apply a downward force on the left end, and find a second force that keeps the board from rotating. There are a number of possibilities.

Discussion after observing the results: How is this situation different from Demonstration 5? Is the board in balance when F does not act on it? Why is there now a torque caused by the gravitational force on the board? For each case where the board is in equilibrium, what is the net torque?

__Demonstration 7:__ **Stable and unstable equilibrium.** You can use the block from Demonstrations 1-3. After the prediction and discussion steps, tilt the block to several angles, and show that up to a certain angle it tends to return to equilibrium. Show that beyond that angle, it tends to fall down.

Discussion after observing the results: When you tilt the block, why does it tend to rotate? What is the cause of the torque on it? What point is it rotating about? What determines if the torque tends to rotate it back or tends to cause it to fall over? What is equilibrium? What is *stable* equilibrium? What is unstable equilibrium?

STATICS (STAT)
TEACHER PRESENTATION NOTES

Demonstration 1: Opposite forces along the same line through center. At first pull with equal forces, then with a stronger force on the right and finally with a stronger force on the left.

- With the equal and opposite forces does the block translate? Does it rotate? With the stronger force on the right does it translate? Rotate? With the stronger force on the left?
- Is there a net force in any of these cases? Is there a net torque about the center of mass? What are the two conditions for equilibrium?

Demonstration 2: Two forces acting at the ends in opposite directions. At first pull with equal forces, then with a stronger force on the top, and then with a stronger force on the bottom.

- With the equal and opposite forces does the block translate? Does it rotate? With the stronger force on the top does it translate? Rotate? With the stronger force on the bottom?
- Is there a net force in any of these cases? Is there a net torque about the center of mass?

Demonstration 3: Two forces acting at the ends in the same direction. At first pull with equal forces, then with a stronger force on the top, and then with a stronger force on the bottom.

- With the equal and opposite forces does the block translate? Does it rotate? With the stronger force on the top does it translate? Rotate? With the stronger force on the bottom?
- Is there a net force in any of these cases? Is there a net torque about the center of mass?

Demonstration 4: Vertical board with axis through center. The force 2F should be applied so that its moment arm is half that of the force F (1/4 the length of the board). It could act on the top half of the board to the right or on the bottom half toward the left.

- How are torques calculated? Why is the net torques zero in these cases?
- Why is there no torque caused by the gravitational force on the board?

Demonstration 5: Horizontal board with axis through center. There are a number of possibilities: force F downward on the right end, force 2F downward midway between the center and the right end, 2F upward midway between the center and the left end, F upward at the left end.

- What is the net torque in each of these cases? What are the details of calculating the torques?
- Why is there no torque caused by the gravitational force Mg? Is the board in balance when F does not act on it?

Demonstration 6: Horizontal board with axis off-center. There are a number of possibilities.

- How is this situation different from Demonstration 5? Is the board in balance when F does not act on it? Why is there now a torque caused by the gravitational force on the board?
- For each case where the board is in equilibrium, what is the net torque?

Demonstration 7: Stable and unstable equilibrium. Beyond a certain angle, the block falls down.

- What is the cause of the torque on the block when you tilt it? What point is it rotating about?
- What determines if the torque tends to rotate it back or tends to cause it to fall over? What is equilibrium? What is *stable* equilibrium? What is *unstable* equilibrium?

Fluid Statics (FLUS)

Hand in this sheet **Name**______________________________

INTERACTIVE LECTURE DEMONSTRATION
PREDICTION SHEET—**FLUID STATICS**

Directions: This sheet will be collected. Write your name at the top to record your presence and participation in these demonstrations. Follow your instructor's directions. You may write whatever you wish on the attached Results Sheet and take it with you.

Demonstration 1: A cylinder of density greater than the density of water ($\rho_{cylinder} > \rho_{water}$) is hung from a force probe with a rigid rod. It is lowered slowly into a container of water. On the axes on the right, sketch your prediction for the force probe reading as a function of time. (Assume that a pull on the force probe is a positive net force, and a push is a negative net force.) Be sure to include the initial reading before the cylinder touches the water, and also the reading when the cylinder is completely submerged.

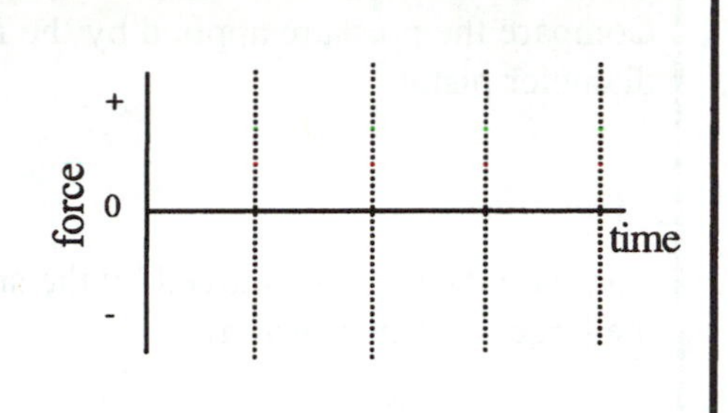

Demonstration 2: The cylinder is replaced by one that is the same size (volume) as the one in Demonstration 1, but has density equal to the density of water($\rho_{cylinder} = \rho_{water}$). As in Demonstration 1, it is hung from a rod and lowered slowly into a container of water. On the axes on the right, sketch your prediction for the force probe reading as a function of time. Be sure to include the initial reading before the cylinder touches the water, and also the reading when the cylinder is completely submerged. Also pay attention to the difference between these two values.

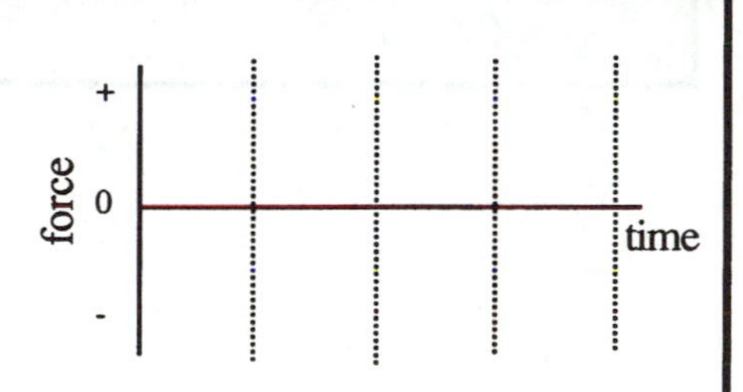

Demonstration 3: The cylinder is replaced by one that is the same size (volume) as the one in Demonstrations 1 and 2, but has density less than the density of water($\rho_{cylinder} < \rho_{water}$). As in Demonstrations 1 and 2, it is hung from a rod and lowered slowly into a container of water. On the axes on the right, ketch your prediction for the force probe reading as a function of time. Be sure to include the initial reading before the cylinder touches the water, and also the reading when the cylinder is completely submerged. Also pay attention to the difference between these two values.

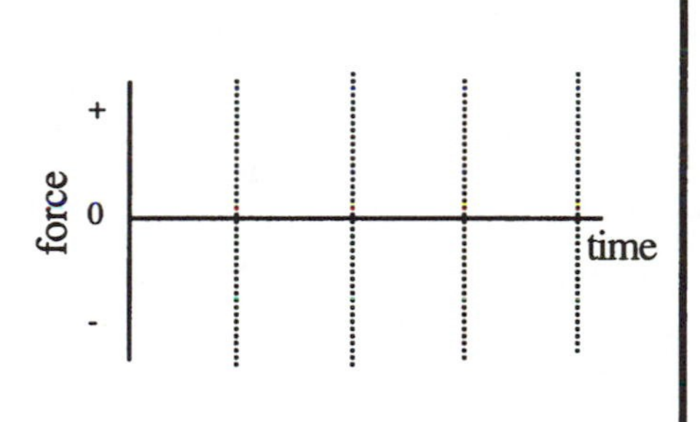

Demonstration 4: Suppose that the cylinders in Demonstrations 1, 2 and 3 were hung from a string instead of from a rigid rod. Would there be any difference in the force vs. time? For each demonstration, either use a dotted line to indicate any differences in the graph above, or write SAME next to the graph. In each case, is there a difference in the maximum buoyant force exerted on the cylinder? Explain.

Demonstration 5: A cylindrical weight is placed on the larger diameter piston (constant force). When the smaller diameter piston is pushed down at a steady rate with a constant force, the weight rises at a steady rate.

Compare the force applied to the smaller diameter piston to the weight of the cylinder.

Compare the pressure applied by the fluid to the smaller diameter piston to that applied to the larger diameter piston.

Compare the distance moved by the smaller diameter piston to the distance moved by the weight (and the larger diameter piston).

Keep this sheet

INTERACTIVE LECTURE DEMONSTRATION
RESULTS SHEET—**FLUID STATICS**

You may write whatever you wish on this sheet and take it with you.

Demonstration 1: A cylinder of density greater than the density of water ($\rho_{cylinder} > \rho_{water}$) is hung from a force probe with a rigid rod. It is lowered slowly into a container of water. On the axes on the right, sketch your prediction for the force probe reading as a function of time. (Assume that a pull on the force probe is a positive net force, and a push is a negative net force.) Be sure to include the initial reading before the cylinder touches the water, and also the reading when the cylinder is completely submerged.

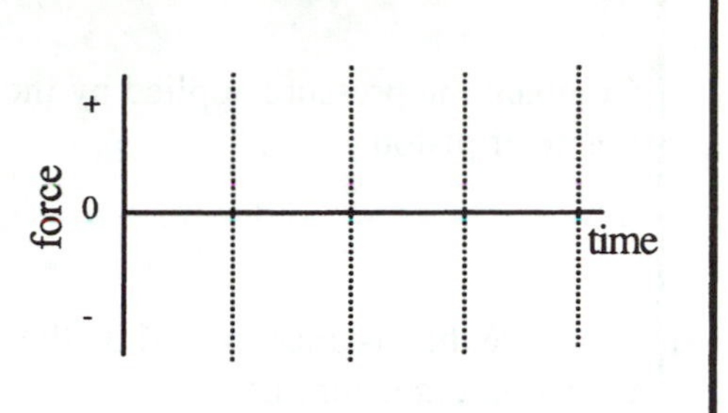

Demonstration 2: The cylinder is replaced by one that is the same size (volume) as the one in Demonstration 1, but has density equal to the density of water($\rho_{cylinder} = \rho_{water}$). As in Demonstration 1, it is hung from a rod and lowered slowly into a container of water. On the axes on the right, sketch your prediction for the force probe reading as a function of time. Be sure to include the initial reading before the cylinder touches the water, and also the reading when the cylinder is completely submerged. Also pay attention to the difference between these two values.

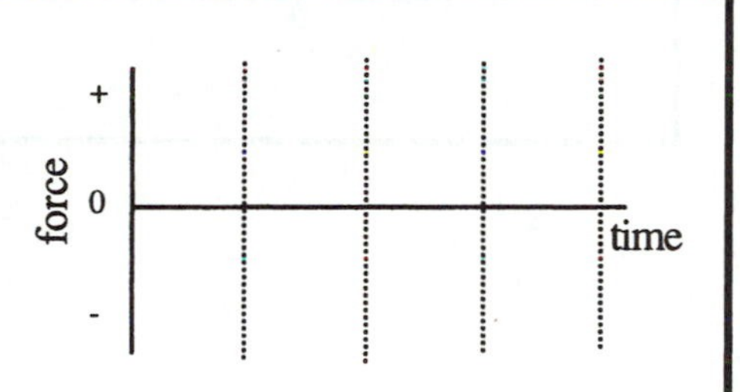

Demonstration 3: The cylinder is replaced by one that is the same size (volume) as the one in Demonstrations 1 and 2, but has density less than the density of water($\rho_{cylinder} < \rho_{water}$). As in Demonstrations 1 and 2, it is hung from a rod and lowered slowly into a container of water. On the axes on the right, ketch your prediction for the force probe reading as a function of time. Be sure to include the initial reading before the cylinder touches the water, and also the reading when the cylinder is completely submerged. Also pay attention to the difference between these two values.

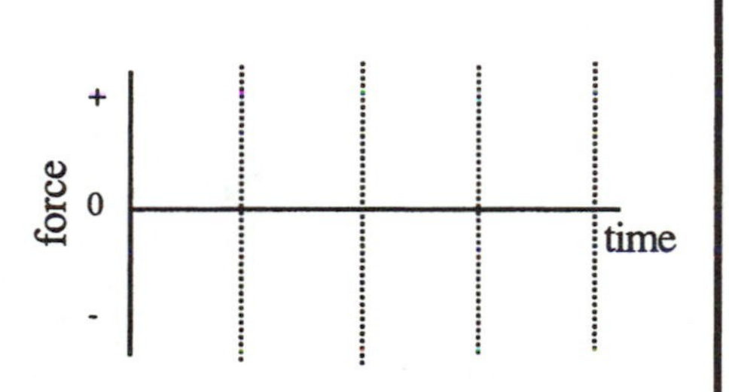

Demonstration 4: Suppose that the cylinders in Demonstrations 1, 2 and 3 were hung from a string instead of from a rigid rod. Would there be any difference in the force vs. time? For each demonstration, either use a dotted line to indicate any differences in the graph above, or write SAME next to the graph. In each case, is there a difference in the maximum buoyant force exerted on the cylinder? Explain.

Demonstration 5: A cylindrical weight is placed on the larger diameter piston (constant force). When the smaller diameter piston is pushed down at a steady rate with a constant force, the weight rises at a steady rate.

Compare the force applied to the smaller diameter piston to the weight of the cylinder.

Compare the pressure applied by the fluid to the smaller diameter piston to that applied to the larger diameter piston.

Compare the distance moved by the smaller diameter piston to the distance moved by the weight (and the larger diameter piston).

FLUID STATICS (FLUS)
TEACHER'S GUIDE

Prerequisites:

Students should already have studied Pascal's and Archimedes' principles either in lecture or in their texts. However, this *ILD* sequence could also be used as an introduction to these topics. Students should have been introduced to the terms pressure, buoyant force, Archimedes' principle and Pascal's principle.

Equipment:

force probe with rod (See below.)
stand and support rod
beaker filled with water
lab jack
35 mm film container filled with water (See below.)
aluminum (or any metal) cylinder the same size as the film container
wood cylinder the same size as the film container
hydraulic press apparatus (See below.)
200 g cylindrical mass

General Notes on Preparation and Equipment:

Force probe with rod:
Either the Vernier (www.vernier.com) Dual-Range Force Sensor (DFS-BTA), PASCO (www.pasco.com) Force Sensor (CI-6537) or PASCO Economy Force Sensor (CI-6746) will work well for these demonstrations. The force probe should be mounted on the stand and support rod, and fitted with a rod to which the cylinders can be rigidly attached. The setup is shown in Figure II-32.

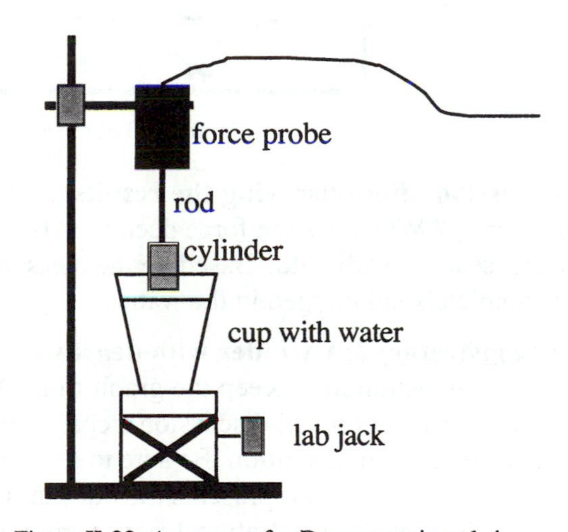

Figure II-32: Apparatus for Demonstrations 1-4.

35 mm film container:
A film container filled with water will have nearly the same average density as water. The container should have a screw-eye attached to the top so that it can be fastened rigidly to the rod with tape.

Hydraulic press apparatus: This apparatus can be constructed from two glass syringes of different sizes supported with stands and test tube clamps and connected by tubing. Glass syringes of various sizes (10 ml, 20 ml, 30 ml, and 50 ml and larger) are available from PGC Scientifics Corporation (www.pgcsci.com). If the syringes are kept very clean, they have very little friction. The pistons can be cleaned with detergent and rinsed well with distilled water.

PASCO makes a hydraulic system (SE-8764), although an additional support for the second syringe will be needed.

Demonstrations and Sample Results:

Demonstration 1: Metal cylinder with density greater than water. (Use experiment configuration file **FLUSD1** to display force vs. time axes.) Zero the force probe with the rod attached to it, but without the cylinder. Attach the metal cylinder rigidly to the bottom of the rod. After the prediction and discussion steps, begin graphing, and slowly but uniformly raise the beaker until the cylinder is completely submerged. Figure II-33 shows typical graphs for Demonstrations 1 and 2. Note the starting and ending force probe readings, and their difference. The latter is the buoyant force with the cylinder completely submerged. Keep this graph persistently displayed on the screen for comparison in Demonstration 2.

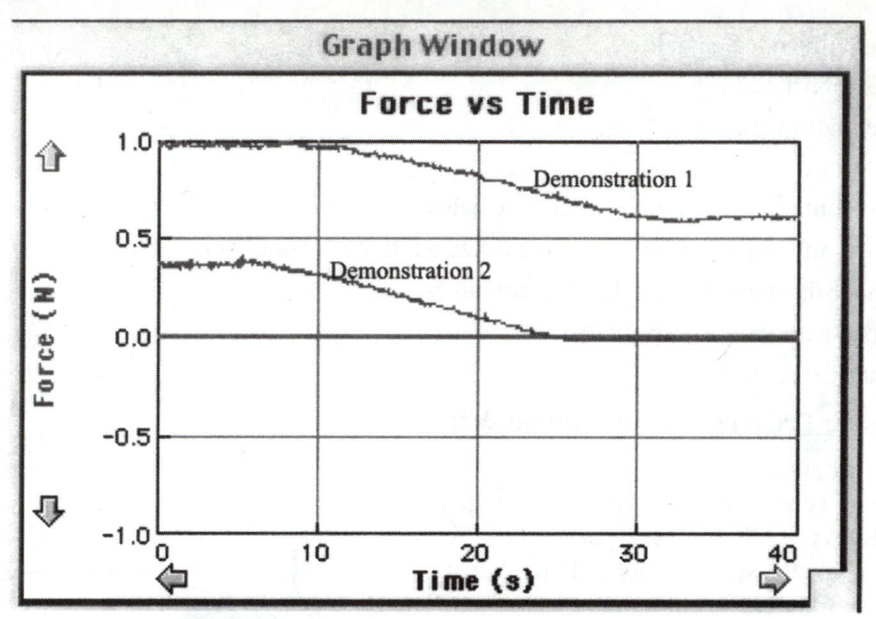

Figure II-33 Force vs. time graphs for Demonstrations 1 and 2.

Discussion after observing the results: Ask students to describe the graph. What is the force probe measuring? Why does the force decrease? How does Archimedes' principle explain this? Why does it decrease at a steady rate? Based on the measurements, how large is the buoyant force when the cylinder is completely submerged in the water?

Demonstration 2: Cylinder with density equal to water. (Use the same experiment configuration file as in Demonstration 1. Keep the graph from Demonstration 1 persistently displayed on the screen.) After the prediction and discussion steps, graph as in Demonstration 1. Figure II-33 shows a typical graph for this demonstration. Be sure to record the beginning force and the ending force, and their difference. This again represents the buoyant force for the cylinder completely submerged. Keep this and the graph from Demonstration 1 persistently displayed for comparison in Demonstration 3.

Discussion after observing the results: Ask students to describe the graph. Why does the shape of the graph look similar to Demonstration 1? What is the meaning of the starting force? What is the final force with the cylinder completely submerged? Why is it zero? How does the buoyant force with the cylinder completely submerged compare to that in Demonstration 1? How does Archimedes' principle explain this? Does the buoyant force depend on what the cylinder is made from? What does it depend on?

Demonstration 3: Wood cylinder with density less than water. (Use the same experiment configuration file as in Demonstrations 1 and 2. Keep the graphs from Demonstrations 1 and 2 persistently displayed on the screen.) After the prediction and discussion steps, graph as in Demonstration 1. Figure II-34 shows a typical graph for this demonstration. Be sure to record the beginning force and the ending force, and their difference. This again represents the buoyant force for the cylinder completely submerged. Keep this and the graphs from Demonstrations 1 and 2 persistently displayed for comparison in Demonstration 4.

Discussion after observing the results: Ask students to describe the graph. Why does the shape of the graph look similar to Demonstrations 1 and 2? Why is the final force with the cylinder completely submerged negative? How does the buoyant force with the cylinder completely submerged compare to that in Demonstrations 1 and 2? How does Archimedes' principle explain this?

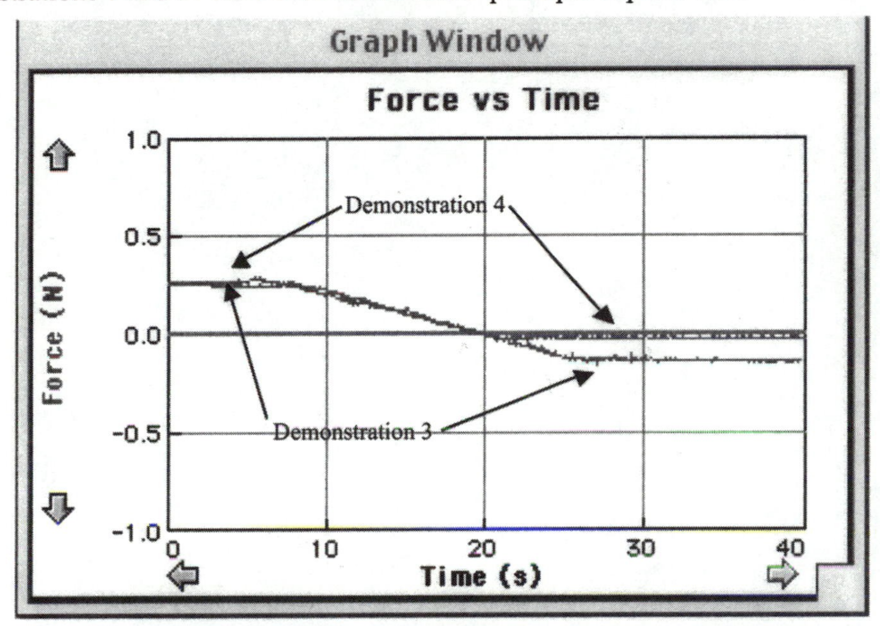

Figure II-34 Force vs. time graphs for Demonstrations 3 and 4.

Demonstration 4: Wood cylinder hanging from string. (Use the same experiment configuration file as in Demonstrations 1, 2 and 3. Keep the graphs from Demonstrations 1, 2 and 3 persistently displayed on the screen.) It will be best to have a screw eye coming out of the side of the wood cylinder to suspend it from the string. The cylinder will be more stable floating in the water on its side. After the prediction and discussion steps, graph as in Demonstration 1. Figure II-34 shows a typical graph for this demonstration. Be sure to record the beginning force and the ending force, and their difference. This again represents the buoyant force for the cylinder completely submerged.

Discussion after observing the results: Ask students to describe the graph. Why is the final force zero this time rather than negative? Is the cylinder completely submerged? Why not? How does the final buoyant force compare to that in Demonstrations 1, 2 and 3? How does Archimedes' principle explain this?

Demonstration 5: Hydraulic press. Have the hydraulic press apparatus set up. Place the 200 g cylindrical mass on top of the larger diameter piston. You can push down on the smaller piston either with the mass that is needed to just raise the 200 g mass, or with the force probe. In either case, record the force that must be applied to just raise the mass, and also record the weight of the cylindrical mass

[$F_g = mg = (0.200)(9.8)$] = 2.0 N. You can also measure the diameters of the two pistons, and calculate the cross-sectional areas.

Discussion after observing the results: How does the applied force compare to the weight of the cylindrical mass? Can you raise the cylinder with a force smaller than its weight? How does Pascal's principle explain this? How does the pressure against the bottom of the smaller syringe compare to the pressure against the bottom of the larger syringe? How does the force on the smaller syringe compare to the force on the larger syringe? If you calculated the cross-sectional areas, how does their ratio compare to the ratio of the forces? How does the distance moved by the smaller piston compare to the distance moved by the larger piston?

FLUID STATICS (FLUS)
TEACHER PRESENTATION NOTES

Demonstration 1: Metal cylinder with density greater than water. Use experiment configuration file **FLUSD1**. Zero the force probe with the rod, but without the cylinder. Attach the metal cylinder rigidly to the bottom of the rod and graph. Note the starting and ending force probe readings, and their difference. Keep this graph persistently displayed.

- What is the force probe measuring? Why does the force decrease? Why does it decrease at a steady rate?
- How does Archimedes' principle explain this?
- How large is the buoyant force when the cylinder is completely submerged in the water?

Demonstration 2: Cylinder with density equal to water. Use the same experiment configuration file. Keep the graph from Demonstration 1 persistently displayed . Record the beginning and the ending forces, and their difference. Keep the graphs persistently displayed.

- Why does the shape of the graph look similar to Demonstration 1? What is the meaning of the starting force?
- What is the final force with the cylinder completely submerged? Why is it zero?
- How does the buoyant force compare to Demonstration 1? How does Archimedes' principle explain this? Does the buoyant force depend on cylinder material? What does it depend on?

Demonstration 3: Wood cylinder with density less than water. Use the same experiment configuration file. Keep the graphs from Demonstrations 1 and 2 persistently displayed. Record the beginning and the ending forces, and their difference. Keep the graphs persistently displayed.

- Why does the shape of the graph look similar to Demonstrations 1 and 2? Why is the final force with the cylinder completely submerged negative?
- How does the buoyant force with the cylinder completely submerged compare to that in Demonstrations 1 and 2? How does Archimedes' principle explain this?

Demonstration 4: Wood cylinder hanging from string. Use the same experiment configuration file. Keep the graphs persistently displayed. Record the forces, and their difference.

- Why is the final force zero this time? Is the cylinder completely submerged? Why not?
- How does the final buoyant force compare to that in Demonstrations 1, 2 and 3? How does Archimedes' principle explain this?

Demonstration 5: Hydraulic press. Place the 200 g cylindrical mass on top of the larger diameter piston. Record the force to raise the mass, and also the weight of the cylindrical mass [$F_g = mg = (0.200)(9.8)$] = 2.0 N.

- How does the applied force compare to the weight of the cylindrical mass? Can you raise the cylinder with a force smaller than its weight?
- How does Pascal's principle explain this? How does the pressure against the bottom of the smaller syringe compare to the pressure against the bottom of the larger syringe? How does the force on the smaller syringe compare to the force on the larger syringe? How does the ratio of the cross-sectional areas compare to the ratio of the forces?
- How does the distance moved by the smaller piston compare to the distance moved by the larger piston?

Section III: Interactive Lecture Demonstrations in Oscillations and Waves

SIMPLE HARMONIC MOTION (SHM)

Hand in this sheet **Name**______________________________

INTERACTIVE LECTURE DEMONSTRATION
PREDICTION SHEET—**SIMPLE HARMONIC MOTION**

Directions: This sheet will be collected. Write your name at the top to record your presence and participation in these demonstrations. Follow your instructor's directions. You may write whatever you wish on the attached Results Sheet and take it with you.

In all cases we consider a mass hanging on a spring. The mass is started in motion by pulling it down a small distance below the equilibrium position and releasing it. The displacement is zero whenever the mass is at its equilibrium position.

The force on the mass is proportional to the displacement--the distance from the equilibrium point.

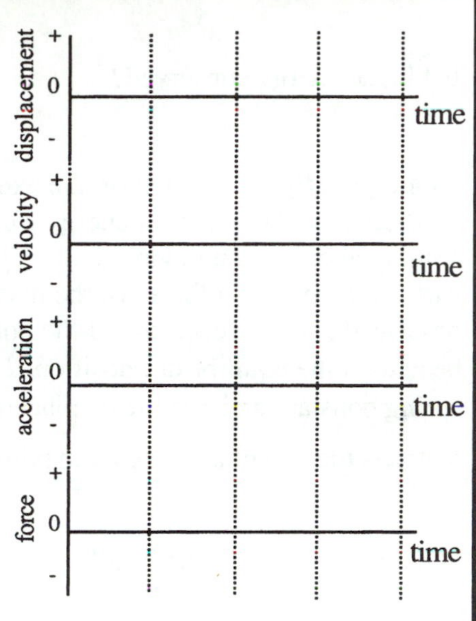

Demonstration 1: Sketch your prediction of the graph of displacement vs. time from watching the motion. Draw it on the displacement axes to the right.

Demonstration 2: From the displacement (position)–time graph and from watching the motion, predict the velocity-time and acceleration-time graphs, and sketch them on the axes to the right.

When is the velocity zero? A maximum?

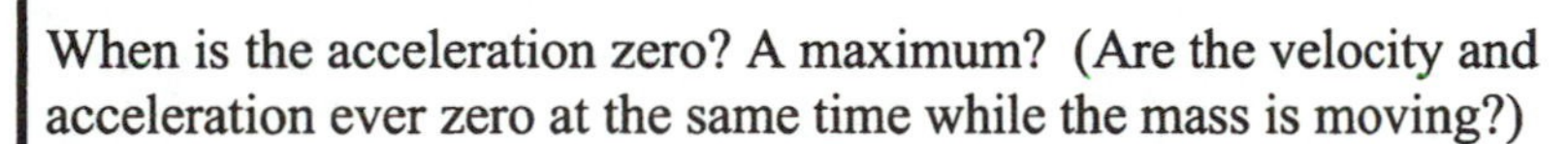

When is the acceleration zero? A maximum? (Are the velocity and acceleration ever zero at the same time while the mass is moving?)

Demonstration 3: From the acceleration-time graph and from watching the motion, predict the force-time graph. Sketch your prediction on the axes to the right.

Compare the force-time graph to the acceleration-time graph. How are they similar? How different?

Is the force ever zero? If so, at what point in the motion?

Demonstration 4: Suppose that you wanted to model the motion (displacement) of the mass hanging from a spring, and write down a mathematical expression for displacement vs. time. What mathematical expression would you use? What additional measurements would you need to make so that the expression would exactly represent the motion?

Demonstration 5: Sketch on the axes to the right your prediction of the kinetic energy (the energy due to motion) of the mass vs. time as it moves as described above. Recall that the kinetic energy, $K = \frac{1}{2}mv^2$, where m is the mass of the moving object, and v is its velocity.

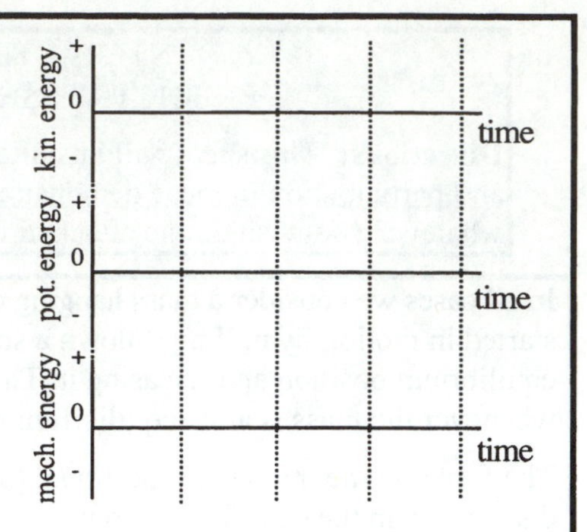

When is the kinetic energy zero? Maximum?

Is kinetic energy conserved?

Demonstration 6: Sketch on the axes to the right your prediction of the potential energy (the energy due to the combined effects on the mass of the gravitational force of the earth and the spring force) of the mass vs. time as it moves as described above. Recall that if the potential energy is defined to be zero at the equilibrium position, $U = \frac{1}{2}kx^2$, where k is the spring constant and x is the displacement.

When is the potential energy a maximum?

Is potential energy conserved?

Demonstration 7: Sketch on the axes to the right your prediction of the mechanical energy, $E^{mech} = K + U$, of the mass over time as it moves as described above.

Is the mechanical energy ever zero?

Is mechanical energy conserved? Explain what conserved means.

Where does the mass get its initial energy?

Keep this sheet

INTERACTIVE LECTURE DEMONSTRATION
RESULTS SHEET—**SIMPLE HARMONIC MOTION**

You may write whatever you wish on this sheet and take it with you.

In all cases we consider a mass hanging on a spring. The mass is started in motion by pulling it down a small distance below the equilibrium position and releasing it. The displacement is zero whenever the mass is at its equilibrium position.

The force on the mass is proportional to the displacement--the distance from the equilibrium point.

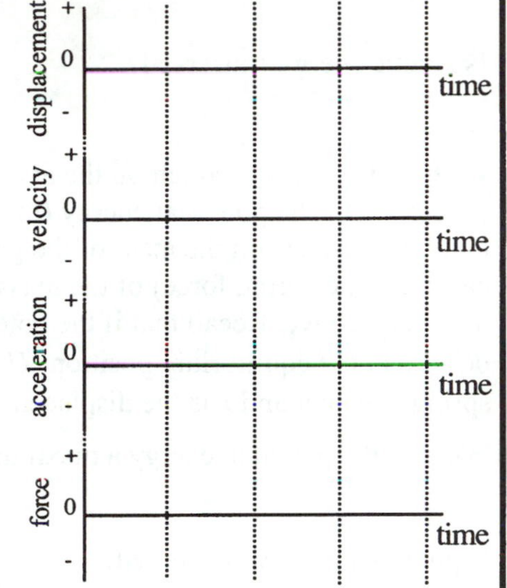

Demonstration 1: Sketch your prediction of the graph of displacement vs. time from watching the motion. Draw it on the displacement axes to the right.

Demonstration 2: From the displacement (position)–time graph and from watching the motion, predict the velocity-time and acceleration-time graphs, and sketch them on the axes to the right.

When is the velocity zero? A maximum?

When is the acceleration zero? A maximum? (Are the velocity and acceleration ever zero at the same time while the mass is moving?)

Demonstration 3: From the acceleration-time graph and from watching the motion, predict the force-time graph. Sketch your prediction on the axes to the right.

Compare the force-time graph to the acceleration-time graph. How are they similar? How different?

Is the force ever zero? If so, at what point in the motion?

Demonstration 4: Suppose that you wanted to model the motion (displacement) of the mass hanging from a spring, and write down a mathematical expression for displacement vs. time. What mathematical expression would you use? What additional measurements would you need to make so that the expression would exactly represent the motion?

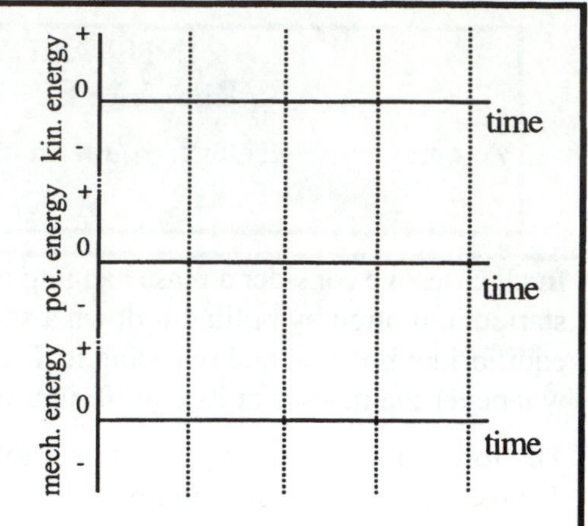

Demonstration 5: Sketch on the axes to the right your prediction of the kinetic energy (the energy due to motion) of the mass vs. time as it moves as described above. Recall that the kinetic energy, $K = \frac{1}{2}mv^2$, where m is the mass of the moving object, and v is its velocity.

When is the kinetic energy zero? Maximum?

Is kinetic energy conserved?

Demonstration 6: Sketch on the axes to the right your prediction of the potential energy (the energy due to the combined effects on the mass of the gravitational force of the earth and the spring force) of the mass vs. time as it moves as described above. Recall that if the potential energy is defined to be zero at the equilibrium position, $U = \frac{1}{2}kx^2$, where k is the spring constant and x is the displacement.

When is the potential energy a maximum?

Is potential energy conserved?

Demonstration 7: Sketch on the axes to the right your prediction of the mechanical energy, $E^{mech} = K + U$, of the mass over time as it moves as described above.

Is the mechanical energy ever zero?

Is mechanical energy conserved? Explain what conserved means.

Where does the mass get its initial energy?

SIMPLE HARMONIC MOTION (SHM)
TEACHER'S GUIDE

Prerequisites:

Students should have already done the *Kinematics 1—Human Motion, Kinematics 2—Motion of Carts, Newton's* 1st *&* 2nd *Laws* and *Energy of a Cart on a Ramp ILD* sequences. If they have done *RealTime Physics Mechanics* Labs 1-5 and 11-12 or *Tools for Scientific Thinking Motion and Force* Labs 1-4, you can skip the prerequisite *ILDs.*

Equipment:

computer-based laboratory system
motion detector (See below.)
force probe (See below.)
ILD experiment configuration files
0.5 kg hanging mass
brass spring (See below.)
base, support rod (ring stand) and cross rod (See below.)
c-clamp or table clamp (See below.)
screen for motion detector (Optional—see below.)

General Notes on Preparation and Equipment:

The motion detector, force probe and screen:
See the *Kinematics 1—Human Motion* Teacher's Guide for information on motion detectors, and the *Kinematics 2—Motion of Carts* Teacher's Guide for information on force probes. It is advisable to protect the motion detector with a screen fabricated from course, stiff wire mesh.

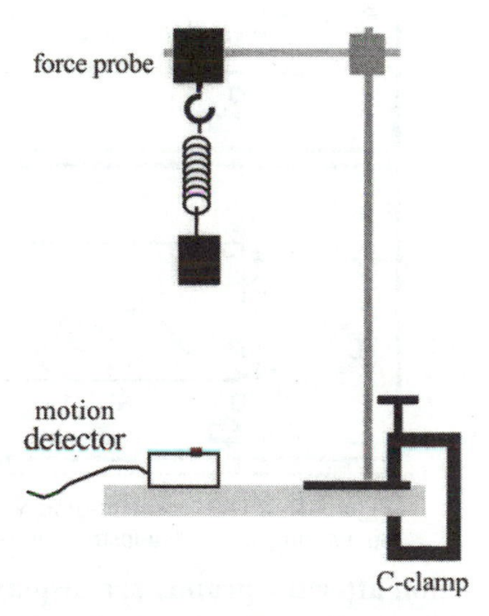

Hanging mass, brass spring and base and support:
In order for students to see the motion clearly, the frequency of the oscillations should be 1 Hz or less. A 0.5 kg hanging mass and a brass spring (Sargent Welch CP75490-00 with spring constant 9.6 N/m) (http://www.sargentwelch.com) are ideal. The base, support rod and cross rod should be as rigid as possible to avoid complex coupled oscillations. It is suggested that you clamp the base to the table with a c-clamp, or use a table clamp to support the rod.

If you are using a motion detector with a minimum distance of 0.5 m, the force probe, spring and mass should be mounted so that the mass doesn't get closer than this at the lowest point in its oscillations.

Equilibrium position:
The oscillations of the mass are most easily analyzed relative to the new equilibrium position after the mass is hung from the spring. This distance from the motion detector can easily be measured with the motion detector. The experiment configuration file is set up to display measurements relative to this new equilibrium position when the value is entered into the data column used to calculate displacement.

Demonstrations and Sample Graphs:

Demonstration 1: Displacement-time graph. (Use experiment configuration file **SHMD1**.) Open window 1 with axes for the displacement vs. time. Be sure to enter the equilibrium distance from the motion detector so that displacements relative to this position are displayed. Remind the students that displacement means "position relative to the equilibrium position."

Try to make the mass oscillate up and down without swinging back and forth. Remember that the mass should not come closer than 0.5 m from the motion detector. A typical displacement-time graph is shown in Figure III-1.

You need only take one set of data for all of the demonstrations in this sequence. Save the data so that the graph is persistently displayed on the screen.

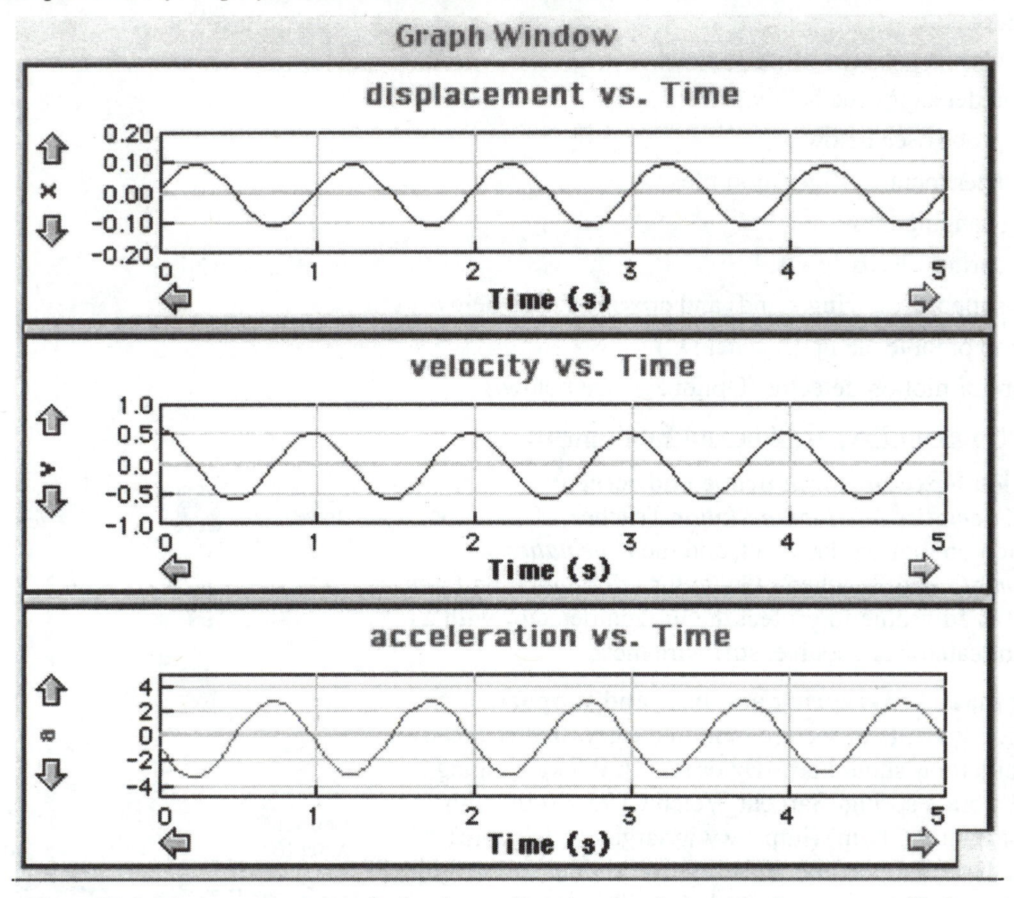

Figure III-1: Displacement-time, velocity-time and acceleration-time graphs for a mass oscillating on a spring as in Demonstrations 1-3.

Discussion after the graphs are displayed: Ask students to describe the displacement-time graph, and explain how it represents the motion of the oscillating mass. Ask them the significance of the initial displacement on the graph at t=0, and point out that there is an arbitrary phase (ϕ) determined by when the motion detector starts measuring. You might also point out the definitions of amplitude (A) and frequency (f), and ask how they could be determined from the graph. (This will be useful for Demonstration 4.)

Demonstration 2: Velocity-time and acceleration-time graphs. (Use the experiment configuration file as in Demonstration 1.) You can use the data from Demonstration 1. Open window 2 with axes for displacement-time, velocity-time and acceleration-time graphs. The velocity-time and acceleration-time graphs are initially hidden. Show the mass oscillating again before students make their predictions. See Figure III-1 for typical graphs.

Discussion after the graphs are displayed: Show the velocity-time and acceleration-time graphs. Use the examine feature of the software to compare the locations of the peaks of displacement, velocity and acceleration. Ask students what is the displacement when velocity is zero? When velocity is maximum? What is the displacement when acceleration is zero? When acceleration is maximum? What are velocity and acceleration when the mass is at its equilibrium position (displacement zero)? What are the velocity and acceleration when the mass is at its maximum position?

Demonstration 3: Force-time graph. (Use the same experiment configuration file.) You can use the data from Demonstration 1. Open window 3 that has axes for acceleration-time and force-time graphs. The force-time graph is initially hidden. See Figure III-2 for typical graphs.

Discussion after the graphs are displayed: Show the force-time graph. Use the examine feature of the software to compare the locations of the peaks of acceleration and force. Ask students what the relationship is? Do the peaks occur at the same times? Are the values identical? What is the physical law that defines the relationship between acceleration and force? (Newton's 2nd Law, $F^{net} = ma$.)

Demonstration 4: Model of the motion. (Use the same experiment configuration file.) Open window 4 that displays only the displacement-time graph.

Discussion after the graphs are displayed: Ask students to describe the functional relationship between displacement and time. It is a sinusoidal one. Ask them what are the parameters that would go into the equation. You should show them the relationship $x = A\sin(2\pi ft + \phi)$, and ask them what each term represents. How are these mathematical terms related to physical attributes of the mass hanging from the spring, and how could you measure them? (A is the amplitude, the maximum displacement. This can be measured from the value of the peak displacement on the graph. f is the frequency, the number of oscillations per second. It is also related to the mass and spring constant by $f = \frac{1}{2\pi}\sqrt{\frac{k}{m}}$. The spring constant, k, can be measured using $F = kx$, finding the displacement x for a given force, $F = mg$, applied to the spring. t is the time, and $t = 0$ is when the motion detector began measuring. ϕ is a phase angle related to where the mass was in its motion at $t = 0$.

Experiment configuration file **SHMD1** has a model for x as a function of t incorporated into it that you can display along with the graph of the actual data. You can enter the measured values for A, k, m and f into the model, and adjust ϕ to get the graph of the model to match the actual data.

Demonstration 5: Kinetic energy of the oscillator. (Use the same experiment configuration file.) You can use the data from Demonstration 1. Open window 5 with axes for velocity vs. time and kinetic energy vs. time. Initially the kinetic energy graph is hidden. A typical graph for kinetic energy vs. time is shown in Figure III-3.

Discussion after the graphs are displayed: Show the kinetic energy-time graph, and show students the equation in the data column used to calculate kinetic energy, e.g. (0.5*1*"vel"^2), with the actual mass entered instead of 1. (Depending on the spring you are using, you may need to add some fraction of the spring's mass to the hanging mass, since part of the spring is also oscillating. Typically for brass, conical springs, this is about one third of the spring's mass.)

Use the examine feature to locate the zeros and maximums of the kinetic energy. Where in the actual motion of the spring do these occur? Ask the meaning of "conserved." Is kinetic energy conserved?

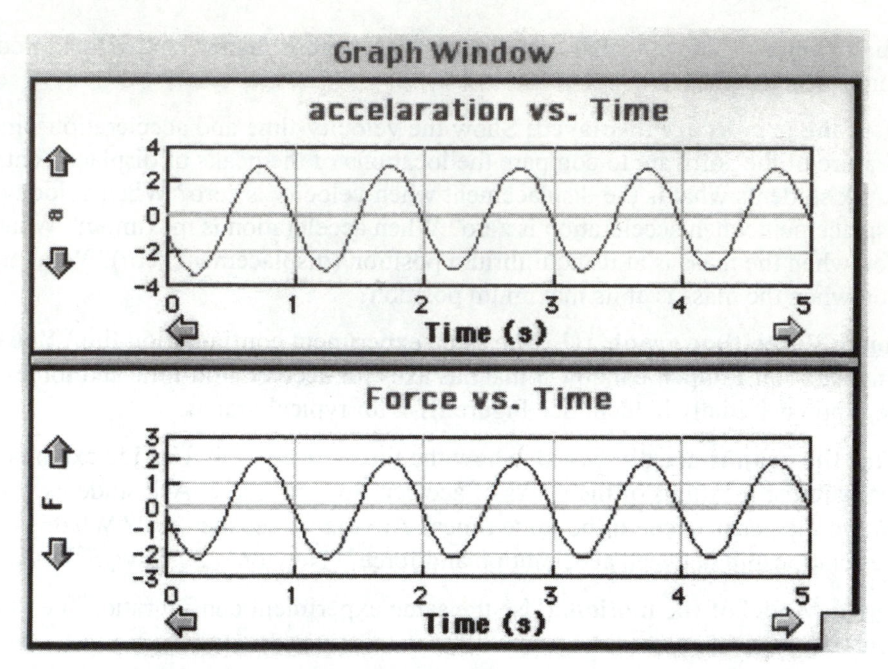

Figure III-2: Acceleration-time and force-time graphs for a mass oscillating on a spring as in Demonstrations 1-3.

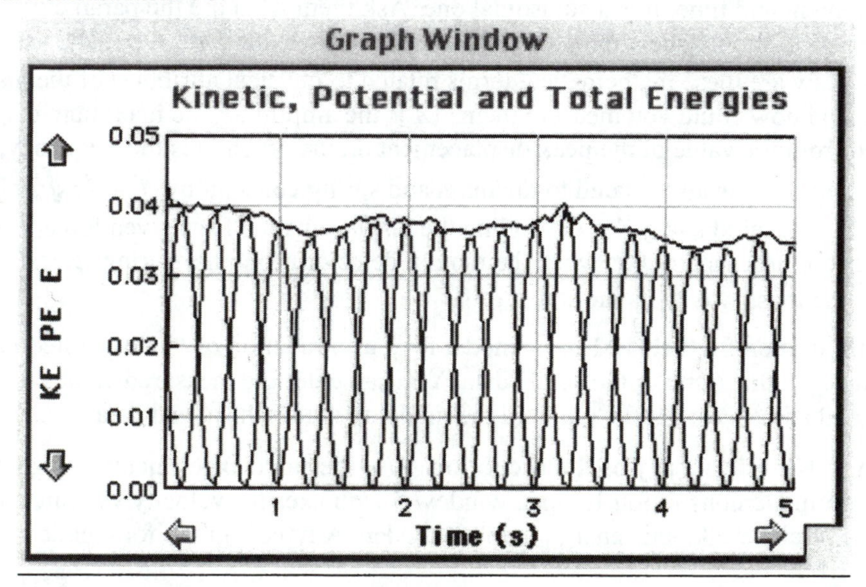

Figure III-3: Kinetic energy-time, potential energy-time and total mechanical energy-time graphs for a mass oscillating on a spring as in Demonstrations 1-3.

Demonstration 6: Potential energy of the oscillator. (Use the same experiment configuration file.) You can use the data from Demonstration 1. Open window 6 with axes for displacement vs. time and potential energy vs. time. Initially potential energy vs. time is hidden. A typical graph for potential energy vs. time is shown in Figure III-3.

Discussion after the graphs are displayed: Show the potential energy graph, and show students the equation in the data column used to calculate potential energy, e.g. (0.5*1*"displ"^2), with the actual spring constant entered instead of 1.

Use the examine feature to locate the zeros and maximums of the potential energy. Where in the actual motion of the spring do these occur? Is potential energy conserved? Why is it only necessary to consider the elastic potential energy of the spring, and not the gravitational potential energy of the mass? (Examining motion around the new equilibrium position already takes into account and cancels out the gravitational force and potential energy.)

Demonstration 7: Mechanical energy of the oscillator. (Use the same experiment configuration file.) You can use the data from Demonstration 1. Open window 7 with axes for kinetic energy, potential energy and mechanical energy vs. time. Initially only kinetic and potential energies are displayed. A typical graph for mechanical energy vs. time is shown in Figure III-3.

Discussion after the graphs are displayed: Show the mechanical energy graph. Display the mechanical energy, and show students the equation in the data column used to calculate mechanical energy, e.g. ($E=K+U$). Is the mechanical energy conserved? Why?

SIMPLE HARMONIC MOTION (SHM)
TEACHER PRESENTATION NOTES

Classroom introduction to *Simple Harmonic Motion:*
Show the students the mass hanging from the spring, and show them the equilibrium position. Show them the location of the motion detector, and tell them that it is measuring the displacement from the equilibrium position.

Demonstration 1: Displacement-time graph. Use experiment configuration file **SHMD1**. Open window 1 with only displacement-time displayed. Pull the mass down, and release it.

- After showing the graph, save the data so that they are persistently displayed on the screen.
- Ask the students for a description of the graph in terms of the motion.
- Ask for the significance of the displacement at t=0.
- Point out the definitions of A, f and ϕ.

Demonstration 2: Velocity-time graph. Use the same experiment configuration file. Use the data from Demonstration 1. Open window 2 with displacement-time shown and velocity-time and acceleration-time hidden. Show the mass oscillating again before students make their predictions.

- Ask students what is the displacement when velocity is zero? When velocity is maximum?
- What is the displacement when acceleration is zero? When acceleration is maximum?
- What are velocity and acceleration when the mass is at its equilibrium position (displacement zero)?
- What are the velocity and acceleration when the mass is at its maximum position?

Demonstration 3: Force-time graph. Use the same experiment configuration. Use the data from Demonstration 1. Open window 3 with acceleration-time displayed and force-time hidden. Show the mass oscillating again before students make their predictions.

- Ask students what the relationship is between a and F? Do the peaks occur at the same times? Are the values identical?
- What is the physical law that defines the relationship? (Newton's 2nd Law, $F^{net}=ma$.)

Demonstration 4: Model of the motion. Use the same experiment configuration file. Open window 4 that displays only the displacement-time graph. Show the mass oscillating again before students make their predictions.

- Ask students to describe the functional relationship between displacement and time. Ask them what are the parameters that would go into the equation.
- Show them the relationship $x = A\sin(2\pi ft + \phi)$, and ask them what each term represents. How are these mathematical terms related to physical attributes of the mass hanging from the spring, and how could you measure them?
- Display the mathematical model for x as a function of t in experiment configuration file **SHMD1** along with the graph of the actual data. Enter the measured values for A, k, m and f into the model, and adjust ϕ to get the graph of the model to match the actual data.

Demonstration 5: Kinetic energy of the oscillator. Use the same experiment configuration file. Use the data from Demonstration 1. Open window 5 with velocity vs. time shown and kinetic energy vs. time hidden. Show the mass oscillating again before students make their predictions.

- Show the kinetic energy-time graph, and show the equation in the data column used to calculate kinetic energy, e.g. (0.5*1*"vel"^2), with the actual mass entered instead of 1.
- Locate the zeros and maximums of the kinetic energy. Where in the motion do these occur?
- Ask the meaning of "conserved." Is kinetic energy conserved?

<u>Demonstration 6:</u> **Potential energy of the oscillator.** Use the same experiment configuration file. Use the data from Demonstration 1. Open window 6 with displacement vs. time displayed and potential energy vs. time hidden. Show the mass oscillating again before students make predictions.

- Show the potential energy graph, and show the equation in the data column used to calculate kinetic energy, e.g. (0.5*1*"displ"^2), with the actual spring constant instead of 1.
- Use the examine feature to locate the zeros and maximums of the potential energy. Where in the actual motion of the spring do these occur?
- Is potential energy conserved?
- Why is it only necessary to consider the elastic potential energy of the spring, and not the gravitational potential energy of the mass?

<u>Demonstration 7:</u> **Mechanical energy of the oscillator.** Use the same experiment configuration file. You can use the data from Demonstration 1. Open window 7 with kinetic energy and potential energy displayed and mechanical energy hidden. Show the mass oscillating again.

- Display the mechanical energy, and show the equation in the data column used to calculate mechanical energy, e.g. (E=K + U).
- Is the mechanical energy conserved? Why?

SOUND (SND)

Hand in this sheet **Name**__________________________

INTERACTIVE LECTURE DEMONSTRATIONS
PREDICTION SHEET—**SOUND**

Directions: This sheet will be collected. Write your name at the top to record your presence and participation in these demonstrations. Follow your instructor's directions. You may write whatever you wish on the attached Results Sheet and take it with you.

Demonstration 1: A sound with *constant pitch* is emitted by a speaker and captured by a microphone.

Predict the shape of the sound wave's sound pressure vs. ***time*** at the location of the microphone.

Sound Pressure

time

Demonstration 2: A sound with a *higher constant pitch* than the sound in Demonstration 1 is emitted by the speaker.

Predict the shape of the sound wave's sound pressure vs. time for several periods.

Sound Pressure

time

Demonstration 3: A *louder sound with the same constant pitch* as the sound in Demonstration 2 is emitted by the speaker.

Predict the shape of the sound wave's sound pressure vs. time for several periods.

Sound Pressure

time

Demonstration 4: A person speaks in a normal tone of voice into a microphone.

Predict the shape of the sound wave's sound pressure vs. time for several periods.

Sound Pressure

time

Demonstration 5: Two sounds with about the same volume but with *slightly different pitches* are played. Describe what you expect to hear:

Predict the shape of the sound pressure vs. time for the wave that represents the combined sound of the two pitches that are played.

Sound Pressure

time

Demonstration 6: Two sounds with about the same volume are played, but one sound is a *whole octave higher* in pitch than the other. Describe what you expect to hear:

Predict the shape of the sound pressure vs. time for the wave that represents the combined sound of the two pitches that are played.

Sound Pressure

time

Demonstration 7: A short tapping sound is made into a long tube that is open at both ends. The sound pulse is recorded by a microphone near the end, then travels down the tube, is reflected back and recorded again by the microphone. Predict the shape of the wave (sound pressure vs. time) recorded by the microphone.

Based on the actual wave that was recorded by the microphone, how could you determine the approximate speed of sound in the tube? For your information (you may or may not need this information), the instructor will supply you with the diameter and length of the tube.

Record your calculation for the speed of sound in the space on the right.

Sound Pressure

time

Keep this sheet

INTERACTIVE LECTURE DEMONSTRATIONS
RESULTS SHEET— **SOUND**

You may write whatever you wish on this sheet and take it with you.

Demonstration 1: A sound with *constant pitch* is emitted by a speaker and captured by a microphone.

Predict the shape of the sound wave's sound pressure vs. ***time*** at the location of the microphone.

Sound Pressure

time

Demonstration 2: A sound with a *higher constant pitch* than the sound in Demonstration 1 is emitted by the speaker.

Predict the shape of the sound wave's sound pressure vs. time for several periods.

Sound Pressure

time

Demonstration 3: A *louder sound with the same constant pitch* as the sound in Demonstration 2 is emitted by the speaker.

Predict the shape of the sound wave's sound pressure vs. time for several periods.

Sound Pressure

time

Demonstration 4: A person speaks in a normal tone of voice into a microphone.

Predict the shape of the sound wave's sound pressure vs. time for several periods.

Sound Pressure

time

Demonstration 5: Two sounds with about the same volume but with *slightly different pitches* are played. Describe what you expect to hear:

Sound Pressure

time

Predict the shape of the sound pressure vs. time for the wave that represents the combined sound of the two pitches that are played.

Demonstration 6: Two sounds with about the same volume are played, but one sound is a *whole octave higher* in pitch than the other. Describe what you expect to hear:

Sound Pressure

time

Predict the shape of the sound pressure vs. time for the wave that represents the combined sound of the two pitches that are played.

Demonstration 7: A short tapping sound is made into a long tube that is open at both ends. The sound pulse is recorded by a microphone near the end, then travels down the tube, is reflected back and recorded again by the microphone. Predict the shape of the wave (sound pressure vs. time) recorded by the microphone.

Sound Pressure

time

Based on the actual wave that was recorded by the microphone, how could you determine the approximate speed of sound in the tube? For your information (you may or may not need this information), the instructor will supply you with the diameter and length of the tube.

Record your calculation for the speed of sound in the space on the right.

SOUND (SND)
TEACHER'S GUIDE

Prerequisites:

This *ILD* sequence could serve as an introduction to sound, or it could be used as a review of sound concepts. Students should have been introduced to the following terms: wave, frequency, wavelength, period, sound pressure, amplitude and speed of propagation.

Equipment:

computer-based laboratory system
ILD experiment configuration files
microphone (See below.)
keyboard (See below.)
long tube (See below.)

General Notes on Preparation and Equipment:

Microphone:
Microphones for computer-based laboratory systems are available from Vernier (www.vernier.com) (MCA-BTA) and PASCO (www.pasco.com) (CI-6506B).

Keyboard:
A variety of inexpensive keyboards are available. It is important for the keyboard to produce pure tones (pure sine waves). Often the purest tone is produced when the keyboard is simulating the sound of a flute. It should be easy to vary the volume of the sound, to play two notes *simultaneously* that are very close together, and to play two notes *simultaneously* that are an octave apart.

Long tube:
The tube is for Demonstration 7, the determination of the speed of sound. It should be at least 2 m in length. A cardboard carpet roller will work well, as will a piece of PVC pipe.

Demonstrations and Sample Graphs:

Demonstration 1: Sound with constant pitch. (Use experiment configuration file **SNDD1**. Axes for sound pressure vs. time will be displayed. The software will be in repeat mode, so that it will continue to collect and display new samples of sound until graphing is stopped.) Connect the microphone to the computer interface. Play a steady note on the keyboard for the students and ask them to make their predictions. After the prediction and discussion steps, play the same note, and display he graph. A sample graph is shown in Figure II-4.

Discussion after the graph is displayed: Ask students to describe the graph. What is characteristic of a constant pitch? Does this sound wave have a constant frequency? Does it have a constant amplitude? What does the graph actually represent?

Demonstration 2: Sound with constant higher pitch. (Use the same experiment configuration file as in Demonstration 1.) Play a higher steady note, and try to make it just as loud as loud as the sound in Demonstration 1. After the prediction and discussion steps, play the same note, and display the graph.

Discussion after the graph is displayed: Ask students to compare this graph to the one in Demonstration 1. What is similar and what is different? Does changing the pitch mean changing the frequency, changing the amplitude or both?

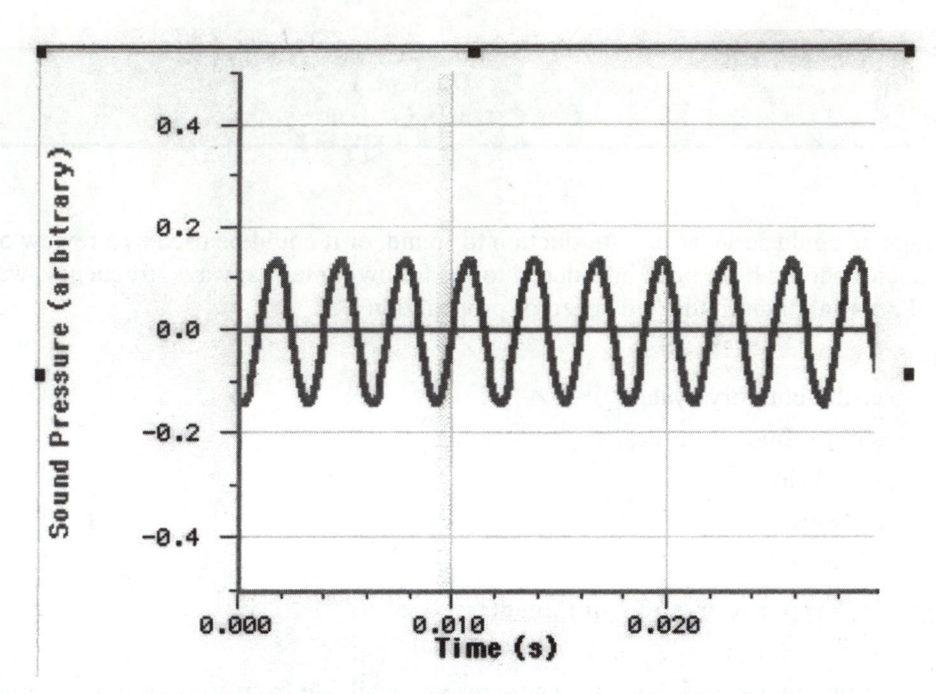

Figure II-4: Sound pressure display for a constant pitch (pure) tone.

Demonstration 3: Louder sound with same pitch. (Use the same experiment configuration file as in Demonstration 1.) Play the same steady note as in Demonstration 2, only make it significantly louder. After the prediction and discussion steps, play the same louder note, and display the graph.

Discussion after the graph is displayed: Ask students to compare this graph to the one in Demonstration 2. What is similar and what is different? Does making a sound louder mean changing the frequency, changing the amplitude or both?

Demonstration 4: Speaking into the microphone. (Use the same experiment configuration file as in Demonstration 1.) After the prediction and discussion steps, speak into the microphone and graph the sound.

Discussion after the graph is displayed: Ask students to compare this graph to the ones in Demonstrations 1-3. What is similar and what is different? Does speaking produce a steady frequency and amplitude? Why not?

Demonstration 5: Two sounds with slightly different pitches. (Use the same experiment configuration file as in Demonstration 1.) It is probably best to break this down into two parts. First ask for predictions of what the resulting sound will sound like. After the prediction and discussion steps, play two adjacent notes from the keyboard simultaneously. An E and F at mid-keyboard work well. Then ask the students to predict the graph. After the prediction and discussion steps, play the two notes simultaneously into the microphone and graph the sound. A sample graph is shown in Figure II-5.

Discussion after the graph is displayed: Ask students to compare this graph to the ones in Demonstrations 1-3. What is similar and what is different? What characteristic of this sound wave results in the characteristic beats that are heard? Is the frequency varying or is the amplitude varying? Why is this happening?

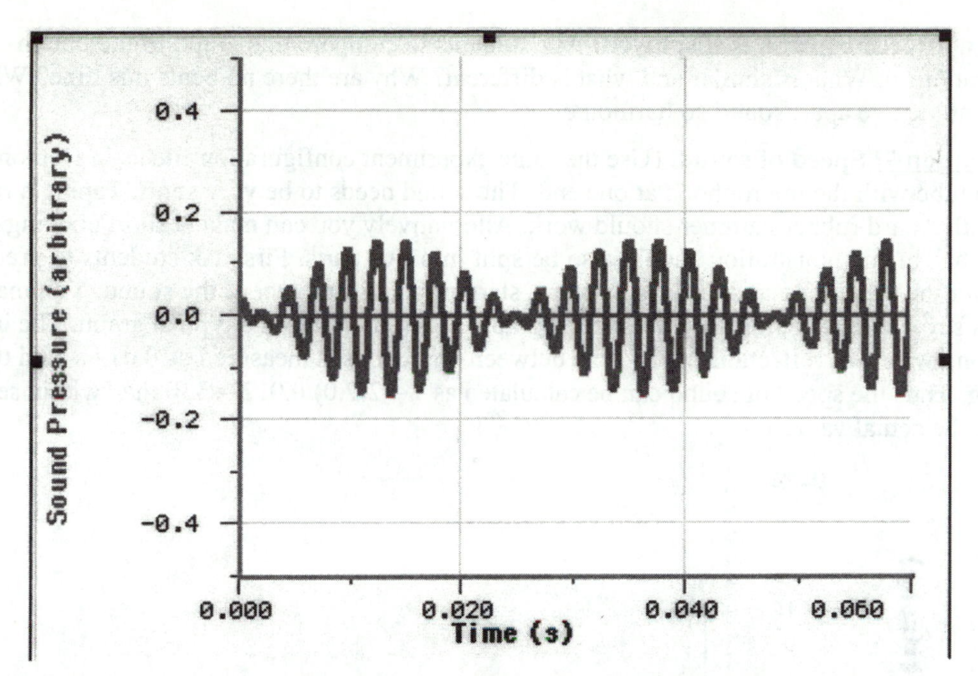

Figure II-5: Sound wave resulting from two notes close together played simultaneously. The amplitude variations correspond to the characteristic beats that are heard.

Demonstration 6: Two sounds an octave apart. (Use the same experiment configuration file as in Demonstration 1.) It is probably best to break this down into two parts. First ask for predictions of what the resulting sound will sound like. After the prediction and discussion steps, play two notes from the keyboard an octave apart simultaneously. Then ask the students to predict the graph. After the prediction and discussion steps, play the two notes simultaneously into the microphone and graph the sound. Figure II-6 shows a typical graph.

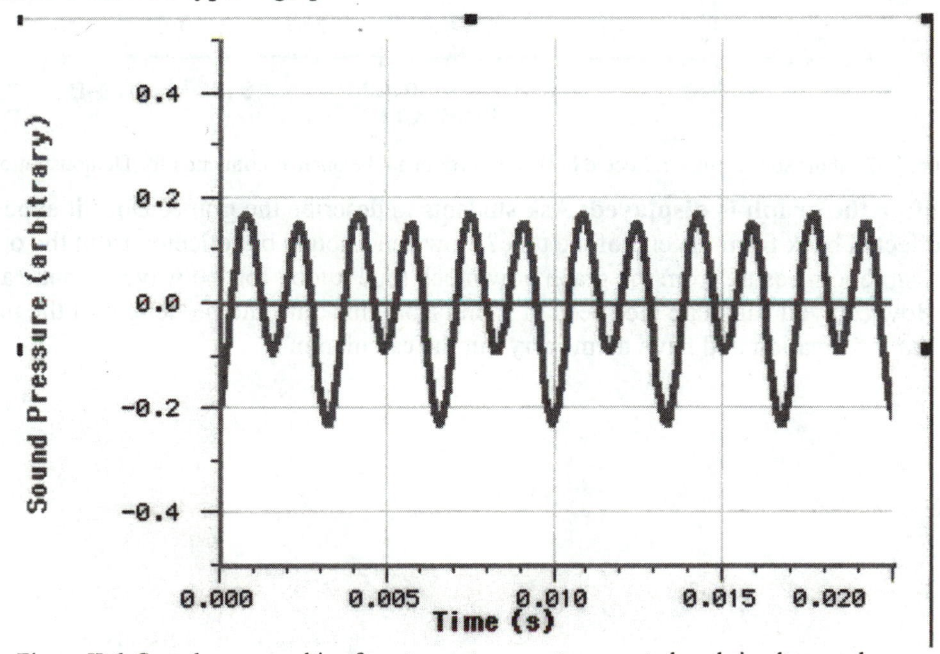

Figure II-6: Sound wave resulting from two notes an octave apart played simultaneously.

Discussion after the graph is displayed: Ask students to compare this graph to the one in Demonstrations 6. What is similar and what is different? Why are there no beats this time? Why do the two notes an octave apart sound so harmonious?

Demonstration 7: Speed of sound. (Use the same experiment configuration file as in Demonstration 1.) Set up the tube with the microphone at one end. The sound needs to be very short. Taping a rubber stopper with a hard rubber hammer should work. Alternatively you can make a short clicking sound with your mouth. This demonstration should also be split into two parts. First ask students to predict the graph. After the prediction and discussion steps, start graphing, and record the sound. You may need to repeat this several times before you get a good graph. Figure II-7 shows a typical graph. The initial pulse is seen along with two reflections. (The time between reflections is measured as 0.012 s, and the tube is 2.0 m long. Thus the speed of sound can be calculated as $v = 2(2.0)/0.012 = 330$ m/s, within several percent of the actual value.)

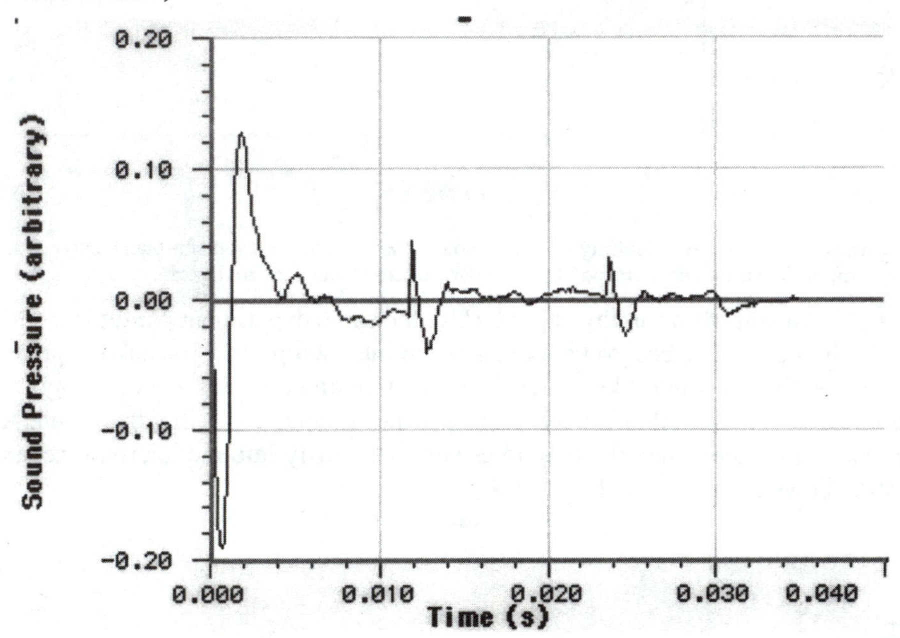

Figure II-7: Short sound pulse reflected back and forth in a tube open at both ends for Demonstration 7.

Discussion after the graph is displayed: Ask students to describe the graph. Does it appear that the sound was reflected back from the end of the tube? How can a sound be reflected from the open end of a tube? Is it possible to measure from the graph how much time it took for the wave to make a round trip in the tube? How can you calculate the speed of sound from this time and the length of the tube? Give the students the information and have them carry out the calculation.

SOUND (SND)
TEACHER PRESENTATION NOTES

Demonstration 1: Sound with constant pitch. Use experiment configuration file **SNDD1**. Play a steady note on the keyboard and ask students to make their predictions. After the prediction and discussion steps, play the same note, and display he graph.

- What is characteristic of a constant pitch? Does this sound wave have a constant frequency? Does it have a constant amplitude?
- What does the graph actually represent?

Demonstration 2: Sound with constant higher pitch. Use the same experiment configuration file. Play a higher steady note, and try to make it just as loud as the sound in Demonstration 1.

- What is similar and what is different compared to Demonstration 1? Does changing the pitch mean changing the frequency, changing the amplitude or both?

Demonstration 3: Louder sound with same pitch. Use the same experiment configuration file. Play the same steady note as in Demonstration 2, only make it significantly louder.

- What is similar and what is different compared to Demonstration 2? Does making a sound louder mean changing the frequency, changing the amplitude or both?

Demonstration 4: Speaking into the microphone. Use the same experiment configuration file. Speak into the microphone and graph the sound.

- What is similar and what is different compared to Demonstrations 1-3? Does speaking produce a steady frequency and amplitude? Why not?

Demonstration 5: Two sounds with slightly different pitches. Use the same experiment configuration file. Break this into two parts. First ask for predictions of what the resulting sound will sound like. Then play two adjacent notes simultaneously, and have students predict the graph.

- What is similar and what is different compared to Demonstrations 1-3?
- What characteristic of this sound wave results in the characteristic beats that are heard? Is the frequency varying or is the amplitude varying? Why is this happening?

Demonstration 6: Two sounds an octave apart. Use the same experiment configuration file. Break this into two parts. First ask for predictions of what the resulting sound will sound like. Then play two notes an octave apart simultaneously, and have students predict the graph.

- What is similar and what is different compared to Demonstration 6? Why are there no beats?
- Why do the two notes an octave apart sound so harmonious?

Demonstration 7: Speed of sound. Use the same experiment configuration file. The sound needs to be very short. Break this into two parts. First ask students to predict the graph.

- Was the sound reflected back from the end of the tube? How can a sound be reflected from the open end of a tube?
- How can you determine the round trip time of the wave from the graph? How can you calculate the speed of sound from this time and the length of the tube? Have students carry out the calculation.

SECTION IV: INTERACTIVE LECTURE DEMONSTRATIONS IN HEAT AND THERMODYNAMICS

Introduction to Heat and Temperature (INHT)

Hand in this sheet **Name**________________________________

INTERACTIVE LECTURE DEMONSTRATIONS

PREDICTION SHEET— **INTRODUCTION TO HEAT & TEMPERATURE**

Directions: This sheet will be collected. Write your name at the top to record your presence and participation in these demonstrations. Follow your instructor's directions. You may write whatever you wish on the attached Results Sheet and take it with you.

Demonstration 1: A small piece of metal has been raised to a high temperature, around 80-90°C. Sketch below your prediction for the temperature-time graph for the piece of metal cooling in the room air. Be sure to carefully sketch the shape of the curve.

What do you think the final temperature of the metal will be? Zero degrees C? Room temperature? Something different?

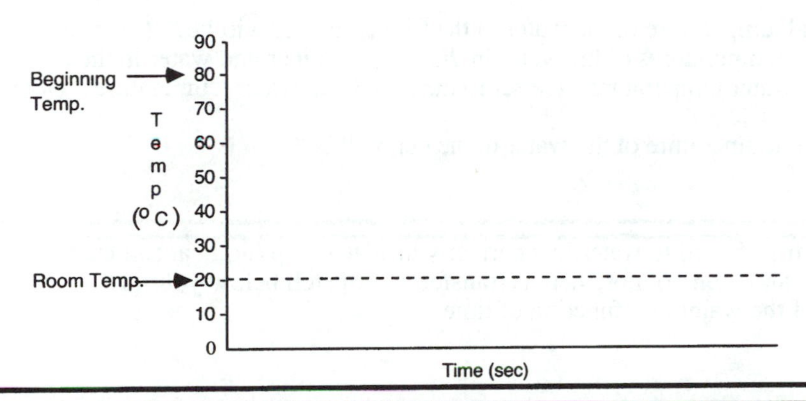

Demonstration 2: Now the same piece of metal at a high temperature (around 80-90°C)) is immersed in a cup filled with cool water (around 20°C). Sketch below your predictions for the temperature-time graphs of the piece of metal and the water in the cup. Be sure to carefully sketch the shapes of the curves.

What do you think the final temperature of the metal will be? (Zero degrees C? Midway between the initial temperatures of the water and the metal? Closer to the initial water temperature? Closer to the initial metal temperature? Other?)

What do you think the final temperature of the water in the cup will be?

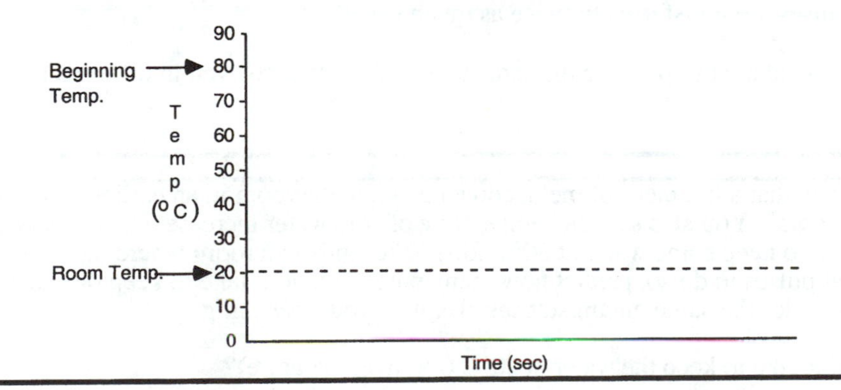

Demonstration 3: Now a small film container filled with water at a high temperature (around 80-90°C) is immersed in a cup filled with room temperature water (around 20°C). Sketch below your predictions for the temperature-time graphs of the film container of hot water and the water in the cup. Be sure to carefully sketch the shapes of the curves.

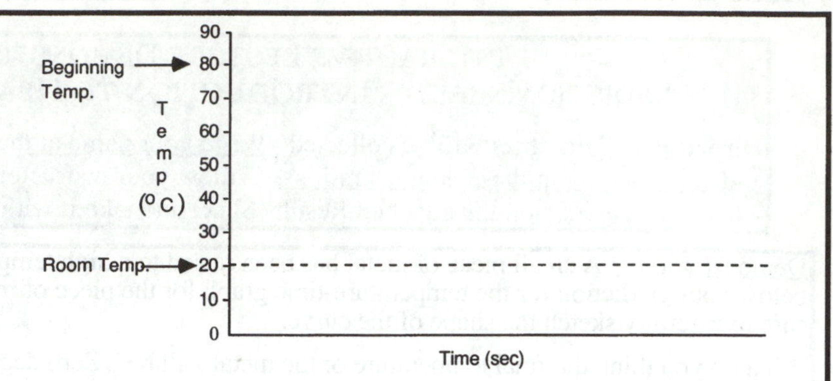

Compare these temperatures to those in Demonstration 2.

What do you think the final temperature of the water in the film container will be? (Zero degrees C? Midway between the initial temperatures of the water in the film container and water in the cup? Closer to the initial film container water temperature? Closer to the initial cup water temperature? Other?)

What do you think the final temperature of the water in the cup will be?

Demonstration 4: Heat is transferred to water in a perfectly insulated cup (no heat can leak in or out) at a steady rate for 80 seconds, and then no more heat is transferred. Sketch below your prediction for the graph of the temperature of the water as a function of time.

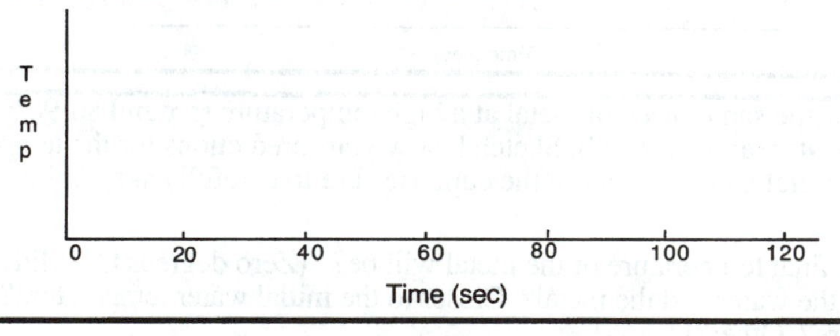

Demonstration 5: A heat pulser can transfer a fixed amount of heat into water for each pulse. The temperature of a small amount of water increases by 8°C when 3 pulses are delivered.
What is the temperature change when 6 pulses of heat are transferred to the water?

What happens when 3 pulses are transferred to twice as much water?

Does the same amount of heat always produce the same temperature change even in different amounts of water?

Demonstration 6: You saw that a hot piece of metal cooled down in the room in an earlier demonstration. Hot water would do the same. You also saw the temperature of cold water increase when heat was transferred to it. We want to keep some water at 80°C for 100 seconds in a room where the temperature is 20°C. If it took 12 heat pulses to do so, predict how many pulses would it take to keep the same water at 50°C for 100 seconds under the same circumstances. Explain your reasoning.

How many pulses would it take to keep the water at 20°C (room temperature)?

Keep this sheet

INTERACTIVE LECTURE DEMONSTRATIONS
RESULTS SHEET— **INTRODUCTION TO HEAT & TEMPERATURE**

You may write whatever you wish on this sheet and take it with you.

Demonstration 1: A small piece of metal has been raised to a high temperature, around 80-90°C. Sketch below your prediction for the temperature-time graph for the piece of metal cooling in the room air. Be sure to carefully sketch the shape of the curve.

What do you think the final temperature of the metal will be? Zero degrees C? Room temperature? Something different?

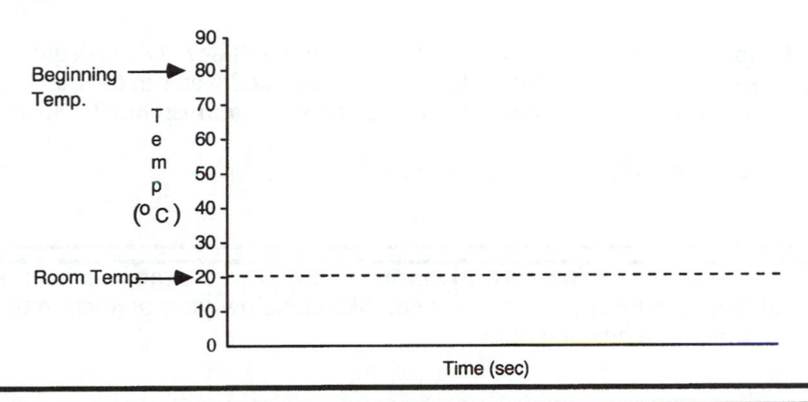

Demonstration 2: Now the same piece of metal at a high temperature (around 80-90°C)) is immersed in a cup filled with room temperature water (around 20°C). Sketch below your predictions for the temperature-time graphs of the piece of metal and the water in the cup. Be sure to carefully sketch the shapes of the curves.

What do you think the final temperature of the metal will be? (Zero degrees C? Midway between the initial temperatures of the water and the metal? Closer to the initial water temperature? Closer to the initial metal temperature? Other?)

What do you think the final temperature of the water in the cup will be?

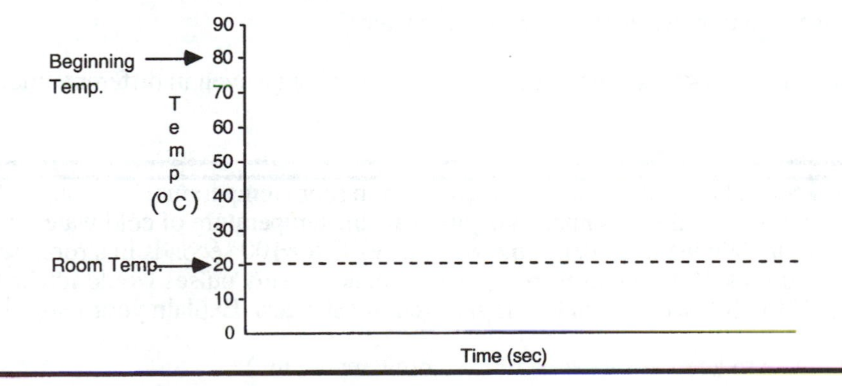

Demonstration 3: Now a small film container filled with water at a high temperature (around 80-90°C) is immersed in a cup filled with room temperature water (around 20°C). Sketch below your predictions for the temperature-time graphs of the film container of hot water and the water in the cup. Be sure to carefully sketch the shapes of the curves.

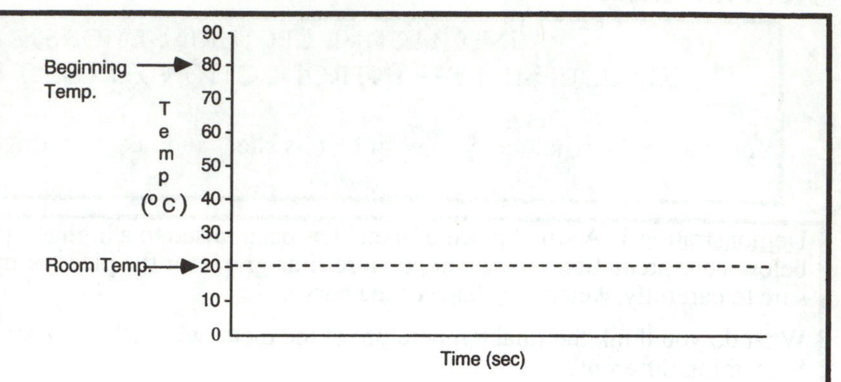

Compare these temperatures to those in Demonstration 2.

What do you think the final temperature of the water in the film container will be? (Zero degrees C? Midway between the initial temperatures of the water in the film container and water in the cup? Closer to the initial film container water temperature? Closer to the initial cup water temperature? Other?)

What do you think the final temperature of the water in the cup will be?

Demonstration 4: Heat is transferred to water in a perfectly insulated cup (no heat can leak in or out) at a steady rate for 80 seconds, and then no more heat is transferred. Sketch below your prediction for the graph of the temperature of the water as a function of time.

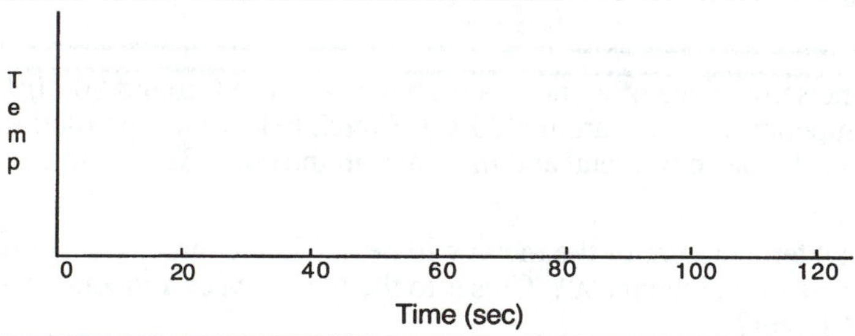

Demonstration 5: A heat pulser can transfer a fixed amount of heat into water for each pulse. The temperature of a small amount of water increases by 5°C when 3 pulses are delivered.
What is the temperature change when 6 pulses of heat are transferred to the water?

What happens when 3 pulses are transferred to twice as much water?

Does the same amount of heat always produce the same temperature change even in different amounts of water?

Demonstration 6: You saw that a hot piece of metal cooled down in room temperature air in an earlier demonstration. Hot water would also do the same. You also saw the temperature of cold water increase when heat was transferred to it. We want to keep some water at 80°C for 100 seconds in a room where the temperature is 20°C. If it took 12 heat pulses to do so, predict how many pulses would it take to keep the same water at 50°C for 100 seconds under the same circumstances. Explain your reasoning.

How many pulses would it take to keep the water at 20°C (room temperature)?

INTRODUCTION TO HEAT & TEMPERATURE (INHT)
TEACHER'S GUIDE

Prerequisites:

There are no pre-requisites the *Introduction to Heat & Temperature ILDs.*

Equipment:

computer-based laboratory system
ILD experiment configuration files
two calibrated temperature probes (See below.)
heat pulser (relay box and at least 200 W immersion heater) (See below.)
small glass beaker
small insulated container (e.g., Styrofoam cup)
thermos bottle with 80 to 90° water
room temperature water
small cylinder of metal with a hole to fit over temperature probe (See below.)
film canister with lid (plastic) with press fit hole for temperature sensor.
Magic tape

General Notes on Preparation and Equipment:

Temperature probes:
Vernier Software and Technology (www.vernier.com) sells a variety of temperature probes to work with their *Lab Pro* interface and *Logger Pro* software. We prefer the Stainless Steel Temperature Probes (TPM-BTA). PASCO Scientific (www.pasco.com) also sells a variety of temperature probes, e.g., the CI-6505B for the *Science Workshop/Data Studio* system, and the PS-2125 for the *PASSPORT* system.

Heat pulser:
Vernier sells a Heat Pulser that plugs into the interface and is controlled by the computer. Do not pulse the heater while it is out of the water or you will burn it out. You want a continuous supply of heat. It is convenient to use the heat pulser but you could also plug an immersion heater into a power strip with a switch and turn it on and off. Some software (Vernier) allows you to "stack" pulses so that if you pre-set the pulser length to 10 seconds and then activate the heat pulser 6 times you will get 60 seconds worth of heat. You can continuously transfer heat for an arbitrary amount of time by activating the pulser once more before the current pulse runs out.

Metal cylinder for cooling experiments:
Figure IV-1 shows the geometry of the cylinder. Use 3/8" brass rod about 5/8" long. Drill a hole in the center that will allow a press fit of the temperature sensor. It would of course be possible to let the mass in the temperature sensor itself cool. We have found it confusing to students initially to have the sensor be the measuring device and also the cooling object.

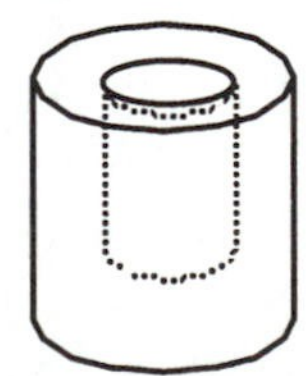

Figure IV-1: Metal cylinder for cooling experiment (Demonstration 2).

Film canister:
Demonstration 3 uses a standard film canister with a lid. Use a drill or an awl to make a hole in the lid as close as possible to the diameter of the temperature sensor. Use tape to secure the sensor to the lid so

that you can use the film canister to stir the water. Care must be taken in filling the film canister with 80 °C water to avoid burning fingers.

Experimental setup:
If you are using a Styrofoam cup, you may want to place it inside a glass beaker to keep it from tipping.

Demonstrations and Sample Graphs:

We suggest you work through the demonstrations before you do them for the class making use of the suggestions below and comparing your results to those shown. Work out any difficulties and be sure the experiment configuration files display well with your equipment.

Demonstration 1: Small piece of metal surrounding the temperature probe cools from 80 or 90°C in air. (Use experiment configuration file **INHTD1**.) Have a previously made graph of the entire cooling process saved and hidden. Leave the second probe suspended nearby to measure room temperature. Heat the metal cylinder in 80 or 90°C water, and dry quickly before letting it cool. After 20 to 30 seconds of cooling, switch to the saved graph explaining you made this graph before class to save time. Figure IV-2 shows a typical cooling curve.

Discussion after graph is displayed: Elicit from the class that the temperature of the metal gradually approaches room temperature. During discussion you can bring up the fact that the curve is exponential (same % change in any fixed time period). Also discuss the fact that heat energy transfer is proportional to the temperature difference, the greater the temperature difference, and the faster the heat transfer, the faster the temperature changes. Point out that when the metal and the air reach the same temperature, they are at thermal equilibrium where the metal transfers as much heat energy to the air as it receives from the air. While the metal is cooling, it transfers more heat energy to the air than it receives from the air. After discussion save data so that the graph is persistently displayed, and then hide the graph.

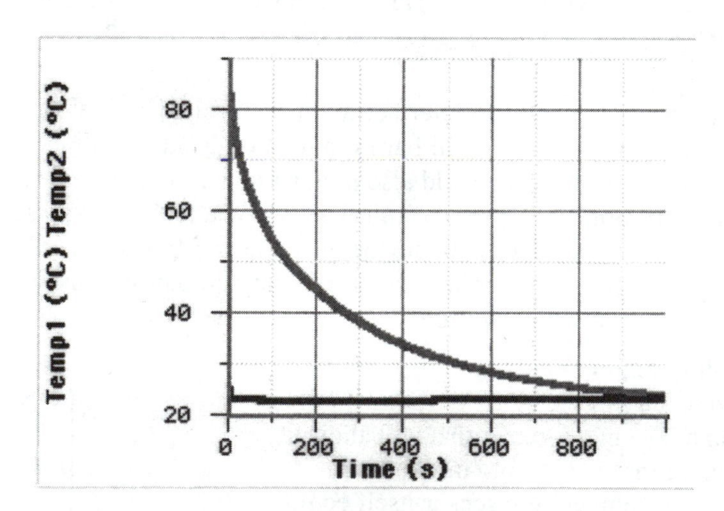

Figure IV-2: Cooling curve for a small metal cylinder cooling in air. The bottom graph shows room temperature. Note that it took almost 1000 seconds to come to equilibrium. This is why it is best to switch to a pre-saved graph.

Demonstration 2: Small piece of metal surrounding the temperature probe cools from 80-90°C in room temperature water. (Use the same experiment configuration file as in Demonstration 1.) Use about 50 ml of water in an insulated cup. You should be able to stop collection well before the time runs out. The second probe should be held midway in the water if possible. Stir constantly. Figure IV-3 shows a typical graph.

Discussion after graph is displayed: Elicit from the class that the temperature of the metal gradually approaches the temperature of the water. Note that the water temperature changes very little. During discussion you can bring up the fact that the curve is essentially exponential (same % change in any fixed time period). Point out that when the metal and the water reach the same temperature, they are in thermal equilibrium where the metal transfers as much heat energy to the water as it receives from the water. While the metal is cooling, it transfers more heat energy to the water than it receives from the water. Compare to Demonstration 1 by showing stored data. Ask students what the difference is. Explain that the water transfers heat energy more effectively than air can so that the materials come to thermal equilibrium more rapidly.

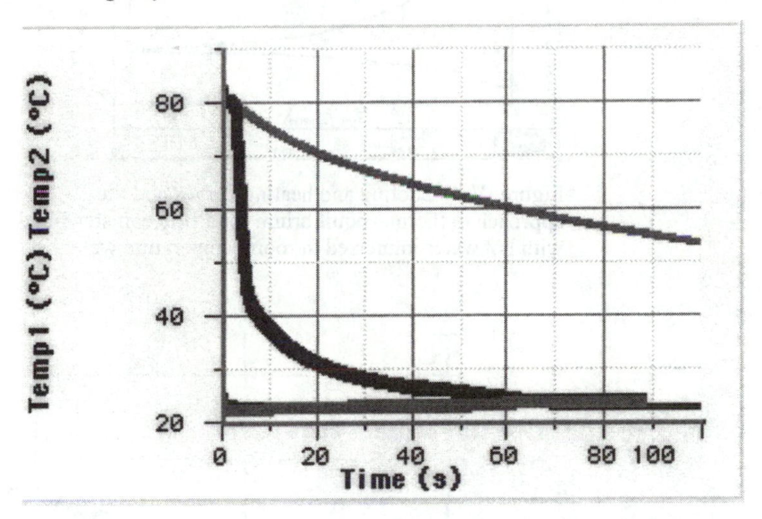

Figure IV-3: Cooling curves for a small metal cylinder in air (top graph from Demonstration 1) and in room temperature water (black graph). Note how quickly the cylinder cools in water, and that the water temperature only increases by 1°C while the metal temperature decreases 60 °C.

<u>**Demonstration 3:**</u> **A small film container filled with water at a high temperature (around 80-90°C) is immersed in a cup filled with room temperature water (around 20°C).** (Use experiment configuration file **INHTD3**.) Use only enough water in the insulated cup to cover the film canister comfortably as you stir. (Do not lift the canister out of the water as you stir.) Mount the second temperature probe about midway in the container if possible. Stir constantly. Figure IV-4 shows a typical graph.

Discussion after the graph is displayed: Elicit from the class that the temperature of the water in the film canister gradually approaches the temperature of the water in the insulated cup. Note that the change in water temperature in the cup is less than the change in the temperature of the water in the film canister. The canister and water come to thermal equilibrium even though there is no transfer of mass. Note that the shapes of the graphs are the same as in Demonstration 2. Ask if the final temperature of the water would be thc same if the experiment were repeated by dumping the water directly into the beaker and stirring. (The amount of heat energy in the canister is very small.) Would the water come to equilibrium slower or faster than before?

Demonstration 4: Heat is transferred to water in a perfectly insulated cup for 80 seconds. (Use experiment configuration file **INHTD4.**) The water should be in an insulated cup. Stir constantly even after you cease heating. Figure IV-5 shows a typical graph.

Discussion after the graph is displayed: Elicit from the class that the temperature-time graph is a straight line during the time heat energy is being transferred at a constant rate. Contrast this to the results of Demonstrations 1-3.

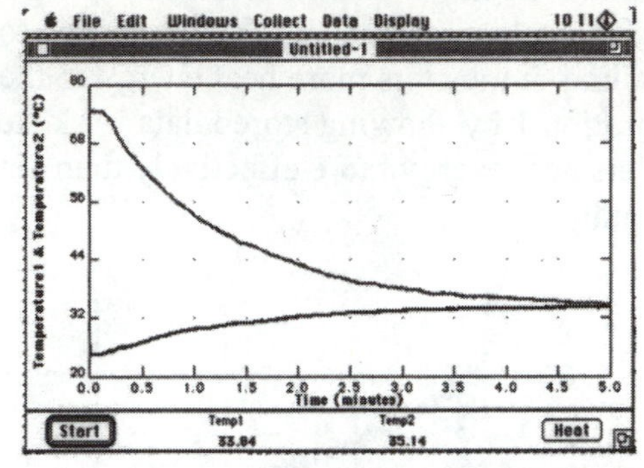

Figure IV-4: Cooling and heating curves and the approach to thermal equilibrium for a film canister filled with hot water immersed in room temperature water.

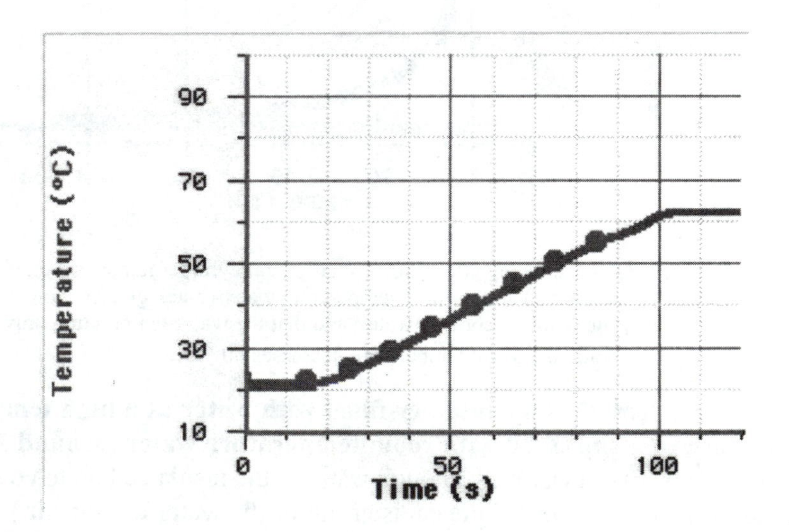

Figure IV-5: Heating curve for heat being steadily transferred to 75 ml of water for 80 seconds. The circles mark the beginnings of 8 continuous 10-second heat pulses.

Demonstration 5: The temperature of a small amount of water increases by 8°C when 3 pulses are transferred. What is the temperature change when 6 pulses of heat are transferred to the water? (Use experiment configuration file **INHTD5.**) Transfer three pulses rapidly to the water in an insulated cup stirring constantly. (75 ml of water works well with 5 second pulses and a 200 watt heater.) Wait until the temperature stops changing, then transfer three more pulses, stirring constantly. (The temperature change for the first three pulses does not need to be exactly 8°C.) A typical graph is shown in Figure IV-6. Using the examine feature of your software show that the temperature change for all six pulses is twice that for three pulses. Save and hide the graph.

Transfer three pulses to twice as much water. A typical graph is shown in Figure IV-7.

Discussion after the graphs are displayed: Emphasize that transferring twice as much heat into the same amount of water gives twice the temperature *change*.The amount of the temperature change for a given amount of heat energy transfer depends directly on the amount of water. Twice as much water results in half the temperature change.

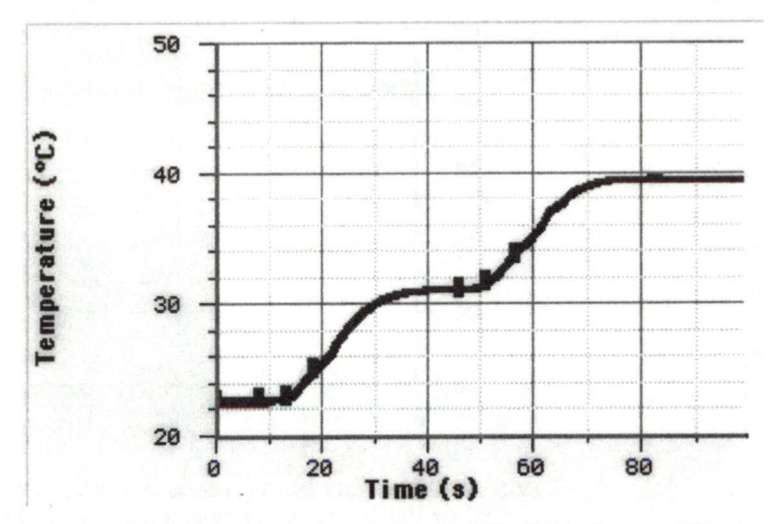

Figure IV-6: Three 5 s heat pulses are transferred to 75ml of water. After the temperature stabilizes, three more pulses are transferred. The temperature rises 8.4 °C in each case. Twice as many pulses result in twice the temperature change.

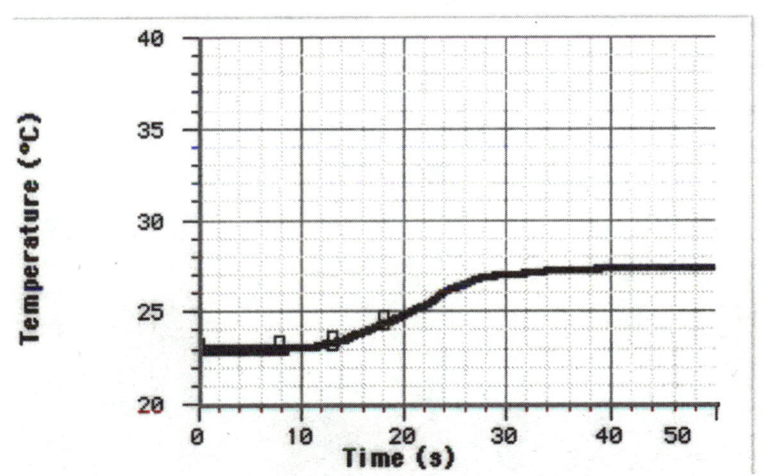

Figure IV-7: Three 5 s heat pulses are transferred to 150 ml of water. The temperature rises 4.2 °C compared to 8.4 °C rise with 75 ml of water, shown in Figure IV-6.

Demonstration 6: Keep some water at 80°C for 100 seconds in a room where the temperature is 20°C. (Use experiment configuration file **INHTD6.**) The water should be in an *un-insulated* cup. Keep stirring and add a pulse of heat when the temperature begins to drop. About 100 ml of water works well. Use 2 second heat pulses. A typical graph is shown in Figure IV-8. Before starting you can use multiple pulses to bring the water to the temperature you want after pouring it from the thermos. Save the results and repeat for water at 50°C. Do not use an insulated cup.

Discussion after the graph is displayed: Elicit from the class that the heat energy transferred from the water in the cup to the air is proportional to the temperature difference between the air and the water in the cup (ΔT). Since the temperature difference is 60°C in the first case, and only 30°C in the second, there are approximately 1/2 as many pulses required to replace the heat transferred out of the cup in the second case. Discuss the fact that heat energy transfer is proportional to the temperature difference, the less the temperature difference, the slower the heat transfer.

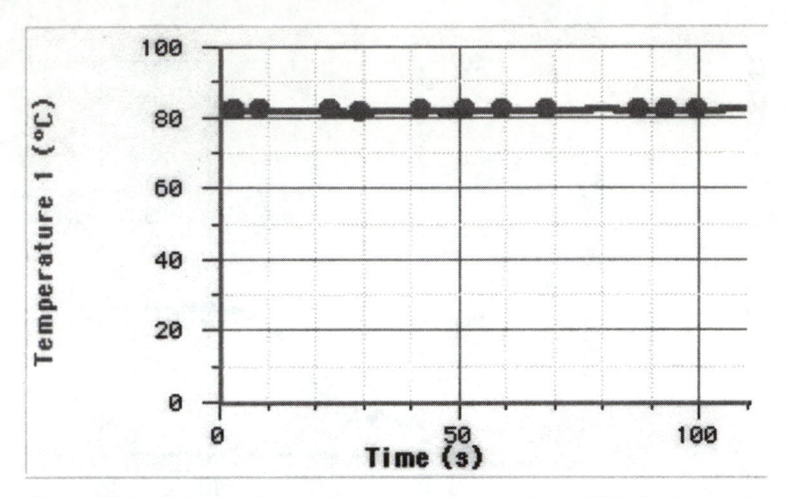

Figure IV-8: Eleven 2 s heat pulses are necessary to keep 100ml of water in a glass beaker at 83°C for 100 seconds with a room temperature of 23 °C (ΔT=60 °C.) When the same amount of water was kept at 53 °C (which is 1/2 the ΔT) only 5 pulses were required (approximately 1/2). Energy transfer is proportional to ΔT.

INTRODUCTION TO HEAT & TEMPERATURE (INHT)
TEACHER PRESENTATION NOTES

Classroom introduction to the *Introduction to Heat & Temperature ILDs*:
Show students the temperature probe and plot a quick temperature vs. time graph where you submerge the probe first in cold water and then in warm water. Remind the students that the probe measures only temperature, not heat energy.

Demonstration 1: Small piece of metal surrounding the temperature probe cools from 80-90°C in air. Use experiment configuration file **INHTD1**. Use *insulated* cup. Use the second probe to measure room temperature. After 20 to 30 seconds, switch to the saved graph. Save the data so that the graph is persistently displayed, and then hide the graph.

- Elicit from the class that the temperature of the metal gradually approaches room temperature.
- The curve is exponential (same % change in any fixed time period). Heat energy transfer is proportional to the temperature difference--the greater the temperature difference, the faster the heat transfer, the faster the temperature change.
- When the metal and the air reach the same temperature, they are in thermal equilibrium.

Demonstration 2: Small piece of metal surrounding the temperature probe cools from 80-90°C in room temperature water. Use the same experiment configuration file. Use *insulated* cup. The second probe should be held midway in the water if possible. Stir constantly.

- Temperature of the metal gradually approaches the temperature of the water. Water temperature changes very little.
- When the metal and the water reach the same temperature, they are in thermal equilibrium. While the metal is cooling, it transfers more heat energy to the water than it receives from the water.
- Compare to Demonstration 1 by showing stored data. What is the difference? Water transfers heat energy more quickly than air so the materials reach thermal equilibrium more rapidly.

Demonstration 3: A small film container filled with water at a high temperature (around 80-90°C) is immersed in a cup filled with room temperature water (around 20°C). Use experiment configuration file **INHTD3**. Use *insulated* cup. Mount second probe about midway in the container.

- Temperature of the water in the film canister gradually approaches the temperature of the water in the cup. Note that the change in water temperature in the cup is less than the change in water temperature in the film canister. Shape of the graph is the same as in Demonstration 2.
- Would the final temperature of the water be the same if hot water dumped directly into the cup and stirred? Would the system come to equilibrium slower or faster than before?

Demonstration 4: Heat is transferred to water in a perfectly insulated cup for 80 seconds. Use experiment configuration file **INHTD4**. Mount probe about midway in the container. Stir constantly even after you cease heating.

- Elicit from the class that the temperature-time graph is a straight line during the time heat energy is being transferred at a constant rate. Contrast this to the results of Demonstrations 1-3.

Demonstration 5: The temperature of a small amount of water increases by 8°C when 3 pulses are transferred. What is the temperature change when 6 pulses of heat are transferred to the water? Use experiment configuration file **INHTD5**. Transfer three pulses rapidly to the water stirring constantly. Wait until the temperature stops changing, then transfer three more stirring constantly. Show that the temperature change for all six pulses is twice that for three pulses. Save and hide the graph. Transfer three pulses to twice as much water.

- Transferring twice as much heat to the same amount of water gives twice the temperature *change*.
- The amount of the temperature change for a given amount of heat energy transfer depends directly on the amount of water. Twice as much water, results in half the temperature change.

<u>Demonstration 6:</u> Keep some water at 80°C for 100 seconds in a room where the temperature is 20°C. Use experiment configuration file **INHTD6**. Use an *un-insulated* cup. Keep stirring and add a pulse of heat when the temperature begins to drop. Save the results and repeat for water at 50°C

- The heat energy transferred to the air is proportional to the temperature difference or ΔT.
- Since the temperature difference is 60°C in the first case, and only 30°C in the second, there are approximately 1/2 as many pulses required to replace the heat transferred out in the second case.
- If the water is at room temperature, no pulses are required since the system is at thermal equilibrium.

Specific Heat (SPHT)

Hand in this sheet **Name**________________________________

INTERACTIVE LECTURE DEMONSTRATIONS
PREDICTION SHEET— **SPECIFIC HEAT**

Directions: This sheet will be collected. Write your name at the top to record your presence and participation in these demonstrations. Follow your instructor's directions. You may write whatever you wish on the attached Results Sheet and take it with you.

Demonstration 1: 100 grams of water is initially at room temperature in a perfectly insulated cup (no heat can leak in or out). Heat energy is transferred to the water at a steady rate for 90 seconds, and then no more heat is transferred. Sketch below your prediction for the graph of the temperature of the water as a function of time.

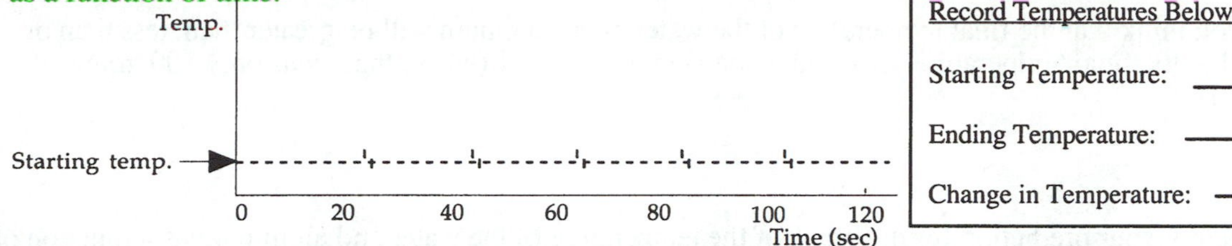

Record Temperatures Below:

Starting Temperature: ________

Ending Temperature: ________

Change in Temperature: ________

Calculation 1: Using the actual starting and ending temperatures of the water, determine the amount of heat energy that was transferred to the water in this process. The specific heat of water is 4186 J/kg C°.

Heat Energy Transferred = Q_1 =

Demonstration 2: 50 grams of aluminum is *added* to 100 grams of water. The aluminum and the water are both initially at room temperature in a perfectly insulated cup. Heat energy is transferred to the system at the same steady rate as in the previous demonstration for 90 seconds, and then no more heat energy is transferred.

Discussion Question:

Do you think that the final temperature of the water plus aluminum will be greater than, less than or equal to the final temperature reached in the previous demonstration?

Sketch below your prediction for the graph of the temperature of the water and aluminum as a function of time.

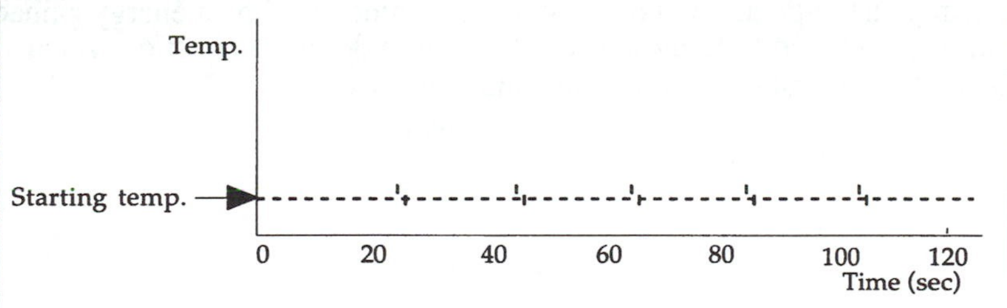

Record Temperatures Below:

Starting Temperature: ________

Ending Temperature: ________

Change in Temperature: ________

Calculation 2: Using the actual starting and ending temperatures of the system, determine the amount of heat that was transferred to the water and aluminum in this process. The specific heat of water is 4186 J/(kg C°) and the specific heat of aluminum is 900 J/(kg C°).

Heat Transferred = Q_2 =

Demonstration 3: Now 50 grams of aluminum are used to *replace* 50 grams of water, so that we start with 50 grams of water and 50 grams of aluminum, both initially at room temperature in a perfectly insulated cup. Heat energy is transferred to the water and aluminum at the same steady rate as in the previous demonstrations for 90 seconds, and then no more heat energy is transferred. Again, the specific heat of water is 4186 J/(kg C°) and the specific heat of aluminum is 900 J/(kg C°).

Discussion Questions:

1. Do you think that the final temperature of the water plus aluminum will be greater than, less than or equal to the final temperature achieved in the Demonstration 2 (where there was 100 grams water with 50 grams aluminum)?

2. Do you think that the final temperature of the water plus aluminum will be greater than, less than or equal to the final temperature achieved in the Demonstration 1 (where there was only 100 grams of water)?

Sketch below your prediction for the graph of the temperature of the water and aluminum as a function of time.

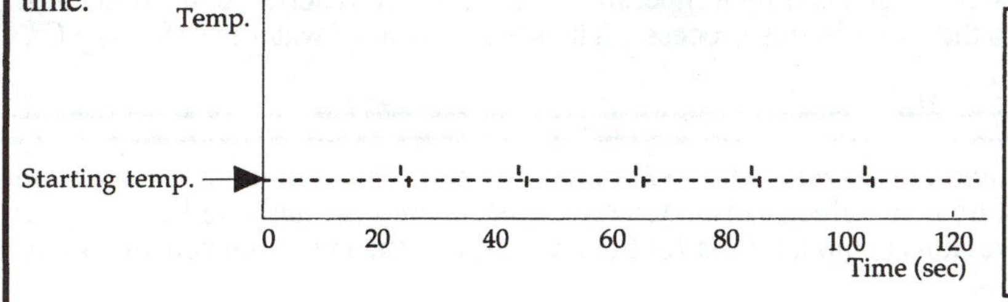

Record Temperatures Below:

Starting Temperature: ________

Ending Temperature: ________

Change in Temperature: ________

Calculation 3: Now let's assume that the same amount of heat energy will be transferred to this system as was transferred in the previous two demonstrations. Since you probably found that these amounts were slightly different, first take the average of the two amounts of heat:

$$\text{Heat Transferred} = Q_3 = \tfrac{1}{2}(Q_1 + Q_2) =$$

Since this heat will be transferred to the water and the aluminum, we can write down an equation that will allow us to predict the theoretical change in temperature of our system. The amount of heat energy gained by each part of the system will be of the form $mc\Delta T$. In the space below, write down an equation which sets Q_3 equal to the heat energy gained by the water and aluminum, and then solve it for ΔT:

$$Q_3 =$$

$$\Rightarrow \quad \Delta T =$$

How does the theoretical numerical answer you just calculated compare with the actual change in temperature of the system? Can you think of a possible explanation for any discrepancy?

Keep this sheet

INTERACTIVE LECTURE DEMONSTRATIONS
RESULTS SHEET— **SPECIFIC HEAT**

You may write whatever you wish on this sheet and take it with you.

Demonstration 1: 100 grams of water is initially at room temperature in a perfectly insulated cup (no heat can leak in or out). Heat energy is transferred to the water at a steady rate for 90 seconds, and then no more heat is transferred. Sketch below your prediction for the graph of the temperature of the water as a function of time.

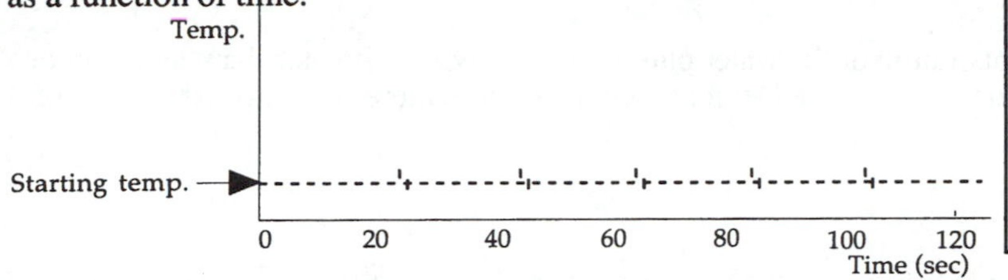

Record Temperatures Below:

Starting Temperature: ________

Ending Temperature: ________

Change in Temperature: ________

Calculation 1: Using the actual starting and ending temperatures of the water, determine the amount of heat energy that was transferred to the water in this process. The specific heat of water is 4186 J/kg C°.

Heat Energy Transferred = Q_1 =

Demonstration 2: 50 grams of aluminum is *added* to 100 grams of water. The aluminum and the water are both initially at room temperature in a perfectly insulated cup. Heat energy is transferred to the system at the same steady rate as in the previous demonstration for 90 seconds, and then no more heat energy is transferred.

Discussion Question:

Do you think that the final temperature of the water plus aluminum will be greater than, less than or equal to the final temperature reached in the previous demonstration?

Sketch below your prediction for the graph of the temperature of the water and aluminum as a function of time.

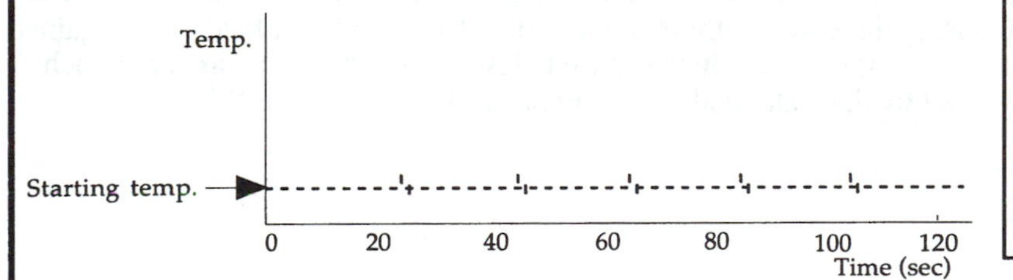

Record Temperatures Below:

Starting Temperature: ________

Ending Temperature: ________

Change in Temperature: ________

Calculation 2: Using the actual starting and ending temperatures of the system, determine the amount of heat that was transferred to the water and aluminum in this process. The specific heat of water is 4186 J/(kg C°) and the specific heat of aluminum is 900 J/(kg C°).

Heat Transferred = Q_2 =

Demonstration 3: Now 50 grams of aluminum are used to *replace* 50 grams of water, so that we start with 50 grams of water and 50 grams of aluminum, both initially at room temperature in a perfectly insulated cup. Heat energy is transferred to the water and aluminum at the same steady rate as in the previous demonstrations for 90 seconds, and then no more heat energy is transferred. Again, the specific heat of water is 4186 J/(kg C°) and the specific heat of aluminum is 900 J/(kg C°).

Discussion Questions:

1. Do you think that the final temperature of the water plus aluminum will be greater than, less than or equal to the final temperature achieved in the Demonstration 2 (where there was 100 grams water with 50 grams aluminum)?

2. Do you think that the final temperature of the water plus aluminum will be greater than, less than or equal to the final temperature achieved in the Demonstration 1 (where there was only 100 grams of water)?

Sketch below your prediction for the graph of the temperature of the water and aluminum as a function of time.

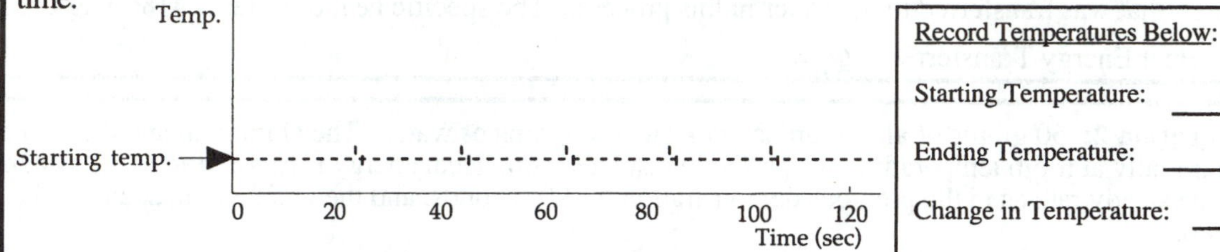

Record Temperatures Below:

Starting Temperature: ________

Ending Temperature: ________

Change in Temperature: ________

Calculation 3: Now let's assume that the same amount of heat energy will be transferred to this system as was transferred in the previous two demonstrations. Since you probably found that these amounts were slightly different, first take the average of the two amounts of heat:

$$\text{Heat Transferred} = Q_3 = \tfrac{1}{2}(Q_1 + Q_2) =$$

Since this heat will be transferred to the water and the aluminum, we can write down an equation that will allow us to predict the theoretical change in temperature of our system. The amount of heat energy gained by each part of the system will be of the form $mc\Delta T$. In the space below, write down an equation which sets Q_3 equal to the heat energy gained by the water and aluminum, and then solve it for ΔT:

$$Q_3 =$$

$$\Rightarrow \quad \Delta T =$$

How does the theoretical numerical answer you just calculated compare with the actual change in temperature of the system? Can you think of a possible explanation for any discrepancy?

SPECIFIC HEAT (SPHT)
TEACHER'S GUIDE

Prerequisites:

The *Introduction to Heat & Temperature ILDs* are prerequisite to this sequence of *ILDs*.

Equipment:

computer-based laboratory system
ILD experiment configuration files
calibrated temperature probe (See below.)
heat pulser (relay box and at least 200 W immersion heater)
room temperature water
50 gm block of aluminum (2 would be more convenient)
small insulated container (Styrofoam cup will work)
beaker marked in ml for measurement

General Notes on Preparation and Equipment:

In this sequence of *ILDs* you will be asking the students to predict the graphs but not predict the calculations. After doing the experiment, students will do calculations individually, and then compare results with their neighbors. Following that you should help students reach a consensus.

Temperature probes:
See the Teacher's Guide for the *Introduction to Heat and Temperature ILDs* for information on temperature probes.

Heat pulser:
See the Teacher's Guide for the *Introduction to Heat and Temperature ILDs* for information on the heat pulser. Do not pulse the heater while it is out of the water or you will burn it out. You want a continuous supply of heat. It is convenient to use the heat pulser but you could also plug an immersion heater into a power strip with a switch and turn it on and off.

Demonstrations and Sample Graphs:

Demonstration 1: Heat is transferred to 100 grams of water at a steady rate for 90 seconds. (Use experiment configuration file **SPHTD1.**) Use an *insulated* cup. Stir constantly even after you cease heating until temperature stabilizes. Find the initial and final temperatures by using the examine feature of your software. Figure IV-9 shows a typical graph. The calculation is straightforward using $Q_1 = m_w c_w \Delta T$.

Discussion after the graph is displayed: Remind students that constant rate of energy transfer results in a constant rate of temperature change. Point out that if the change in temperature is known for a material, the total heat transferred can be calculated.

Demonstration 2: Heat is transferred to 100 grams of water and 50 grams of aluminum at a steady rate for 90 seconds. (Use experiment configuration file **SPHTD2.**) Use an insulated cup. Stir constantly even after you cease heating until temperature stabilizes. Point out that the water and the metal stay in thermal equilibrium because both are so conductive. The calculation is straightforward using $Q_2 = m_w c_w \Delta T + m_{Al} c_{Al} \Delta T$.

Discussion after the graph is displayed: After doing the experiment, point out that transferring the same amount of heat energy to more mass logically results in a smaller temperature change. Note that aluminum has a much smaller heat capacity than water for the same mass (smaller specific heat). This may be counter-intuitive to some students. Since the same amount of heat is transferred in Demonstrations 1 and 2, and some of the heat must be transferred to the aluminum, the amount of heat transferred to the 100 grams of water is less here. Therefore the temperature change is smaller

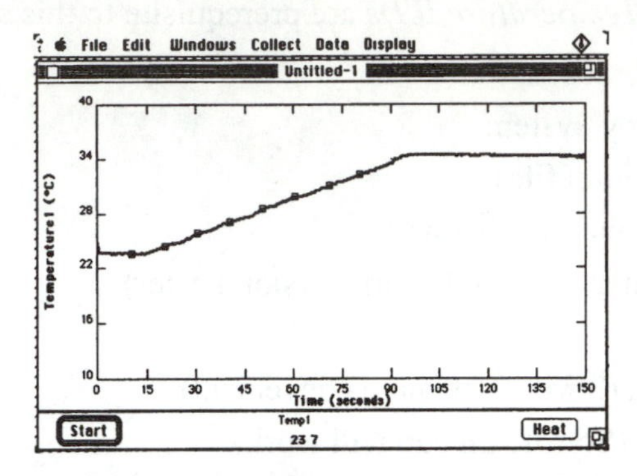

Figure IV-9: All graphs in this sequence will look something like this graph. The slope will change as the material being heated changes. In all cases heat will be transferred for 90 seconds.

Demonstration 3: Heat is transferred to 50 grams of water and 50 grams of aluminum at a steady rate for 90 seconds. (Use experiment configuration file **SPHTD3.**) Use an insulated cup. Stir constantly even after you cease heating until temperature stabilizes. The calculation is straightforward. For any of these experiments, $Q = m_w c_w \Delta T + m_{Al} c_{Al} \Delta T$, and, therefore, $\Delta T = Q/(m_w c_w + m_{Al} c_{Al})$. Since Q is the same in each case, ΔT is determined by the denominator, $(m_w c_w + m_{Al} c_{Al})$. This is largest in Demonstration 2, giving the smallest temperature change. It is smallest in Demonstration 3, giving the largest temperature change, and intermediate in Demonstration 1.

Discussion after the graph is displayed: Point out that the total mass being heated is the same as in Demonstration 1. What is different? The average specific heat is smaller (average of 4186 for water and 900 for aluminum). Therefore, if the same amount of heat is transferred, the temperature change should be greater here.

SPECIFIC HEAT (SPHT)
TEACHER PRESENTATION NOTES

Demonstration 1: Heat is transferred to 100 grams of water at a steady rate for 90 seconds. Use experiment configuration file **SPHTD1**. Use an insulated cup. Stir constantly even after you cease heating until temperature stabilizes. Find the initial and final temperatures by using the examine feature of your software. The calculation is straightforward using $Q_1 = m_w c_w \Delta T$.

- Remind students that constant rate of energy transfer results in a constant rate of temperature change. If the change in temperature is known for a material, the total heat transferred can be calculated..

Demonstration 2: Heat is transferred to 100 grams of water and 50 grams of aluminum at a steady rate for 90 seconds. Use the experiment configuration file **SPHTD2**. Stir constantly even after you cease heating until temperature stabilizes. Point out that the water and the metal stay in thermal equilibrium because both are so conductive. The calculation is straightforward using $Q_2 = m_w c_w \Delta T + m_{Al} c_{Al} \Delta T$.

Discussion after the graph is displayed:

- Point out that transferring the same amount of heat energy to more mass logically results in a smaller temperature change.
- Note that aluminum has a much smaller heat capacity than water for the same mass.
- Since the same amount of heat is transferred in Demonstrations 1 and 2, and some of the heat must be transferred to the aluminum, the amount of heat transferred to the 100 grams of water is less here. Therefore the temperature change is smaller

Demonstration 3: Heat is transferred to 50 grams of water and 50 grams of aluminum at a steady rate for 90 seconds. Use experiment configuration file **SPHTD3**. Use an insulated cup. Stir constantly even after you cease heating until temperature stabilizes. The calculation is straightforward. For any of these experiments, $Q = m_w c_w \Delta T + m_{Al} c_{Al} \Delta T$, and, therefore, $\Delta T = Q/(m_w c_w + m_{Al} c_{Al})$.

- The total mass being heated is the same as in Demonstration 1. What is different?
- The average specific heat is smaller (average of 4186 for water and 900 for aluminum). Therefore, if the same amount of heat is transferred, the temperature change should be greater here.

HEAT & PHASE CHANGE (HTPC)

Hand in this sheet **Name**________________________________

INTERACTIVE LECTURE DEMONSTRATIONS
PREDICTION SHEET—**HEAT & PHASE CHANGES**

Directions: This sheet will be collected. Write your name at the top to record your presence and participation in these demonstrations. Follow your instructor's directions. You may write whatever you wish on the attached Results Sheet and take it with you.

Demonstration 1: Water initially at room temperature is in a perfectly insulated cup (no heat can leak in or out). During the first 20 seconds no heat is transferred to the water, and then during the next 60 seconds, heat is transferred at a steady rate. Then no more heat is transferred. Sketch below your prediction for the graph of the temperature of the water as a function of time.

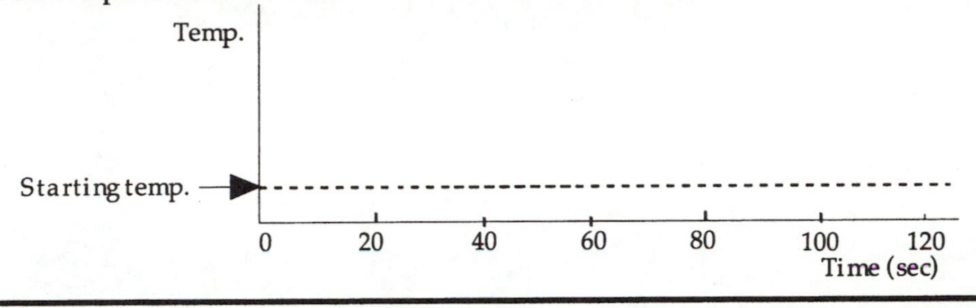

Demonstration 2: Heat is transferred at a steady rate to a mixture of water and ice at 0°C in a perfectly insulated cup (no heat can leak in or out). After the ice has completely melted, heat is still transferred for awhile. Sketch below your prediction for the graph of the temperature as a function of time.

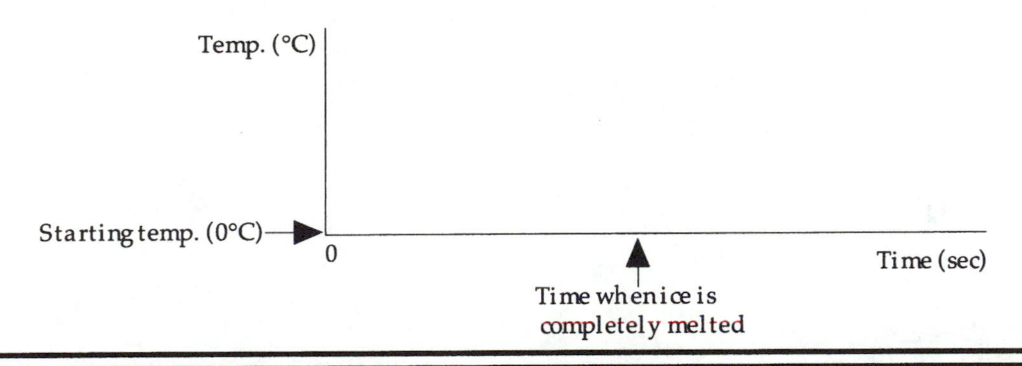

Demonstration 3: Heat is transferred at a steady rate to water initially at 80°C in a perfectly insulated cup (no heat can leak in or out). After the water starts boiling, heat is still transferred for awhile. Sketch below your prediction for the graph of the temperature as a function of time.

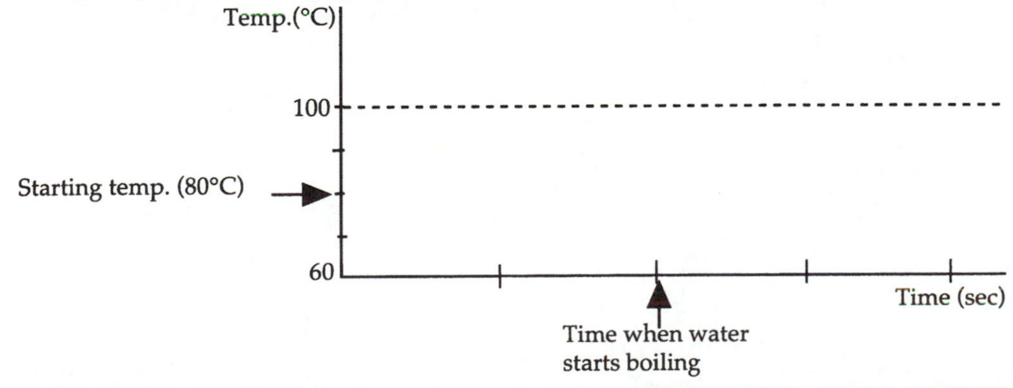

Keep this sheet

INTERACTIVE LECTURE DEMONSTRATIONS
RESULTS SHEET—**HEAT & PHASE CHANGES**

You may write whatever you wish on this sheet and take it with you.

Demonstration 1: Water initially at room temperature is in a perfectly insulated cup (no heat can leak in or out). During the first 20 seconds no heat is transferred to the water, and then during the next 60 seconds, heat is transferred at a steady rate. Then no more heat is transferred. Sketch below your prediction for the graph of the temperature of the water as a function of time.

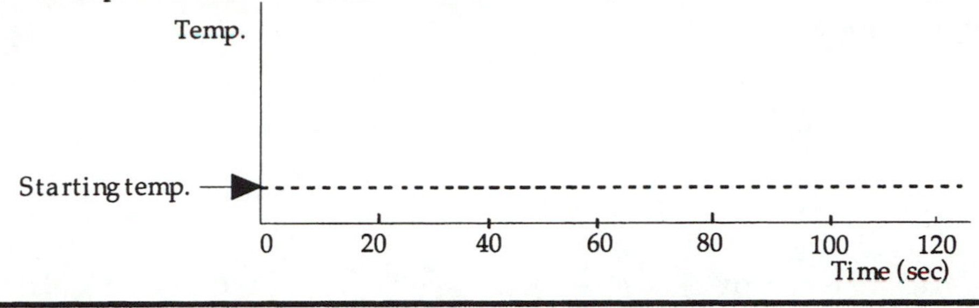

Demonstration 2: Heat is transferred at a steady rate to a mixture of water and ice at 0°C in a perfectly insulated cup (no heat can leak in or out). After the ice has completely melted, heat is still transferred for awhile. Sketch below your prediction for the graph of the temperature as a function of time.

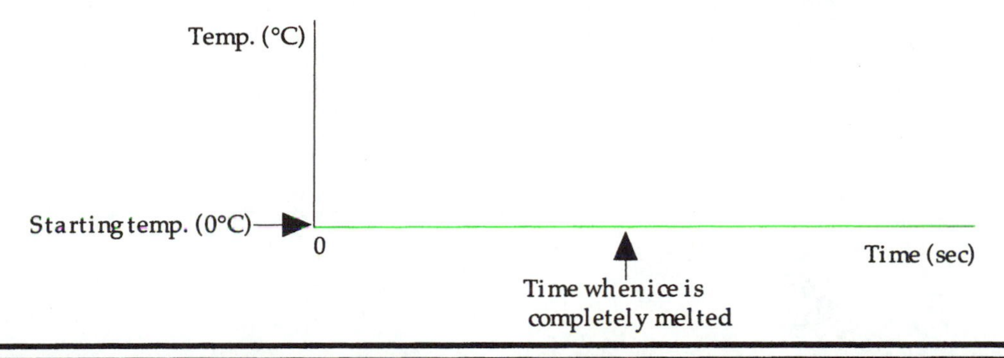

Demonstration 3: Heat is transferred at a steady rate to water initially at 80°C in a perfectly insulated cup (no heat can leak in or out). After the water starts boiling, heat is still transferred for awhile. Sketch below your prediction for the graph of the temperature as a function of time.

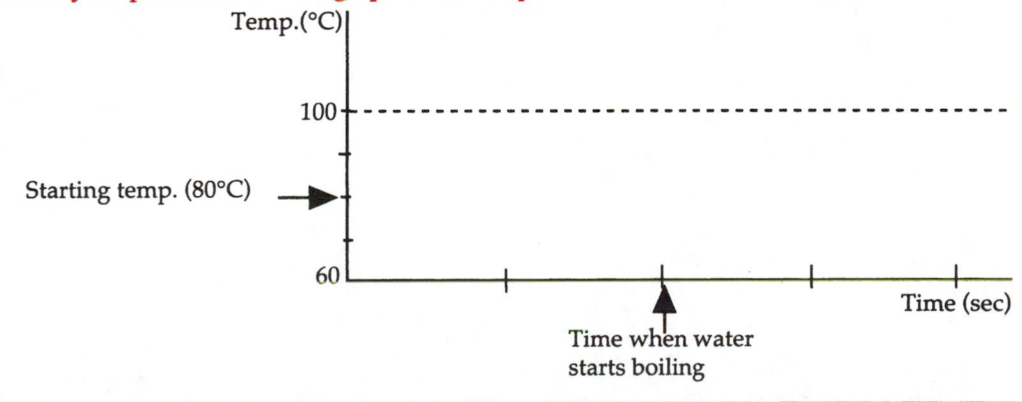

Keep this sheet

HEAT & PHASE CHANGES (HTPC)
TEACHER'S GUIDE

Prerequisites:

The *Heat & Phase Changes ILDs* should be done after the *Introduction to Heat & Temperature* and *Specific Heat ILD* sequences.

Equipment:

computer-based laboratory system
ILD experiment configuration files
temperature probe (see below.)
heat pulser (relay box and at least 200 W immersion heater) (See below.)
room temperature water
crushed ice and cold water in insulated (Styrofoam) cup
water at 80°C (stored in thermos bottle)
small Styrofoam cup in a larger glass beaker
stirring rod

General Notes on Preparation and Equipment:
The key to these experiments is not to use too much water (or water and ice) so that they are relatively quick to do. To just covering the heater coils in a small Styrofoam cup should require about 75 ml or less and should work well. Put the Styrofoam cup inside of a larger glass beaker to keep it from tipping.

Temperature probes:
See the Teacher's Guide for the *Introduction to Heat and Temperature ILDs* for information on temperature probes.

Heat pulser:
See the Teacher's Guide for the ***Introduction to Heat and Temperature ILDs*** for information on the heat pulser. Do not pulse the heater while it is out of the water or you will burn it out. You want a continuous supply of heat. It is convenient to use the heat pulser but you could also plug an immersion heater into a power strip with a switch and turn it on for the length of time required.

Demonstrations and Sample Graphs:

The first demonstration reminds students that transferring heat at a constant rate to a material generally results in a linear temperature rise. The next two demonstrations introduce students to processes in which heat energy transferred to a system brings about a change of state *without a temperature change.* These processes are phase changes. The phenomena of melting, freezing, boiling and condensation are familiar to students, but most students do not understand the temperature histories during a phase change. Many students will be surprised to see that the temperature of water remains *constant* during a phase change. Students who have taken a course in chemistry may have performed similar experiments already, but probably haven't seen the temperature-time graphs displayed in such an accessible manner.

Demonstration 1: During the first 20 seconds no heat is transferred to the water, and then during the next 60 seconds, heat is transferred at a steady rate. (Use experiment configuration file **HTPCD1**.) Use an insulated cup. Keep stirring. Be sure the quantity of water you have chosen will not boil. Figure IV-10 shows a typical graph.

Discussion after graph is displayed: Elicit from the students that transferring heat at a constant rate to a material generally results in a linear temperature rise, as seen in the *Specific Heat ILDs*.

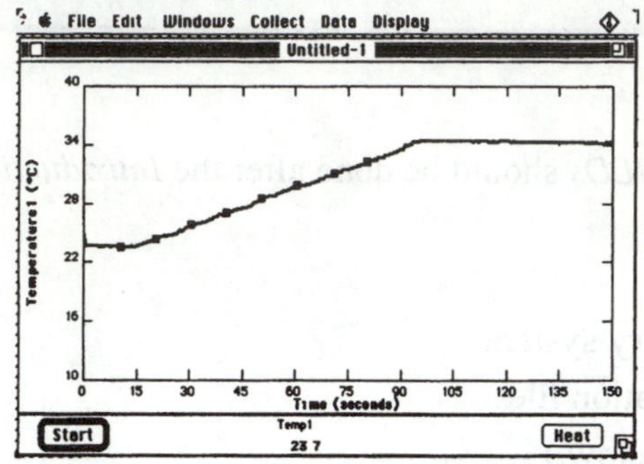

Figure IV-10: The graph will be shaped like this one except the temperature will begin to increase at about 20 seconds and stop changing around 80 seconds.

Demonstration 2: Heat is transferred at a steady rate to a mixture of water and ice at 0°C in a perfectly insulated cup (no heat can leak in or out). (Use experiment configuration file **HTPCD2.**) Use an insulated cup. Be sure the ice is well crushed. Do not use ice cubes. Keep stirring. It is essential to stir constantly and vigorously with the stirring rod. If you don't, the water-ice mixture will not be kept in thermal equilibrium, and the temperature will rise even before the ice melts. Figure IV-11 shows a typical graph.

Discussion after graph is displayed: Ask students where on the graph the ice was entirely melted. Ask the students what the transferred heat energy was doing if it was not raising the temperature.

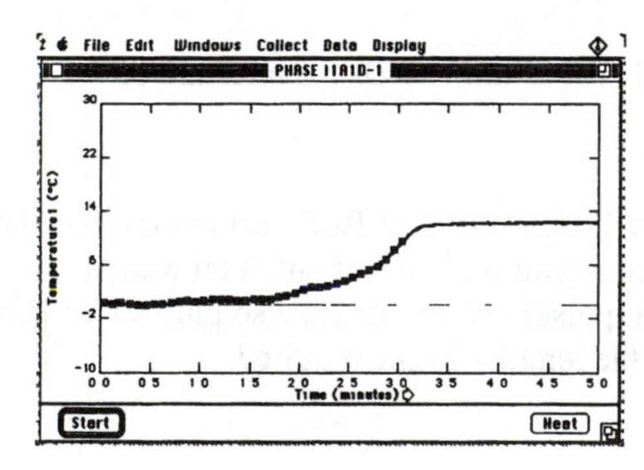

Figure IV-11: The graph shows a typical temperature history for melting ice by transferring heat at a steady rate.

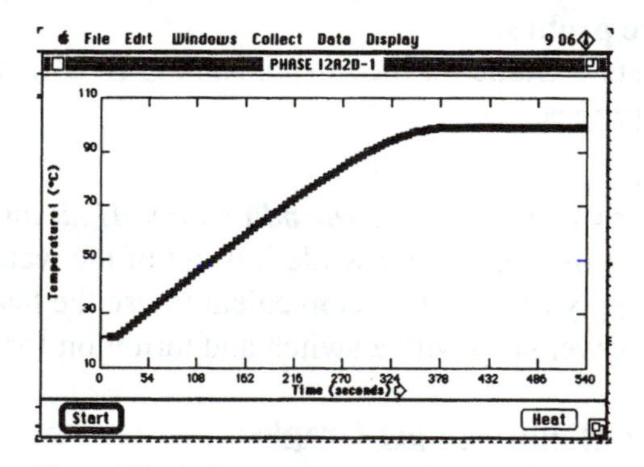

Figure IV-12: Temperature vs. time graph for heat energy transferred to water raising it to the boiling point. To save time, in Demonstration 3 we are start with 80°C water.

Demonstration 3: Heat is transferred at a steady rate to water initially at 80°C in a perfectly insulated cup (no heat can leak in or out). (Use experiment configuration file **HTPCD3.**) Use an insulated cup. Keep transferring heat for awhile after the water begins boiling. If the heater is on the bottom of the cup, you will not need to stir as vigorously. Figure IV-12 shows a typical graph.

Discussion after graph is displayed: After showing the graph, ask students where on the graph the water started to boil. Ask the students what the transferred heat energy was doing if it was not raising the temperature. Introduce the idea of change of state or phase change. Energy is required to rearrange molecules.

HEAT & PHASE CHANGES (HPHC)
TEACHER PRESENTATION NOTES

Classroom introduction to the *Heat Phase Changese ILDs*:
Students should be familiar with the temperature probe and heat pulser from the previous two *ILD* sequences.

Demonstration 1: **During the first 20 seconds no heat is transferred to the water in a perfectly insulated cup (no heat can leak in or out), and then during the next 60 seconds, heat is transferred at a steady rate.** Use experiment configuration file **HTPCD1**. Start with room temperature water in an insulated cup. Keep stirring.

- Elicit from the students that transferring heat at a constant rate generally results in a linear temperature rise.

Demonstration 2: **Heat is transferred at a steady rate to a mixture of water and ice at 0°C in a perfectly insulated cup.** Use experiment configuration file **HTPCD2**. Use an insulated cup. After the ice has completely melted, transfer heat for another 15 seconds. Keep stirring vigorously.

- After showing the graph, ask students where on the graph the ice was entirely melted.
- Ask the students what the transferred heat energy was doing if it was not raising the temperature.

Demonstration 3: **Heat is transferred at a steady rate to water initially at 80°C in a perfectly insulated cup.** Use experiment configuration file **HTPCD3**. After the water starts boiling, heat is still transferred for awhile. If the heater is on the bottom of the cup, you will not need to stir as vigorously.

- After showing the graph, ask students where on the graph the water started to boil.
- Ask the students what the transferred heat energy was doing if it was not raising the temperature.
- Introduce the idea of change of state or phase change. Energy is required to rearrange molecules.

Heat Engine (HENG)

Hand in this sheet **Name**___________________________

INTERACTIVE LECTURE DEMONSTRATIONS
PREDICTION SHEET— **HEAT ENGINE**

Directions: This sheet will be collected. Write your name at the top to record your presence and participation in these demonstrations. Follow your instructor's directions. You may write whatever you wish on the attached Results Sheet and take it with you.

A model of a heat engine consists of a cylinder (a glass syringe) connected by tubing to a pressure probe and a flask filled with air. The flask may be submerged into a HOT reservoir filled with hot water or a COLD reservoir filled with ice water. The job to be done by this engine is to lift a 100 gram mass a certain height. The pressure probe measures the pressure in the syringe and the volume of gas in the syringe can be measured using the markings on the syringe. The axes below may be used to plot a P-V diagram of the cycle of this engine.

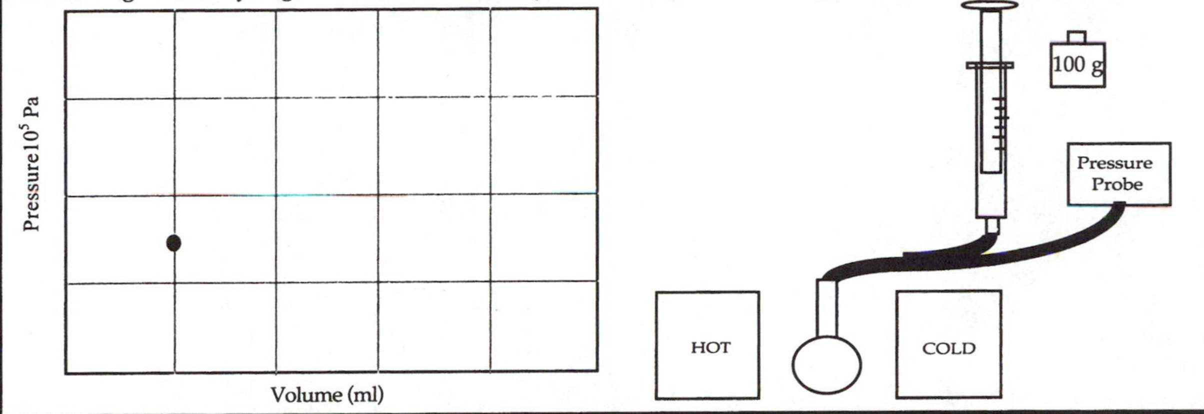

Demonstration 1a: The cycle of the engine begins with the flask in the COLD reservoir, and the 100 gram mass off of the piston. This state is represented by the black dot on the axes above. Sketch on the axes the process that takes place when the mass is quickly put on top of the piston, with the flask left in the COLD reservoir. Describe in words what happens to the pressure and the volume of the air in the syringe.

Demonstration 1b: In the next process of the cycle, the flask is moved from the COLD reservoir to the HOT reservoir, with the mass left on top of the piston. Sketch this process on the axes. Describe in words what happens to the pressure and the volume of the air in the syringe.

Demonstration 1c: In the next process of the cycle, the mass is removed from the top of the piston, with the flask left in the HOT reservoir. Sketch this process on the axes. Describe in words what happens to the pressure and the volume of the air in the syringe.

Demonstration 1d: In the last process of the cycle, the flask is moved from the HOT reservoir back to the COLD reservoir with the mass removed from the top of the piston. Sketch this process on the axes. Describe in words what happens to the pressure and the volume of the air in the syringe.

Demonstration 1e: How would you calculate the work done by the gas during this cycle? Show on your P-V diagram what represents the work done during the cycle.

Keep this sheet

INTERACTIVE LECTURE DEMONSTRATION
RESULTS SHEET—**HEAT ENGINE**

You may write whatever you wish on this sheet and take it with you.

A model of a heat engine consists of a cylinder (a glass syringe) connected by tubing to a pressure probe and a flask filled with air. The flask may be submerged into a HOT reservoir filled with hot water or a COLD reservoir filled with ice water. The job to be done by this engine is to lift a 100 gram mass a certain height. The pressure probe measures the pressure in the syringe and the volume of gas in the syringe can be measured using the markings on the syringe. The axes below may be used to plot a P-V diagram of the cycle of this engine.

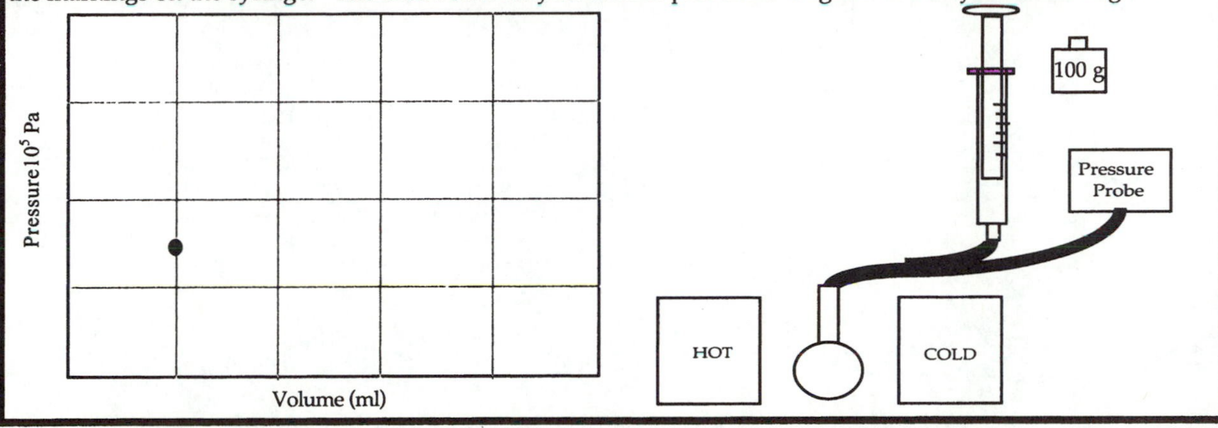

Demonstration 1a: The cycle of the engine begins with the flask in the COLD reservoir, and the 100 gram mass off of the piston. This state is represented by the black dot on the axes above. Sketch on the axes the process that takes place when the mass is quickly put on top of the piston, with the flask left in the COLD reservoir. Describe in words what happens to the pressure and the volume of the air in the syringe.

Demonstration 1b: In the next process of the cycle, the flask is moved from the COLD reservoir to the HOT reservoir, with the mass left on top of the piston. Sketch this process on the axes. Describe in words what happens to the pressure and the volume of the air in the syringe.

Demonstration 1c: In the next process of the cycle, the mass is removed from the top of the piston, with the flask left in the HOT reservoir. Sketch this process on the axes. Describe in words what happens to the pressure and the volume of the air in the syringe.

Demonstration 1d: In the last process of the cycle, the flask is moved from the HOT reservoir back to the COLD reservoir with the mass removed from the top of the piston. Sketch this process on the axes. Describe in words what happens to the pressure and the volume of the air in the syringe.

Demonstration 1e: How would you calculate the work done by the gas during this cycle? Show on your P-V diagram what represents the work done during the cycle.

HEAT ENGINE (HENG)
TEACHER'S GUIDE

Prerequisites:

These *ILDs* should be done after the first three Heat &Temperature *ILD* sequences. It is helpful if the students have been introduced to the following terms: cycle, isobaric, adiabatic, thermodynamic work, heat reservoir and efficiency.

Equipment:

computer-based laboratory system
ILD experiment configuration files
pressure sensor (See below.)
temperature sensor (See below.)
10-cc low friction glass syringe with ring stand support (See below.)
several lengths of Tygon tubing
25-ml flask with one hole rubber stopper
2 insulated (e.g., Styrofoam) containers (to use as reservoirs)
ruler
100-g mass
hot water (about 80-90°C)
ice water

General Notes on Preparation and Equipment:

Pressure sensor:
Since the syringe is near atmospheric pressure, a barometric probe should be used. Vernier Software and Technology (www.vernier.com) sells a barometer probe (BAR-BTA) for use with the *Lab Pro* interface and *Logger Pro* software. While the PASCO (www.pasco.com) absolute pressure sensor (CI-6532A) for the *Science Workshop/Data Studio* system operates over a wider range, it should also work. PASCO also makes absolute pressure (PS-2107) and a barometer sensors (PS-2113) for the *PASSPORT* system.

Temperature probes:
See the Teacher's Guide for the *Introduction to Heat and Temperature ILDs*.

Glass syringe and heat engine:
In this investigation you will be showing students the operation of an actual heat engine. The essential element of this engine is a low friction, 10 cc glass syringe, with a needle luer-lock tip, which is commonly found in chemistry departments, and which may be purchased (around $20) from science equipment supply companies, e.g., PGC Scientifics Corporation (www.pgcsci.com). As long as the plunger and cylinder are clean, the operation of this syringe will be with very low friction. If the plunger starts to stick, clean it in distilled water and detergent and rinse in distilled water. The only drawback of this piece of apparatus is that it is very breakable. Extreme care must be taken that the plunger does not fall out of the syringe. Alternatively, the more expensive PASCO (www.pasco.com) Heat Engine/Gas Law Apparatus (TD-8572) will work very well in this experiment in place of the low friction syringe.

Experimental setup:
The software should be set-up in prompted event mode where it continuously measures pressure. When you decide it is appropriate, you can keep the pressure value and enter a value for the volume of the

syringe, tubing, pressure sensor and flask. Only the syringe changes volume so you can enter an appropriate constant volume for the tubing, pressure sensor and flask into a calculated column before class. As you do the demonstration, you then only need to read and enter the volume of the syringe.

Demonstrations and Sample Graphs:

Have the students make predictions for demonstrations 1a to 1d. For each one they should make an individual predicton, and discuss it with nearest neighbors. You should gather various ideas. After this you can do all four steps of the heat cycle. You should do it rapidly to minimize air leaks in the system and heat loss or gain. After taking all data, you can use the Replay feature of the software to go back over the cycle. Use experiment configuration file **HENGD1**.

Demonstration 1a: The flask is in the COLD reservoir, and the 100 gram mass off of the piston. Let the pressure stabilize. Put the 100 gram mass on top of the piston. The piston moved quickly downward. Since this happened quickly, it is possible it took place without any transfer of heat energy into or out of the syringe, in which case it was an adiabatic change. Let the pressure stabilize.

Demonstration 1b: The flask is moved from the COLD reservoir to the HOT reservoir, with the mass left on top of the piston. The piston and mass rose. Since the pressure in the gas is the force per unit area, it is determined by the force exerted on the gas by the weight of the plunger and 100 gram mass, and by atmospheric pressure exerted on the plunger. Since neither of these changes during the process, the pressure should remain constant. This should be an isobaric (constant pressure) process.

Demonstration 1c: In the next process of the cycle, the mass is removed from the top of the piston, with the flask left in the HOT reservoir. The plunger moved upward. Since this happened quickly, it is possible it was an adiabatic change.

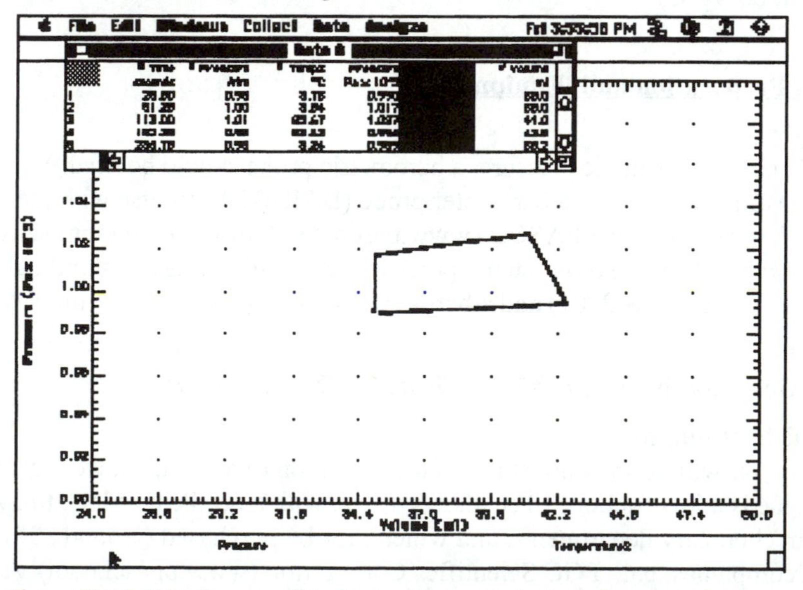

Figure IV-13: P-V diagram of the heat engine cycle in these demonstrations.

Demonstration 1d: In the last process of the cycle, the flask is moved from the HOT reservoir back to the COLD reservoir with the mass removed from the top of the piston. The piston moved downward. Since the pressure is determined by the force exerted on the gas by the weight of the plunger and by atmospheric pressure, this should be an isobaric process. The gas is now back in its original state A typical graph is shown in Figure IV-13. It is clear that the pressure changed somewhat from leakage.

Demonstration 1e: How would you calculate the work done by the gas during this cycle? The area inside the P-V graph is the work done. Use your software to integrate or estimate the area manually.

HEAT ENGINE (HENG)
TEACHER PRESENTATION NOTES

Classroom introduction to the *Heat Engine ILDs*:
Students should be familiar with the temperature probe. Introduce the pressure probe before running the experiment.

Have the students work through making predictions for Demonstrations 1a to 1d. For each one they should make an individual prediction, and discuss it with nearest neighbors. You should gather various ideas for each one before they go on. Call their attention at each stage to how or if the volume or pressure changes. After this you can do the entire demonstration (all four steps of the heat engine cycle). You should do it rapidly to minimize air leaks in the system and heat loss or gain to the system. After taking all data, you can use the Replay feature of the software to go back over the cycle more slowly. Use experiment configuration file **HT04D1**. A description of the effect follows each step in the cycle

<u>**Demonstration 1a**</u>**: The flask is in the COLD reservoir, and the 100 gram mass <u>off of the piston</u>. Let the pressure stabilize.** Put the 100 gram mass on top of the piston. The piston moved quickly downward.

- What kind of process was this? Since this happened quickly, it is possible it took place without any transfer of heat energy into or out of the syringe, in which case it was an adiabatic change. Let the pressure stabilize.

<u>**Demonstration 1b**</u>**: The flask is moved from the COLD reservoir to the HOT reservoir, with the mass left on top of the piston**. The piston and mass rose.

- What kind of process was this? Since the pressure in the gas is the force per unit area, it is determined by the force exerted on the gas by the weight of the plunger and 100 gram mass, and by atmospheric pressure exerted on the plunger.
- Since neither of these changes during the process, the pressure should remain constant. This should be an isobaric (constant pressure) process.

<u>**Demonstration 1c**</u>**: In the next process of the cycle, the mass is removed from the top of the piston, with the flask left in the HOT reservoir.** The plunger moved upward.

- What kind of process was this? Since this happened quickly, it is possible it took place without any transfer of heat energy into or out of the syringe, in which case it was an adiabatic change.

<u>**Demonstration 1d:**</u> **In the last process of the cycle, the flask is moved from the HOT reservoir back to the COLD reservoir with the mass removed from the top of the piston**. The piston moved downward.

- What kind of process was this? Since the pressure in the gas is the force per unit area, it is determined by the force exerted on the gas by the weight of the plunger and by atmospheric pressure exerted on the plunger.
- Since neither of these changes during the process, the pressure should remain constant. This should be an isobaric process. The gas is now back in its original state (except for some possible air leakage).

<u>**Demonstration 1e**</u>**: How would you calculate the work done by the gas during this cycle?** The area inside of the P-V graph will be the work done by the gas. Use your software to integrate or estimate the area manually.

- How is the work in a P-V process calculated? How can it be found from the graph?

SECTION V: INTERACTIVE LECTURE DEMONSTRATIONS IN ELECTRICITY AND MAGNETISM

ELECTRIC FIELD, FORCE AND POTENTIAL (EFFP)

Hand in this sheet **Name**__________________________

INTERACTIVE LECTURE DEMONSTRATIONS

PREDICTION SHEET—**ELECTROSTATIC FIELD, FORCE AND POTENTIAL**

Directions: This sheet will be collected. Write your name at the top to record your presence and participation in these demonstrations. Follow your instructor's directions. You may write whatever you wish on the attached Results Sheet and take it with you.

Demonstration 1: Two identical positive point charges are located as shown in the figure on the right. Draw an arrow to indicate the direction of the electric field at each of the x's.

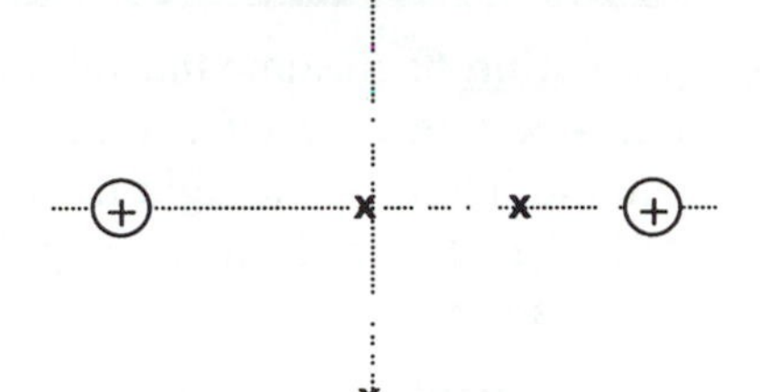

Demonstration 2: Imagine that you place a small positive test charge on one of the x's in the figure on the right. Draw an arrow to indicate the direction of the electrostatic *force* on the test charge. Repeat for each of the other two x's.

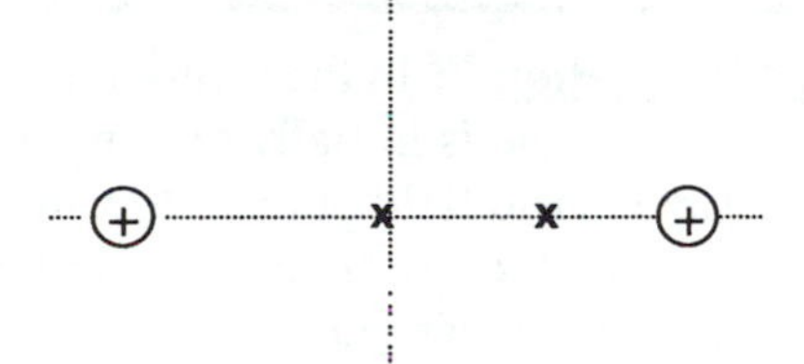

Demonstration 3: Assume the value of the electrostatic potential is zero infinitely far from any point charge. At each x in the figure on the right indicate with a symbol (+,- or 0) whether the electrostatic *potential* is positive, zero, or negative.

At which x would a test charge have the *highest* electrostatic potential energy? How do you know?

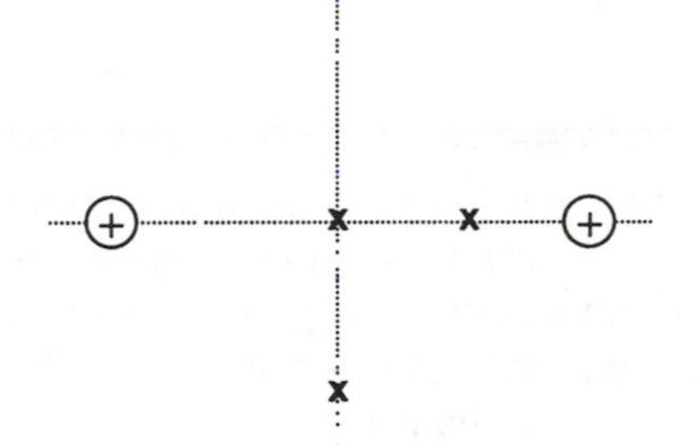

Demonstration 4: A positive and a negative charge of equal strength are located as shown in the figure on the right. Draw an arrow to indicate the direction of the electric field at each of the x's.

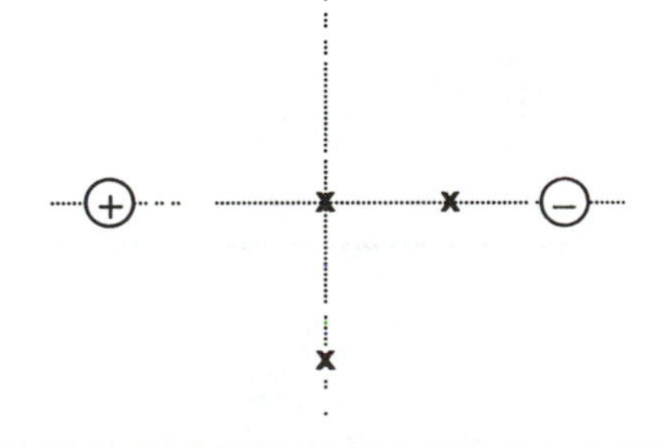

Demonstration 5: Imagine that you place a small positive test charge on one of the x's in the figure on the right. Draw an arrow to indicate the direction of the electrostatic *force* on the test charge. Repeat for each of the other two x's.

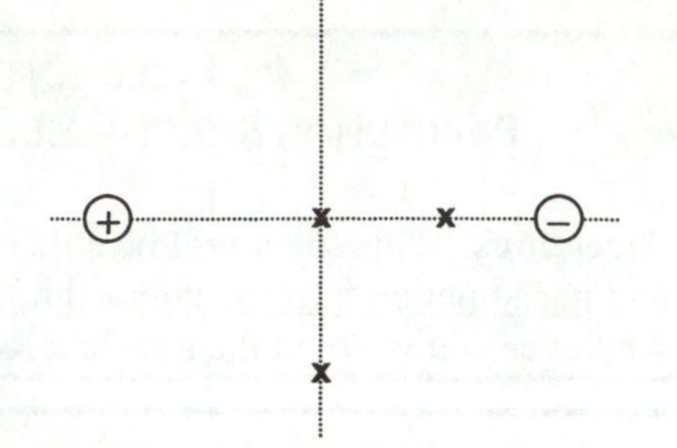

Demonstration 6: Assume the value of the electrostatic potential is zero infinitely far from any point charge. At each x in the figure on the right indicate with a symbol (+,- or 0) whether the electrostatic potential is positive, zero, or negative.

At which x's would a test charge have the *highest* electrostatic potential energy?

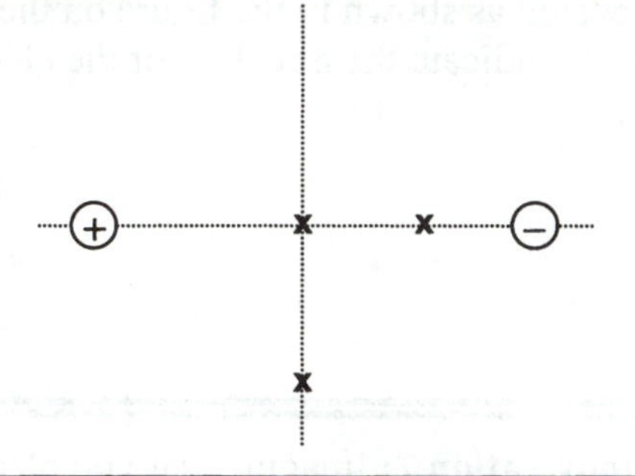

Demonstration 7: In the figure on the right, a *positively* charged particle is initially moving to the right through a uniform electric field pointed towards the top of the page. At this instant, draw an arrow to indicate the direction of the force on the particle.

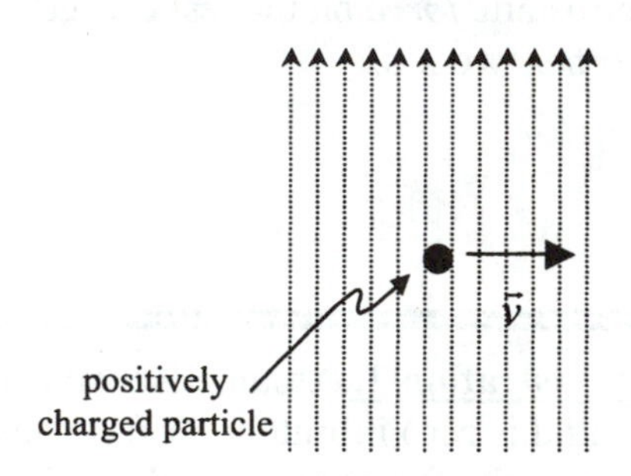

Demonstration 8: In the figure on the right, a *negatively* charged particle is initially moving to the right through a uniform electric field pointed towards the top of the page. At this instant, draw an arrow to indicate the direction of the force on the particle.

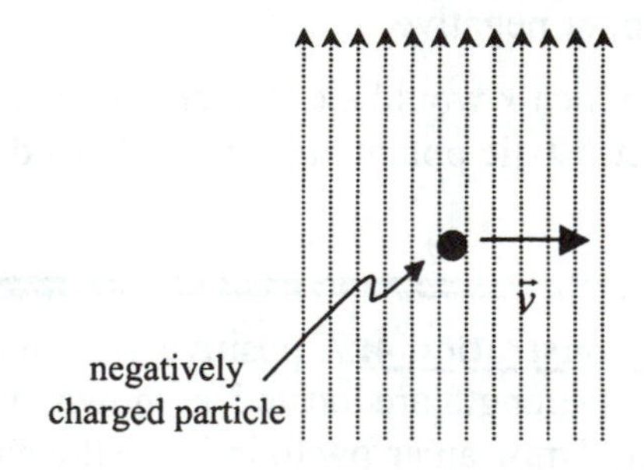

Keep this sheet

INTERACTIVE LECTURE DEMONSTRATIONS

RESULTS SHEET— **ELECTROSTATIC FIELD, FORCE AND POTENTIAL**

You may write whatever you wish on this sheet and take it with you.

Demonstration 1: Two identical positive point charges are located as shown in the figure on the right. Draw an arrow to indicate the direction of the electric field at each of the x's.

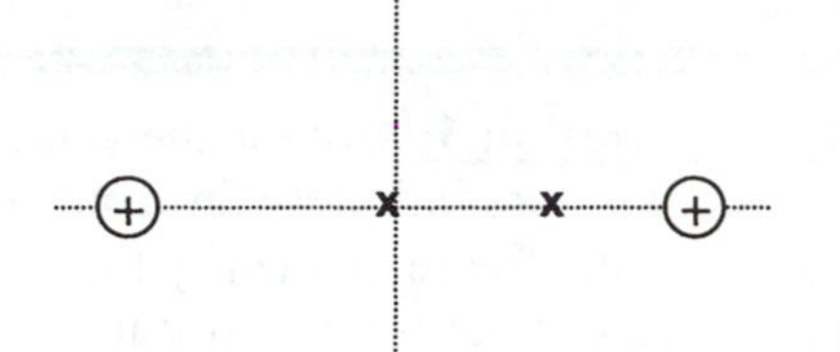

Demonstration 2: Imagine that you place a small positive test charge on one of the x's in the figure on the right. Draw an arrow to indicate the direction of the electrostatic *force* on the test charge. Repeat for each of the other two x's.

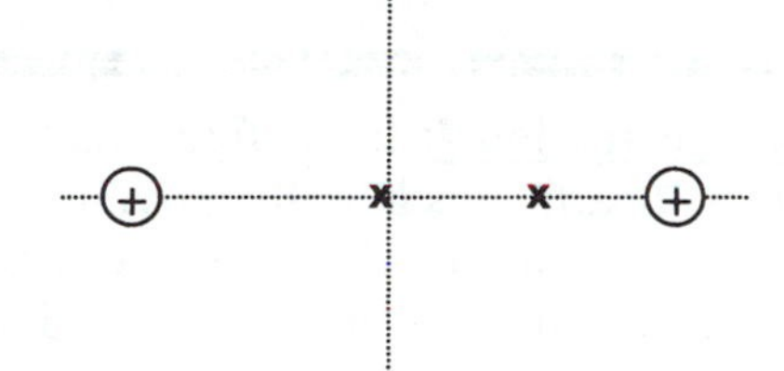

Demonstration 3: Assume the value of the electrostatic potential is zero infinitely far from any point charge. At each x in the figure on the right indicate with a symbol (+,- or 0) whether the electrostatic *potential* is positive, zero, or negative.

At which x would a test charge have the *highest* electrostatic potential energy? How do you know?

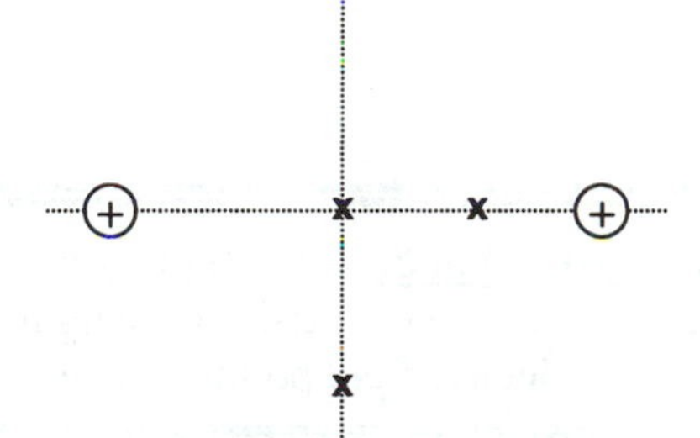

Demonstration 4: A positive and a negative charge of equal strength are located as shown in the figure on the right. Draw an arrow to indicate the direction of the electric field at each of the x's.

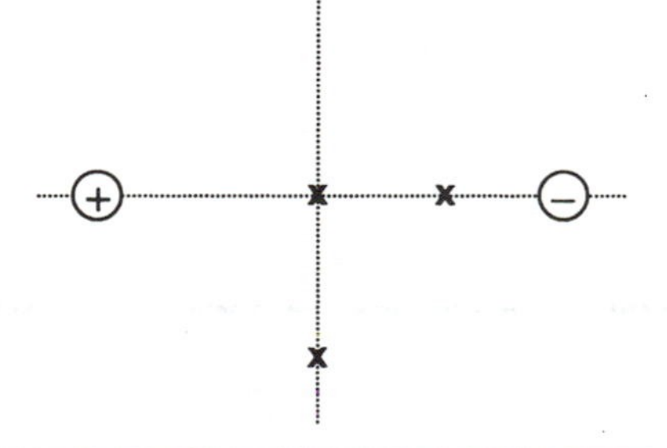

Demonstration 5: Imagine that you place a small positive test charge on one of the x's in the figure on the right. Draw an arrow to indicate the direction of the electrostatic *force* on the test charge. Repeat for each of the other two x's.

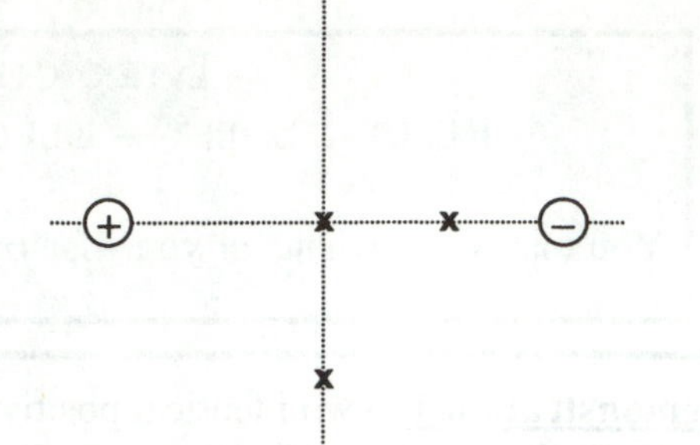

Demonstration 6: Assume the value of the electrostatic potential is zero infinitely far from any point charge. At each x in the figure on the right indicate with a symbol (+,- or 0) whether the electrostatic potential is positive, zero, or negative.

At which x's would a test charge have the *highest* electrostatic potential energy?

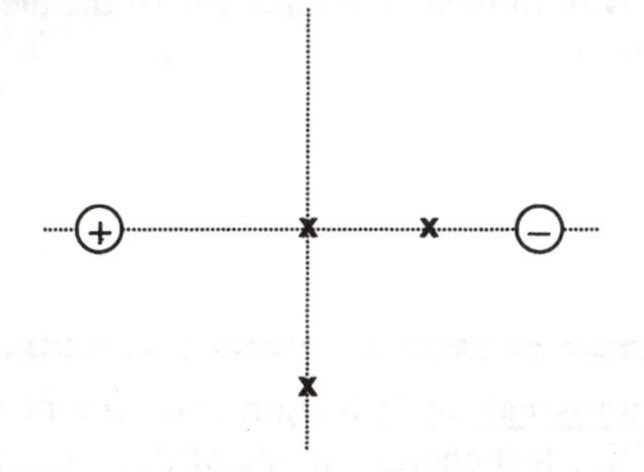

Demonstration 7: In the figure on the right, a *positively* charged particle is initially moving to the right through a uniform electric field pointed towards the top of the page. At this instant, draw an arrow to indicate the direction of the force on the particle.

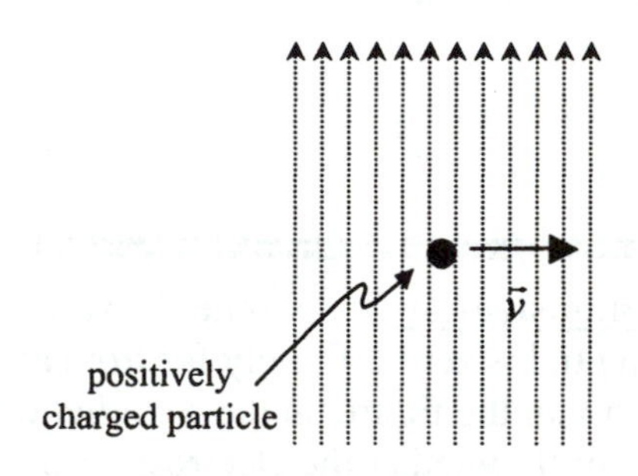

Demonstration 8: In the figure on the right, a *negatively* charged particle is initially moving to the right through a uniform electric field pointed towards the top of the page. At this instant, draw an arrow to indicate the direction of the force on the particle.

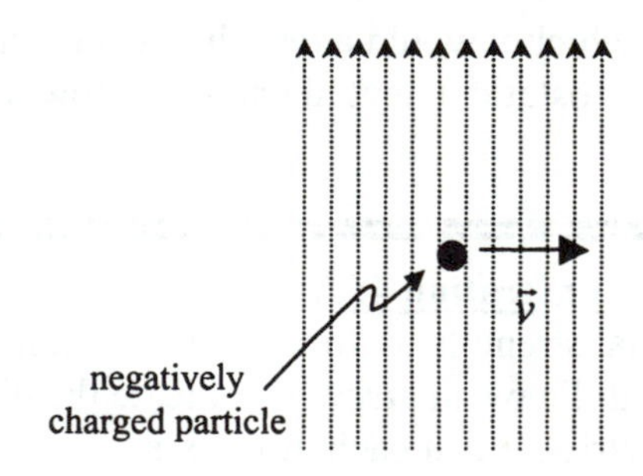

ELECTROSTATIC FIELD, FORCE AND POTENTIAL (EFFP) TEACHER'S GUIDE

Prerequisites:

This *ILD* sequence can be used as a review of electric field, force and potential concepts after they have been covered in lecture and/or text readings. Students should have been introduced to positive and negative charges, Coulomb's law, electric field, test charges and electrostatic potential. Unlike most *ILDs* in this series, this set does not involve actual physical demonstrations.

Equipment:

no equipment is required

Demonstrations and Sample Results:

Demonstration 1: Electric field for two positive point charges. After the prediction and discussion steps are completed, sketch the vectors representing the contribution to the field by each charge at one of the three points. Show how vector addition gives the resultant field at that point. Ask students to describe the process for the other two points.

Discussion after observing the results: How does the electric field depend on the distance from a point charge? How can you determine the direction of the field produced at a point in space by a positive point charge? Why must the field contributions of the two charges be added together by vector addition? What does it mean that there is zero field exactly between the two charges?

Demonstration 2: Force on a positive test charge. After the prediction and discussion steps are completed, sketch the vector representing the force on a positive test charge placed at one of the points.

Discussion after observing the results: What is the relationship between the direction of the electric field at a point in space and the direction of the force on a positive test charge? What is the direction of the force at the other two points? Once you know the electric field at a point, why don't you need to use Coulomb's law to find the force? What would the direction of the force be at each point if the test charge were negative?

Demonstration 3: Electrostatic potential for two positive charges. Describe how the electrostatic potential is calculated as the algebraic (not vector) sum of contributions from the three charges. Show how you can apply this to one of the points. Explain why it is necessary to specify that the potential is zero at an infinite distance from the charges.

Discussion after observing the results: What is the sign of the electrostatic potential at the other two points? Why is the potential positive at all three points? In particular, why is the electrostatic potential positive at the point midway between the two positive charges while the electric field is zero? Is it possible for the electrostatic potential produced by a collection of positive charges to ever be zero anywhere other than infinitely far away from the charges? How can you determine at which point the potential will be largest?

Demonstration 4: Electric field for positive and negative point charges. After the prediction and discussion steps are completed, sketch the vectors representing the contribution to the field by each charge at one of the three points. Show how vector addition gives the resultant field at that point. Ask students to describe the process for the other two points.

Discussion after observing the results: How can you determine the direction of the field produced at a point in space by a negative point charge? How does this change the situation from Demonstration 1? Why is the field no longer zero at the point midway between the two charges? Is there some other point (points) where the field is zero?

<u>Demonstration 5:</u> Force on a positive test charge. After the prediction and discussion steps are completed, sketch the vector representing the force on a positive test charge placed at one of the points.

Discussion after observing the results: Does the relationship between the direction of the electric field at a point in space and the direction of the force on a positive test charge used in Demonstration 2 work here as well? What is the direction of the force at the other two points? What would the direction of the force be at each point if the test charge were negative?

<u>Demonstration 6:</u> Electrostatic potential for positive and negative point charges. Show how you apply the method used in Demonstration 3 to one of the points.

Discussion after observing the results: What is the sign of the electrostatic potential at the other two points? Why is the electrostatic potential now zero at the point midway between the two charges? What does it mean for the potential to be zero? How can you determine at which point the potential will be largest?

<u>Demonstration 7:</u> Force on a moving positive charge. After the prediction and discussion steps are completed, ask for volunteer(s) to describe the force.

Discussion after observing the results: Does the electrostatic force depend on whether or not the charge is moving? Again ask for the relationship between electrostatic force on a positive test charge and electric field at a point in space. How would the charge move?

<u>Demonstration 8:</u> Force on a moving negative charge. After the prediction and discussion steps are completed, ask for volunteer(s) to describe the force.

Discussion after observing the results: What is different when the test charge is negative? How would the charge move?

ELECTROSTATIC FIELD, FORCE AND POTENTIAL (ILDEM01)
TEACHER PRESENTATION NOTES

Demonstration 1: Electric field for two positive point charges. Sketch the vector contributions to the field by each charge at one of the points. Show how vector addition gives the resultant field.

- How does the field depend on distance from a point charge? What is the field direction?
- Why must vector addition be used? What does zero field mean?

Demonstration 2: Force on a positive test charge. Sketch the vector representing the force on a positive test charge placed at one of the points.

- What is the relationship between the direction of the field and the force on a positive charge? What is the direction at the other two points? Why don't you need to use Coulomb's law?
- What would be the direction of the force on a negative test charge?

Demonstration 3: Electrostatic potential for two positive charges. Show how the potential is calculated. Explain why the potential must be zero at an infinite distance from the charges.

- What is the sign of the potential at the other two points? Why is the potential positive at all three points? In particular, why is it positive at the point midway between the charges?
- Can the potential produced by a collection of positive charges ever be zero anywhere?
- How can you determine at which point the potential will be largest?

Demonstration 4: Electric field for positive and negative point charges. Sketch the vector contributions by each charge at one of the points. Show how vector addition gives the resultant field.

- What is the direction of the field produced by a negative point charge? How does this change the situation from Demonstration 1?
- Why is the field non-zero midway between the charges? Is it zero at some other point(s)?

Demonstration 5: Force on a positive test charge. Sketch the vector representing the force on a positive test charge placed at one of the points.

- Does the relationship from Demonstration 2 work here as well? What is the direction of the force at the other two points?
- What would the direction of the force be at each point if the test charge were negative?

Demonstration 6: Electrostatic potential for positive and negative point charges. Show how you apply the method used in Demonstration 3 to one of the points.

- What is the sign of the potential at the other two points? Why is the potential now zero at the point between midway the two charges? What does it mean for the potential to be zero?
- How can you determine at which point the potential will be largest?

Demonstration 7: Force on a moving positive charge. After the prediction and discussion steps are completed, ask for volunteer(s) to describe the force.

- Does the force depend on whether or not the charge is moving? How would the charge move?

Demonstration 8: Force on a moving negative charge. After the prediction and discussion steps are completed, ask for volunteer(s) to describe the force.

- What is different when the test charge is negative? How would the charge move?

Introduction to DC Circuits (INDC)

Hand in this sheet **Name**____________________________

INTERACTIVE LECTURE DEMONSTRATIONS
PREDICTION SHEET—**INTRODUCTION TO DC CIRCUITS**

Directions: This sheet will be collected. Write your name at the top to record your presence and participation in these demonstrations. Follow your instructor's directions. You may write whatever you wish on the attached Results Sheet and take it with you.

Demonstration 1: A resistor (a device that obeys Ohm's law) is connected to a variable source of voltage. Sketch on the right a graph of the current that flows through the resistor as the voltage across the resistor is increased starting from zero.

How would you calculate the resistance of the resistor by reading values from the graph?

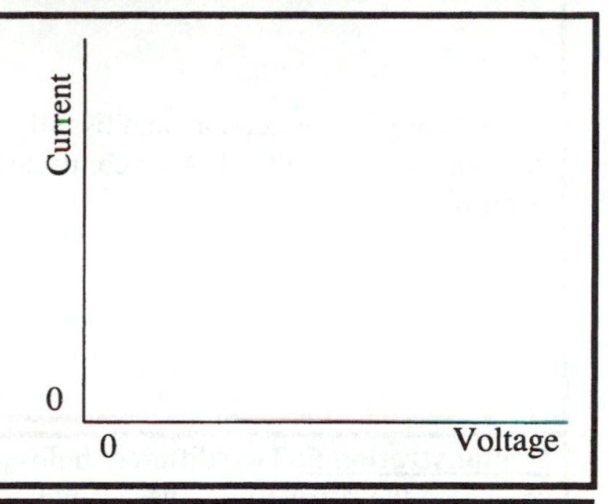

Demonstration 2: The resistance of conductors increases as the temperature increases. As more current flows through the filament of a light bulb, the temperature of the filament gets higher. A light bulb is connected to a variable source of voltage. Sketch on the right a graph of the current that flows through the light bulb as the voltage across the light bulb is increased starting from zero.

Does a light bulb obey Ohm's law?

How would you find the resistance of the light bulb?

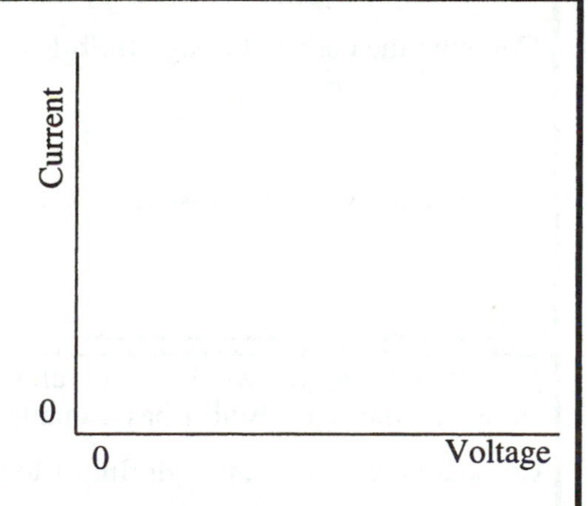

Demonstration 3: Two light bulbs are connected with the same potential difference (voltage) across their terminals. Bulb 1 has a smaller resistance than Bulb 2.

Which bulb has a larger current flowing through it, or do they both have the same current?

How is power related to the voltage across a bulb and the current flowing through the bulb?

Which bulb has the larger power delivered to it, or do they both have the same power?

Which bulb is brighter, or are they both just as bright?

Demonstration 4: Seven bulbs are connected to a battery (B) as shown on the right.

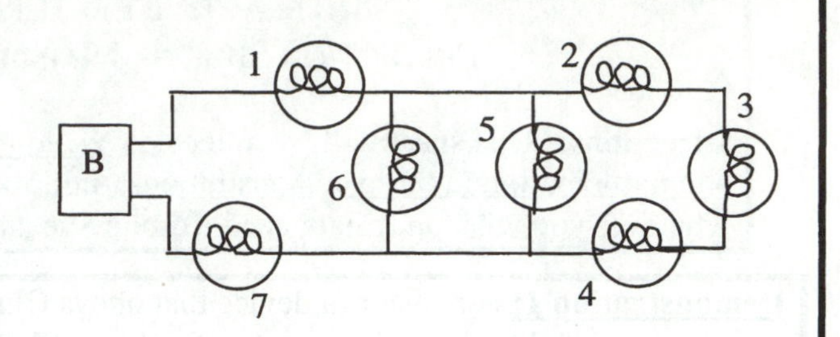

Define series connection, and list all combinations of bulbs that are connected in series.

Define parallel connection, and list all combinations of bulbs that are connected in parallel.

Demonstration 5: Two different bulbs are connected to a battery as shown on the right. Bulb 1 has a smaller resistance than Bulb 2.

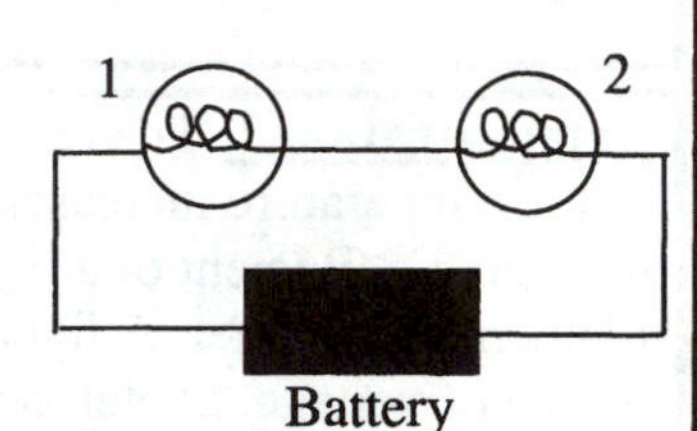

Compare the current through Bulb 1 to the current through Bulb 2.

Compare the voltage across Bulb 1 to the voltage across Bulb 2.

Demonstration 6: Two different bulbs are connected to a battery as shown on the right. Bulb 1 has a smaller resistance than Bulb 2.

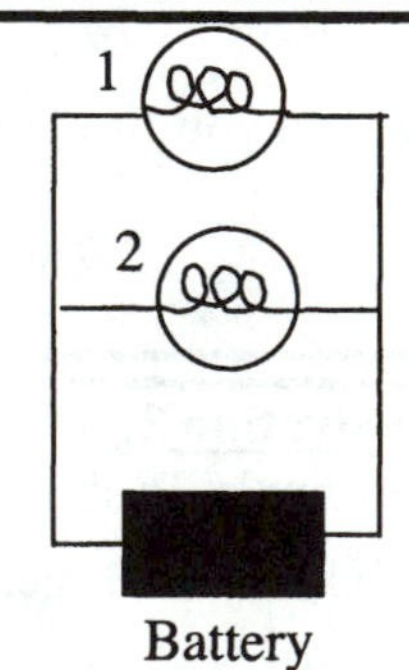

Compare the current through Bulb 1 to the current through Bulb 2.

Compare the voltage across Bulb 1 to the voltage across Bulb 2.

Keep this sheet

INTERACTIVE LECTURE DEMONSTRATIONS
RESULTS SHEET—**INTRODUCTION TO DC CIRCUITS**

You may write whatever you wish on this sheet and take it with you.

Demonstration 1: A resistor (a device that obeys Ohm's law) is connected to a variable source of voltage. Sketch on the right a graph of the current that flows through the resistor as the voltage across the resistor is increased starting from zero.

How would you calculate the resistance of the resistor by reading values from the graph?

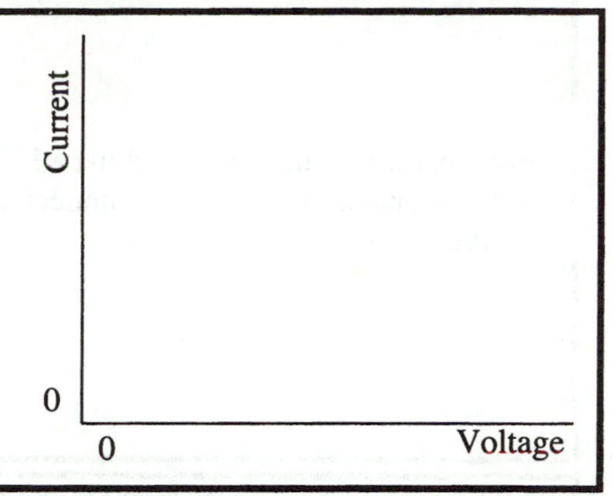

Demonstration 2: The resistance of conductors increases as the temperature increases. As more current flows through the filament of a light bulb, the temperature of the filament gets higher. A light bulb is connected to a variable source of voltage. Sketch on the right a graph of the current that flows through the light bulb as the voltage across the light bulb is increased starting from zero.

Does a light bulb obey Ohm's law?

How would you find the resistance of the light bulb?

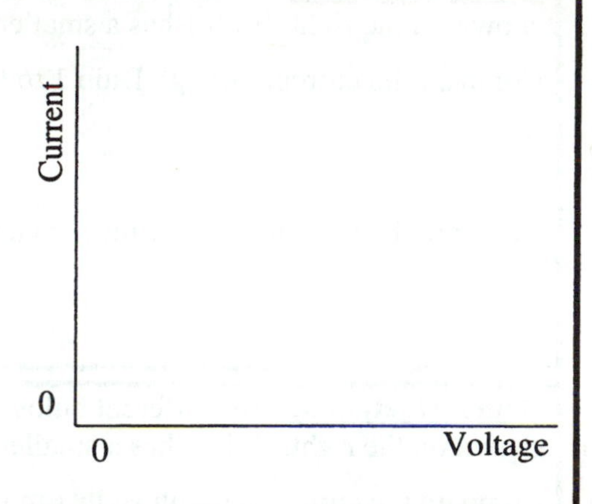

Demonstration 3: Two light bulbs are connected with the same potential difference (voltage) across their terminals. Bulb 1 has a smaller resistance than Bulb 2.

Which bulb has a larger current flowing through it, or do they both have the same current?

How is power related to the voltage across a bulb and the current flowing through the bulb?

Which bulb has the larger power delivered to it, or do they both have the same power?

Which bulb is brighter, or are they both just as bright?

Demonstration 4: Seven bulbs are connected to a battery (B) as shown on the right.

Define series connection, and list all combinations of bulbs that are connected in series.

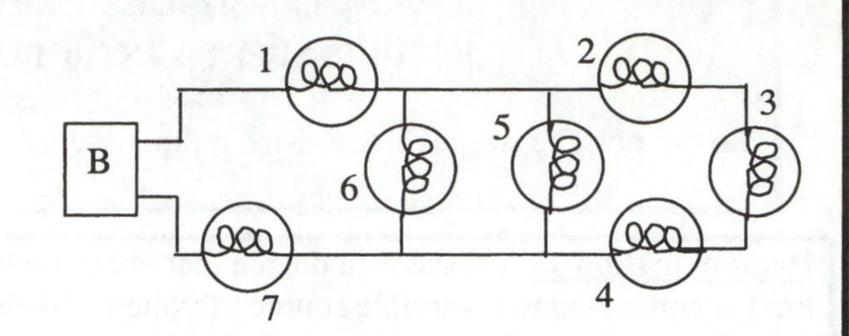

Define parallel connection, and list all combinations of bulbs that are connected in parallel.

Demonstration 5: Two different bulbs are connected to a battery as shown on the right. Bulb 1 has a smaller resistance than Bulb 2.

Compare the current through Bulb 1 to the current through Bulb 2.

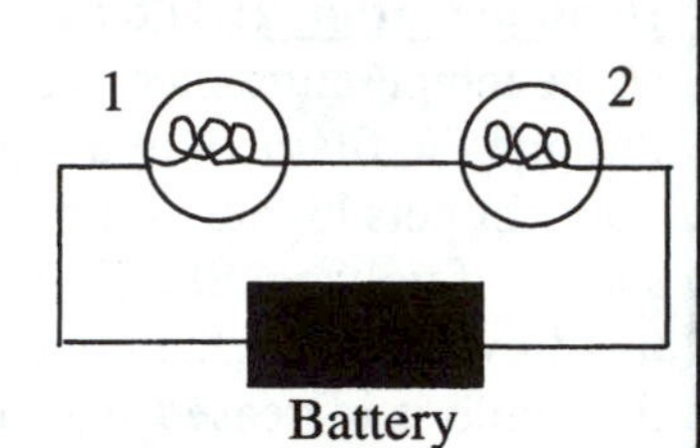

Compare the voltage across Bulb 1 to the voltage across Bulb 2.

Demonstration 6: Two different bulbs are connected to a battery as shown on the right. Bulb 1 has a smaller resistance than Bulb 2.

Compare the current through Bulb 1 to the current through Bulb 2.

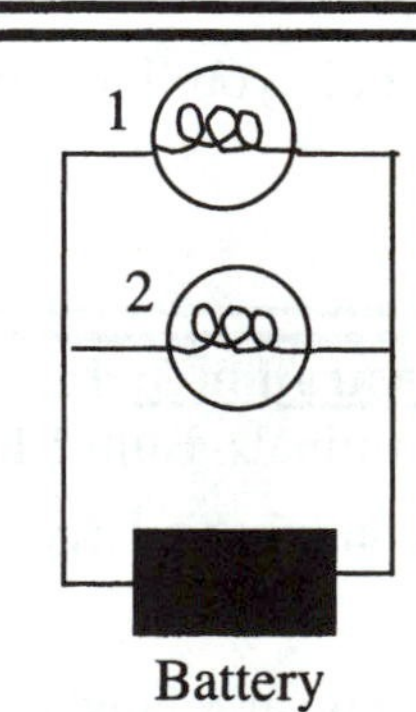

Compare the voltage across Bulb 1 to the voltage across Bulb 2.

Introduction to DC Circuits (INDC)
Teacher's Guide

Prerequisites:

Students should have already heard the definitions of basic circuit quantities such as current, voltage (potential difference), series, parallel, etc.

Equipment:

computer-based laboratory system
2 current probes (See below.)
2 voltage probes (See below.)
ILD experiment configuration files
circuit board with 15 ohm resistor mounted on it (See below.)
circuit board with lantern bulb and socket mounted on it (See below.)
circuit board with two lantern bulbs and sockets connected in series (See below.)
circuit board with two lantern bulbs and sockets connected in parallel (See below.)
two bulbs with different resistances
regulated variable DC power supply (See below.)
6 V lantern battery

General Notes on Preparation and Equipment:

Current and Voltage Probes:
Current and Voltage probes are available from Vernier Software and Technology (www.vernier.com) and from PASCO Scientific (www.pasco.com). The Vernier probes are the Current Probe (DCP-BTA) and the Differential Voltage Probe (DVP-BTA). The PASCO probes are the Current Sensor (CI-6556) and the Voltage Sensor (CI-6503).

Circuit Boards:
The size of the circuit boards should be appropriate to the size of the lecture class. They can be fabricated from poster or foam board. Circuit diagrams identical to those on the Prediction and Results Sheets should be drawn on the board with a thick marker, and the circuit elements should be mounted over the diagram in the appropriate places. The circuits should be wired in front, where students can see the wires, and provision should be made for connecting the probes. It should be possible to use the same board for Demonstrations 1 and 2, if there is provision for disconnecting the resistor and replacing it with the light bulb and socket.

Power Supply:
The DC power supply should be regulated to avoid drift in the current and voltage displays with the probes. It should be able to deliver up to 0.4 amperes at 6 V.

Demonstrations and Sample Graphs:

Demonstration 1: Current-Voltage graph for a resistor. (Use experiment configuration file **INDCD1**.) Two sets of axes will open—current vs. time and voltage vs. time. Connect the power supply to the circuit board. The current probe and voltage probe should be connected to measure the current through and the voltage across the resistor, respectively. When you begin graphing and adjust the power supply from 0 to 6 V, graphs of current vs. time and voltage vs. time will be plotted. After the graphing ends, you can change the horizontal axis on the current vs. time graph to voltage, and display a graph of current vs. voltage. A sample graph is shown in Figure V-1.

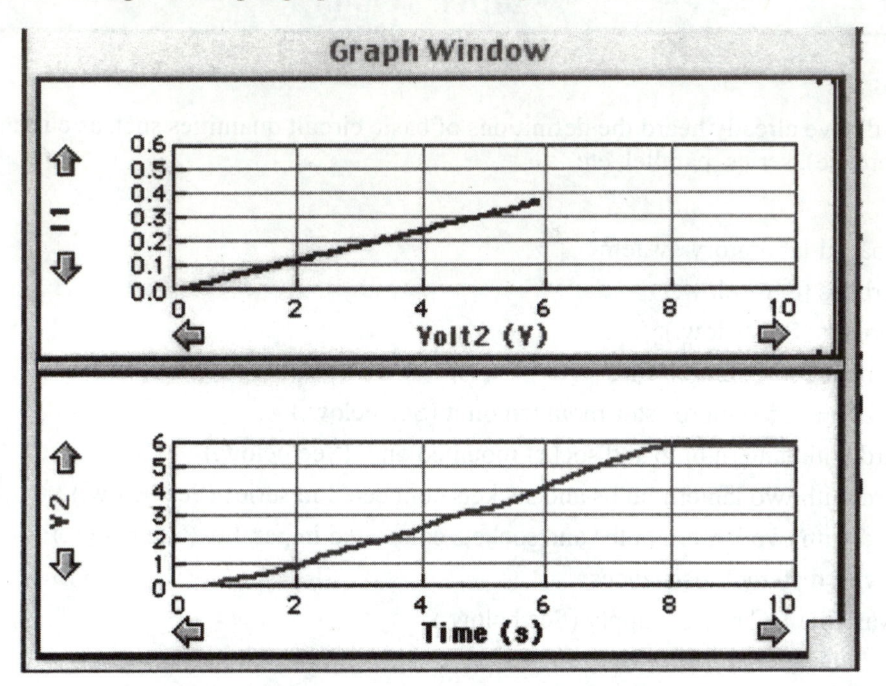

Figure V-1: Current vs. voltage and voltage vs. time graphs for a resistor in Demonstration 1.

Discussion after the graphs are displayed: Ask students to describe the current vs. voltage graph, and what it tells abut the relationship between current and voltage for a resistor. Ask for a volunteer to state Ohm's law: I=V/R. Ask for a volunteer to describe how the graph supports this relationship, and how R could be found from the graph. (Inverse of the slope.)

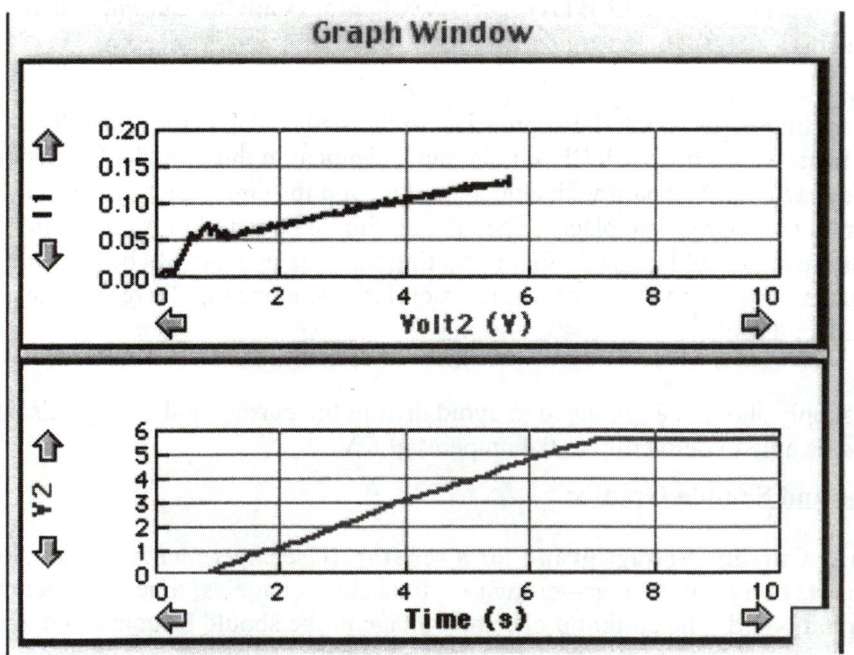

Figure V-2: Current vs. voltage graph for a light bulb in Demonstrations 2.

<u>Demonstration 2:</u> Current-Voltage graph for a light bulb. (Use the same experiment configuration file as in Demonstration 1.) Return the axes to current and voltage vs. time. Replace the resistor with the bulb and socket. Graph as the voltage is increased from zero, and display current vs. voltage as in Demonstration 1. A sample graph is shown in Figure V-2.

Discussion after the graphs are displayed: Ask students to describe the current vs. voltage graph, and what it tells about the relationship between current and voltage for a light bulb. Is it a linear relationship as with the resistor? Does a light bulb obey Ohm's law. (No, it is a non-Ohmic device.) Ask if the resistance can still be defined for a light bulb. (Yes, for a particular operating point, $R = V/I$, but it is not a constant. It changes as the current flowing through the filament changes and heats the filament to a higher temperature.)

<u>Demonstration 3:</u> Bulbs with same potential difference. (Use experiment configuration file **INDCD3**. This will provide a digital display of current from the two current probes.) The probes should be connected to measure the currents through Bulb 1 and Bulb 2. Both bulbs should be connected across the power supply so that the potential differences are the same.

After predictions are made and displayed, turn up the power supply until both bulbs are lighted. Display the two currents.

Discussion after the results are displayed: Which bulb has the larger current? Given that Bulb 2 has a larger resistance, how is this related to Ohm's law, explored in Demonstration 1. What is the power relationship? ($P=IV$) If both bulbs have the same voltage across their terminals, which has the larger power delivered? Which bulb was brighter? How is brightness related to the power delivered to a bulb?

<u>Demonstration 4:</u> Series and parallel connections. This is an exercise in identifying series and parallel connections. Many students are confused by this. For example, many students will think that bulbs 1 and 7 are connected in parallel. Also 2 and 4. Many students will think that bulbs 1, 6 and 7 are in series. Also 2, 3, 4 and 5.

Discussion. Ask for volunteers to define series and parallel connections. Then ask for volunteers to describe how each of the bulbs is connected. Note that 5 and 6 are in parallel, and they are also in parallel with the series combination of 2, 3 and 4. 1 and 7, and 2, 3 and 4 are in series, and 1 and 7 are in series with the combination of 5, 6 and 2, 3, 4.

<u>Demonstration 5:</u> Bulbs in series. (Use experiment configuration file **INDCD5**. This will provide digital displays of either the two current probes or the two voltage probes.) Use the circuit board with the two bulbs in series. First connect the two current probes to measure the currents through the two bulbs. Then connect the voltage probes to measure the voltages across the two bulbs.

Discussion after the results are displayed. Be sure to contrast the results with those from Demonstration 3. Note that the result here will be surprising because of Demonstration 3. Ask for student volunteer(s) to explain why the current is the same through both bulbs. How is this characteristic of a series circuit? After measuring the voltages, ask for a volunteer(s) to explain why one voltage is smaller than the other. Which bulb is having more power delivered to it. Which bulb is brighter?

<u>Demonstration 6:</u> Bulbs in parallel. (Use the experiment configuration file as in Demonstration 5. This will provide digital displays of either the two current probes or the two voltage probes.) Use the circuit board with the two bulbs in parallel. First connect the two current probes to measure the currents through the two bulbs. Then connect the voltage probes to measure the voltages across the two bulbs.

Discussion after the results are displayed. Be sure to point out that theses results are the same as in Demonstration 3. Ask for student volunteer(s) to explain why the voltage is the same across both bulbs.

How is this characteristic of a parallel circuit? After measuring the currents, ask for a volunteer to explain why one current is smaller than the other. Which bulb is having more power delivered to it. Which bulb is brighter?

INTRODUCTION TO DC CIRCUITS (INDC)
TEACHER PRESENTATION NOTES

Demonstration 1: Current-Voltage graph for a resistor. Use experiment configuration file **INDCD1**. Begin with axes for current and voltage vs. time.

- After collecting data, display current vs. voltage. Ask the students for a description of the graph and how it is related to Ohm's law, I=V/R.
- Ask how R can be determined from the graph.

Demonstration 2: Current-Voltage graph for a resistor. Use the same experiment configuration file. Begin with axes for current and voltage vs. time.

- After collecting data, display current vs. voltage. Ask the students for a description of the graph and how it differs from Demonstration 1.
- Is a light bulb Ohmic? How is R defined and determined from the graph?

Demonstration 3: Bulbs with the same potential difference. Use experiment configuration file **INDCD3**. Turn up power supply until both bulbs light, and display the currents.

- Have students observe which bulb has the larger current. How is this related to Ohm's law, if Bulb 2 has the larger resistance?
- What is the power relationship? (P=IV) How is power related to the brightness of bulbs?

Demonstration 4: Series and parallel connections. The common *incorrect* predictions are Parallel: 1 and 7, 2 and 4. Series: 1, 6 and 7, 2, 3, 4 and 5. The *correct* answers: 5 and 6 are in parallel, and they are also in parallel with the series combination of 2, 3 and 4. 1 and 7, and 2, 3 and 4 are in series, and 1 and 7 are in series with the combination of 5, 6 and 2, 3, 4.

- Ask for volunteers to define series and parallel connections.
- Ask students which bulbs are in series and which are in parallel.

Demonstration 5: Bulbs in series. Use experiment configuration file **INDCD5**. First display the two currents, and then the two voltages.

- Ask students to contrast the results with Demonstration 3. Ask for volunteer(s) to explain why the currents are the same here.
- Why is one voltage smaller than the other?
- Which bulb is having more power delivered to it? Which is brighter?

Demonstration 6: Bulbs in parallel. Use the same experiment configuration file as in Demonstration 5. First display the two currents, and then the two voltages.

- Ask students to compare the results with Demonstration 3. Ask for volunteer(s) to explain why the voltages are the same here, as in Demonstration 3.
- Why is one current smaller than the other?
- Which bulb is having more power delivered to it? Which is brighter?

Series and Parallel Circuits (SPC)

Hand in this sheet **Name**______________________________

INTERACTIVE LECTURE DEMONSTRATIONS
PREDICTION SHEET—**SERIES AND PARALLEL CIRCUITS**

Directions: This sheet will be collected. Write your name at the top to record your presence and participation in these demonstrations. Follow your instructor's directions. You may write whatever you wish on the attached Results Sheet and take it with you.

Demonstration 1: In the *top* circuit on the right, the bulb is connected to a perfect battery (with no internal resistance). In the *bottom* circuit, Bulb B (which is identical to Bulb A) is added in series with bulb A, as shown.

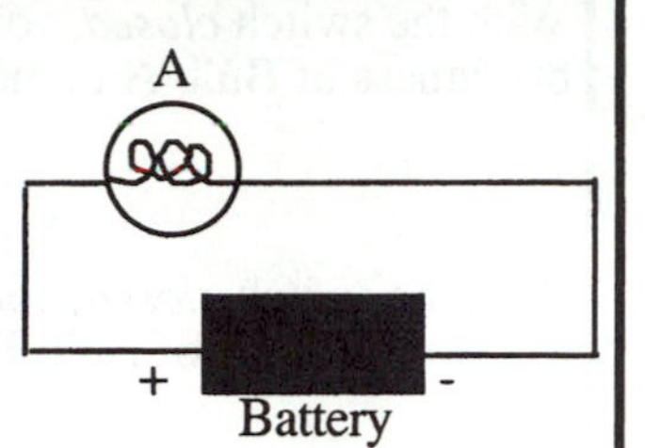

Compare the brightness of Bulb A in the *bottom* circuit to the brightness of Bulb A in the *top* circuit.

In the *bottom* circuit, compare the brightness of Bulb A to the brightness of Bulb B.

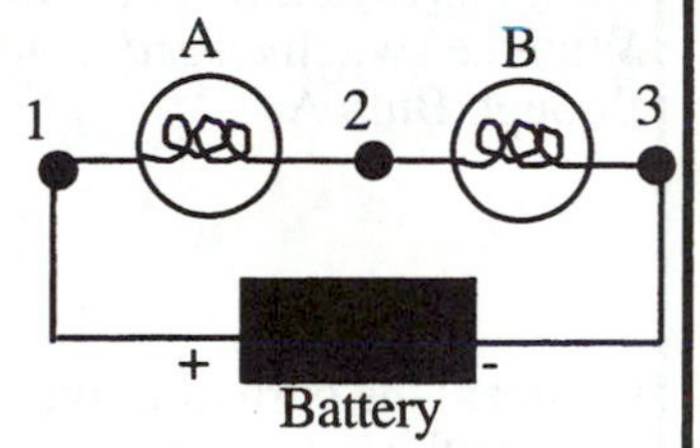

Demonstration 2:
Compare the current at points 1, 2 and 3 in the *bottom* circuit.

Compare the current through Bulb A in the *top* circuit to the current through Bulb A in the *bottom* circuit.

Compare the current through the battery in the *top* circuit to the current through the battery in the *bottom* circuit.

Demonstration 3:
Compare the potential difference (voltage) across Bulb A in the top circuit to the potential difference across Bulb A in the bottom circuit.

In the *bottom* circuit, compare the potential difference across Bulb A to the potential difference across Bulb B.

In the *bottom* circuit, compare the potential difference across Bulb A to the potential difference across the battery.

Demonstration 4: In the *top* circuit on the right, Bulb A is connected to a perfect battery (with no internal resistance). In the *bottom* circuit, Bulb B which is identical to Bulb A is added as shown.

With the switch *open*, Compare the brightness of Bulb A in the *top* circuit to the brightness of Bulb A in the *bottom* circuit.

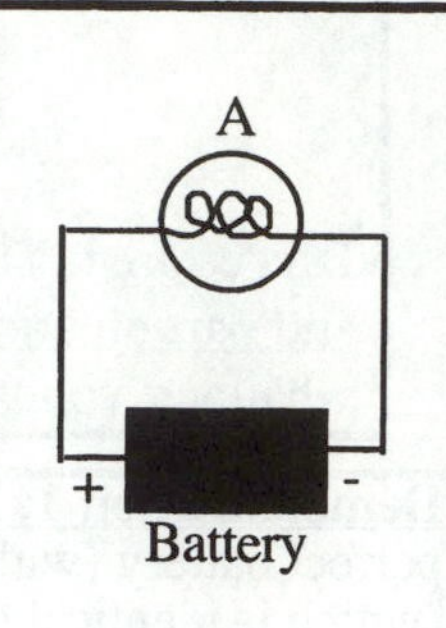

With the switch *closed*, compare the brightness of Bulb A in the *top* circuit to the brightness of Bulb A in the *bottom* circuit.

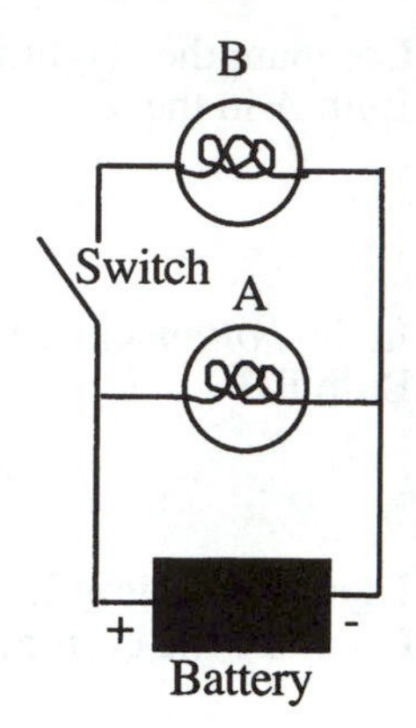

With the switch *closed*, compare the brightness of Bulb A in the bottom circuit to the brightness of Bulb B.

Demonstration 5: In the bottom circuit:
With the switch *closed*, compare the current through Bulb B to the current through Bulb A.

Compare the current through the battery with the switch *closed* to the current through the battery with the switch *open*.

With the switch *closed*, compare the current through the battery to the current through Bulb B.

Demonstration 6: In the bottom circuit:
With the switch *closed*, compare the potential difference (voltage) across Bulb A to the potential difference across Bulb B.

With the switch *closed*, compare the potential difference (voltage) across Bulb B to the potential difference across the *battery*.

Keep this sheet

INTERACTIVE LECTURE DEMONSTRATIONS
RESULTS SHEET—SERIES AND PARALLEL CIRCUITS

You may write whatever you wish on this sheet and take it with you.

Demonstration 1: In the *top* circuit on the right, the bulb is connected to a perfect battery (with no internal resistance). In the *bottom* circuit, Bulb B (which is identical to Bulb A) is added in series with bulb A, as shown.

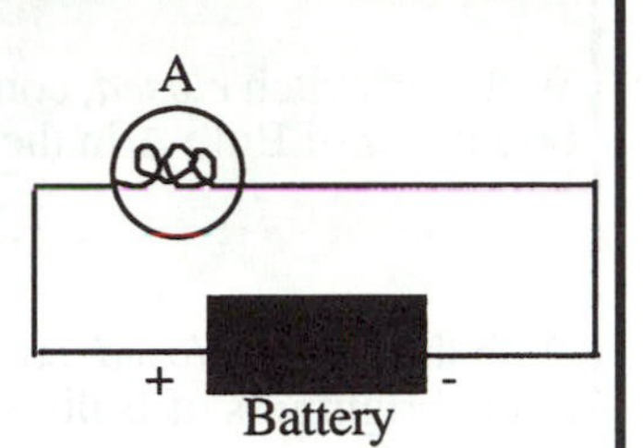

Compare the brightness of Bulb A in the *bottom* circuit to the brightness of Bulb A in the *top* circuit.

In the *bottom* circuit, compare the brightness of Bulb A to the brightness of Bulb B.

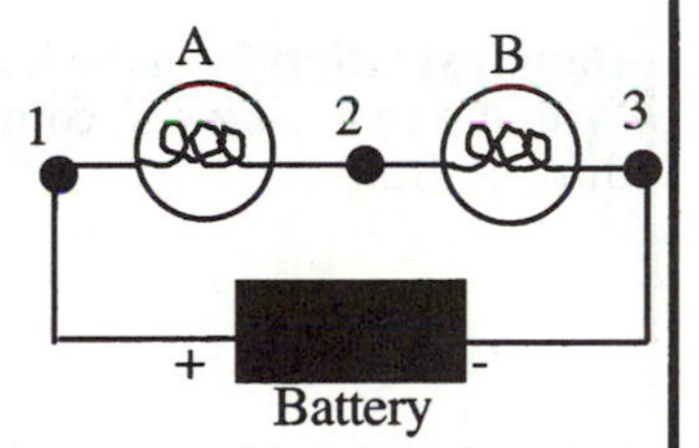

Demonstration 2:
Compare the current at points 1, 2 and 3 in the *bottom* circuit.

Compare the current through Bulb A in the *top* circuit to the current through Bulb A in the *bottom* circuit.

Compare the current through the battery in the *top* circuit to the current through the battery in the *bottom* circuit.

Demonstration 3:
Compare the potential difference (voltage) across Bulb A in the top circuit to the potential difference across Bulb A in the bottom circuit.

In the *bottom* circuit, compare the potential difference across Bulb A to the potential difference across Bulb B.

In the *bottom* circuit, compare the potential difference across Bulb A to the potential difference across the battery.

Demonstration 4: In the *top* circuit on the right, Bulb A is connected to a perfect battery (with no internal resistance). In the *bottom* circuit, Bulb B which is identical to Bulb A is added as shown.

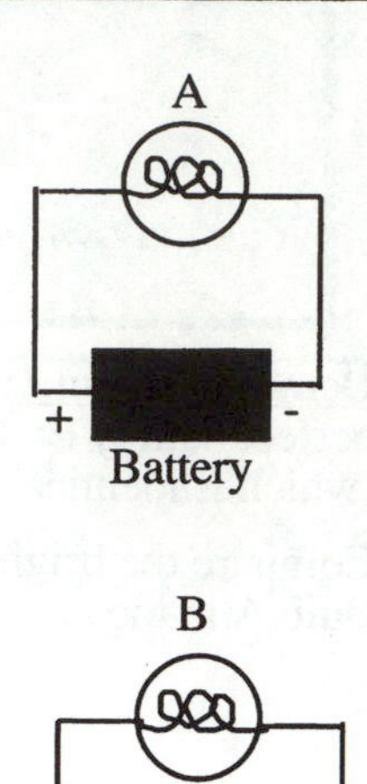

With the switch *open*, Compare the brightness of Bulb A in the *top* circuit to the brightness of Bulb A in the *bottom* circuit.

With the switch *closed*, compare the brightness of Bulb A in the *top* circuit to the brightness of Bulb A in the *bottom* circuit.

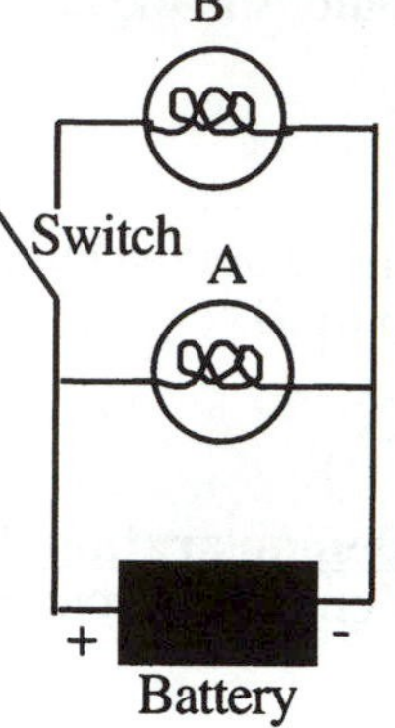

With the switch *closed*, compare the brightness of Bulb A in the bottom circuit to the brightness of Bulb B.

Demonstration 5: In the bottom circuit:
With the switch *closed*, compare the current through Bulb B to the current through Bulb A.

Compare the current through the battery with the switch *closed* to the current through the battery with the switch *open.*

With the switch *closed*, compare the current through the battery to the current through Bulb B.

Demonstration 6: In the bottom circuit:
With the switch *closed*, compare the potential difference (voltage) across Bulb A to the potential difference across Bulb B.

With the switch *closed*, compare the potential difference (voltage) across Bulb B to the potential difference across the *battery.*

SERIES AND PARALLEL CIRCUITS (SPC)
TEACHER'S GUIDE

Prerequisites:

The *Introduction to DC Circuits ILDs* are prerequisite to these. Students should have already been introduced to basic circuit quantities such as current, voltage (potential difference), series, parallel, etc.

Equipment:

computer-based laboratory system

2 current probes (See below.)

2 voltage probes (See below.)

ILD experiment configuration files

circuit board with battery, bulb socket and bulb and provision for second identical bulb in series (See below.)

circuit board with battery, two bulb sockets, two identical bulbs and a switch (See below.)

6 V lantern battery (very fresh)

General Notes on Preparation and Equipment:

Current and Voltage Probes:
Current and Voltage probes are available from Vernier Software and Technology (www.vernier.com) and from PASCO Scientific (www.pasco.com). The Vernier probes are the Current Probe (DCP-BTA) and the Differential Voltage Probe (DVP-BTA). The PASCO probes are the Current Sensor (CI-6556) and the Voltage Sensor (CI-6503).

Circuit Boards:
The size of the circuit boards should be appropriate to the size of the lecture class. They can be fabricated from poster or foam board. Circuit diagrams identical to those on the Prediction and Results Sheets should be drawn on the board with a thick marker, and the circuit elements should be mounted over the diagram in the appropriate places. The circuits should be wired in front, where students can see the wires, and provision should be made for connecting the probes. The board for Demonstrations 1, 2 and 3 should have provision for adding a second bulb in series. Alternatively, it could have both circuits for Demonstrations 1, 2 and 3 on it. The board for Demonstrations 4, 5 and 6 should have a switch that adds a second bulb in parallel.

Demonstrations and Sample Graphs:

Demonstration 1: Bulbs in series--brightness. Light the single bulb in the top circuit. Then add the second identical bulb with the circuit connected to the same battery. Some students will be surprised that the bulbs in the bottom circuit are less bright than the single bulb in the top circuit. They believe that the battery is a source of constant current. Some will be surprised that bulb B is as bright as Bulb A in the bottom circuit. They believe that current "gets used up" by the first bulb that it passes through, leaving less current for the second bulb.

Discussion after displaying results: Ask students to compare the brightness of bulb A in each circuit. Ask for volunteer(s) to explain why the bulb in the bottom circuit is less bright even though the circuit is powered by the same battery. Ask students to compare the brightness of bulbs A and B in the bottom circuit. Ask for volunteer(s) to explain why both of these bulbs have the same brightness.

Demonstration 2: Bulbs in series--current. (Use experiment configuration file **SPCD2**. Two sets of axes will open—Current 1 and Current 3 vs. time.) Begin with the two bulbs in series, and connect the two current probes to measure the currents at points 1 and 3. Be sure that the probes are calibrated. Connect the battery and graph both currents. (Alternatively, you can connect and disconnect the battery several times as you graph.) A sample graph is shown in Figure V-3.

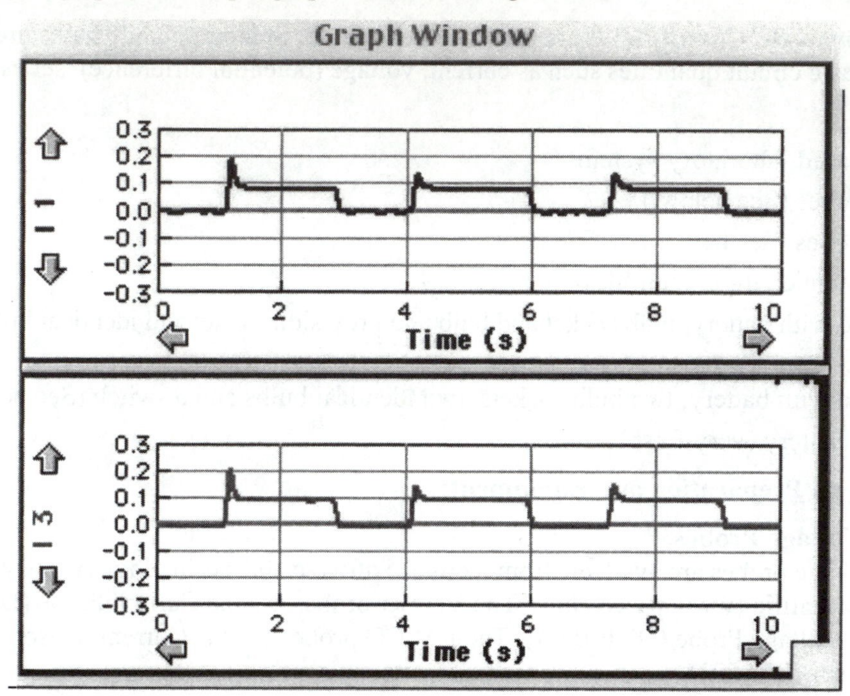

Figure V-3: Current 1 and Current 2 vs. time in Demonstrations 2.

Note that when you close the switch, you get spikes on the current graphs. These are the result of the cold resistance of the filament quickly rising as the filament heats up to a very high temperature. The current, therefore, begins large and quickly decays down to a steady-state value. (Actually, the spikes are larger than observed, since the experiment configuration file has a relatively low data collection rate, and also has some smoothing incorporated.) Most students are not particularly interested in this phenomenon unless it is brought to their attention. Since many students have serious misconceptions about current in a series circuit, this is probably not the time to discuss the origin of the spikes.

Next re-connect the top circuit with two current probes to measure the currents through the battery and bulb. Again connect the battery and graph both currents.

Discussion after the graphs are displayed: Ask students to compare the currents at points 1 and 3. Are they equal in value? What do they think the current at 2 will be? (Hopefully the equal values of the currents at 1 and 3 will convince students that the current also has the same value at 2. Re-connecting a probe to measure the current at 2 would be cumbersome.) Ask for volunteer(s) to explain why the current is the same at all three points. Ask for volunteer(s) to explain why the current through the battery will also have the same value. How is this characteristic of a series circuit? Write down this value.

After the graphs for the top circuit are displayed, ask students to compare the current in the top circuit to that in the bottom circuit. Ask for volunteer(s) to explain why the current is smaller in the bottom circuit, with two bulbs in series. Is a battery a source of constant current? Note: The fact that the current isn't half as large in the bottom circuit may be confusing to students. A brief explanation may be useful at this juncture if students notice. However, the most important feature of these graphs is that connecting two elements in series reduces the current delivered by the battery.

Demonstration 3: Bulbs in series--voltages. (Use experiment configuration file **SPCD3**. Two sets of axes will open—Voltage 1 and Voltage 2 vs. time.) Initially, Voltage Probe 1 should be connected to measure the voltage across Bulb A in the top circuit. Then Voltage Probes 1 and 2 should be connected to measure the voltages across Bulbs A and B in the bottom circuit. Finally, Voltage Probe 2 should be moved so that it measures the voltage across the battery.

Discussion after the graphs are displayed: Ask students to compare the voltages across Bulb A in the top and bottom circuits. Why is the voltage in the bottom circuit half the voltage in the top circuit? A more meaningful discussion can take place after the three voltages in the bottom circuit are measured. What must be true about the sum of the voltages around a series circuit? Is this true in the bottom circuit? How is this related to Kirchhoff's loop rule? Why are the voltages equal for Bulbs A and B?

Demonstration 4: Bulbs in parallel--brightness. Ask students if the top circuit is the same as the bottom circuit with the switch open. Ask volunteer(s) to explain why they are the same. What kind of circuit is this with the switch closed? If the battery is very fresh, then the brightness of Bulb A will not change very much when the switch is closed.

Discussion after the results are displayed: Ask volunteer(s) to explain why the brightness of Bulb A did not change very much when the switch was closed, and to explain why Bulbs A and B are both just as bright. You can remind them of Demonstration 3 in the *Introduction to DC Circuits ILDs*.

Demonstration 5: Bulbs in parallel--currents. (Use experiment configuration file **SPCD5**. Two sets of axes will open—Current 1 and Current 2 vs. time.) The probes should first be connected to measure the currents through Bulbs A and B. Then Current Probe 1 can be moved from Bulb A to the battery. In each case, the switch should be opened and closed several times as you graph. Figure V-4 shows typical graphs for the currents through the battery and Bulb B as the switch is closed and opened.

Discussion after the graphs are displayed: Ask students to describe the currents through Bulbs A and B. Are they equal? Why should they be equal if the bulbs are identical? Is current used up by Bulb A? Why must the current through the battery be the sum of the currents through Bulbs A and B? Try to get volunteer(s) to state the equivalent of Kirchhoff's junction rule.

Demonstration 6: Bulbs in parallel--voltages. (Use experiment configuration file **SPCD6**. Two sets of axes will open—Voltage 1 and Voltage 2 vs. time.) First connect the two probes across the two bulbs. Then move the probe from Bulb A to the battery. In each position, toggle the switch on and off several times as you graph.

Discussion after the graphs are displayed: Ask students to compare the voltages across Bulbs A and B when the switch is closed. Why are they the same? Why are they the same as the voltage across the battery? How is this characteristic of a parallel circuit?

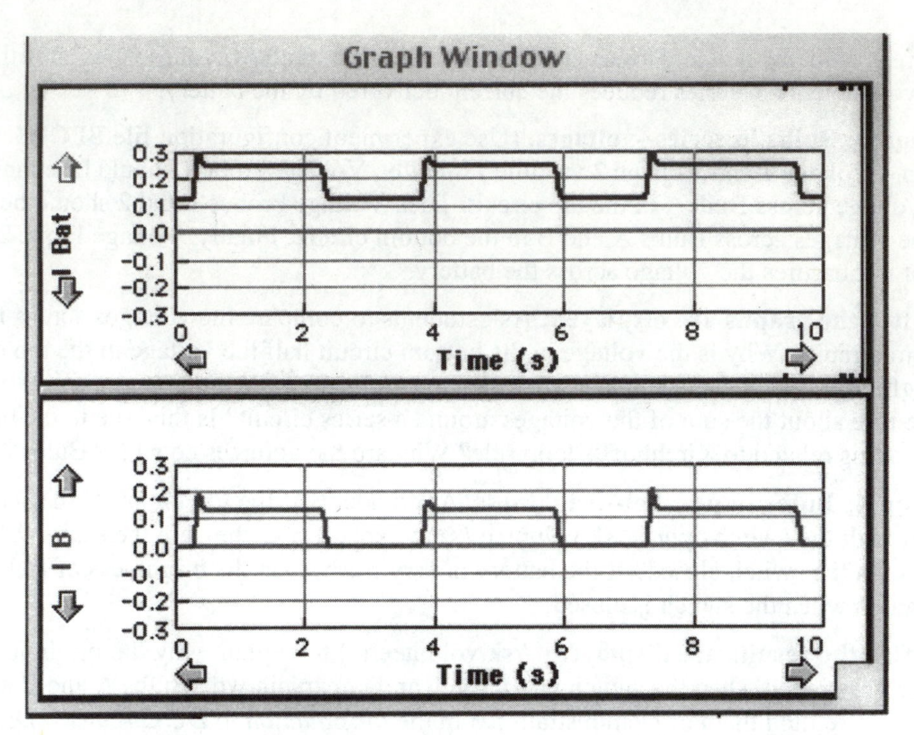

Figure V-4: Current through the battery and current through Bulb B vs. time in Demonstrations 5.

SERIES AND PARALLEL CIRCUITS (SPC)
TEACHER PRESENTATION NOTES

Demonstration 1: Bulbs in series--brightness. After the prediction and discussion steps, light the single bulb, and then light both bulbs connected in series (bottom circuit).

- Why is the bulb in the bottom circuit less bright than the one in the top circuit even when powered by the same battery?
- Why are both bulbs just as bright in the bottom series circuit? Is current used up?

Demonstration 2: Bulbs in series--current. Use experiment configuration file **SPCD2**. First display the currents through the two bulbs (bottom circuit). Write down this value. Then connect the one-bulb circuit (top circuit), and display the currents through the battery and through the bulb.

- Why is the current the same everywhere in the two bulb circuit (bottom circuit)? How is this characteristic of a series circuit?
- Why is the current smaller with two bulbs in series than with one bulb?
- Is a battery a source of constant current?

Demonstration 3: Bulbs in series--voltages. Use experiment configuration file **SPCD3**. First display voltage across the bulb in the single-bulb circuit (top circuit). Then display the voltages across the two bulbs in the two-bulb circuit (bottom circuit). Finally display the voltages across bulb A and the battery (bottom circuit).

- Why is the voltage across Bulb A half as large in the two-bulb circuit (bottom circuit)?
- Why are the voltages across Bulbs A and B equal (bottom circuit)?
- Why is the voltage across the battery equal to the sum of the voltages across the two bulbs? How is this related to Kirchhoff's loop rule?

Demonstration 4: Bulbs in parallel--brightness. Is the top circuit the same as the bottom with the switch open? What type of circuit is this when the switch is closed? Observe the brightness of Bulb A when the switch is closed, and Bulb B lights.

- Why didn't the brightness of Bulb A change?
- Why are Bulbs A and B both just as bright?.

Demonstration 5: Bulbs in parallel--currents. Use experiment configuration file **SPCD5**. First connect probes to measure the currents through Bulbs A and B. Then move Current Probe 1 from Bulb A to the battery.

- Are the currents through Bulbs A and B equal? Why should they be equal? Is current used up in Bulb A?
- Why must the current through the battery equal the sum of the currents through the two bulbs?

Demonstration 6: Bulbs in parallel--voltages. Use experiment configuration file **SPCD6**. First connect probes to measure the voltages across Bulbs A and B. Then move Voltage Probe 1 from Bulb A to the battery.

- Why are the voltages across Bulbs A and B equal, and equal to the voltage across the battery?
- How is this characteristic of a parallel circuit?

RC Circuits (RCC)

Hand in this sheet **Name**________________________

INTERACTIVE LECTURE DEMONSTRATIONS
PREDICTION SHEET—**RC CIRCUITS**

Directions: This sheet will be collected. Write your name at the top to record your presence and participation in these demonstrations. Follow your instructor's directions. You may write whatever you wish on the attached Results Sheet and take it with you.

Demonstration 1: The circuit on the right consists of capacitor C in series with a bulb of resistance R. The capacitor is initially charged with +Q on the top plate and -Q on the bottom plate.

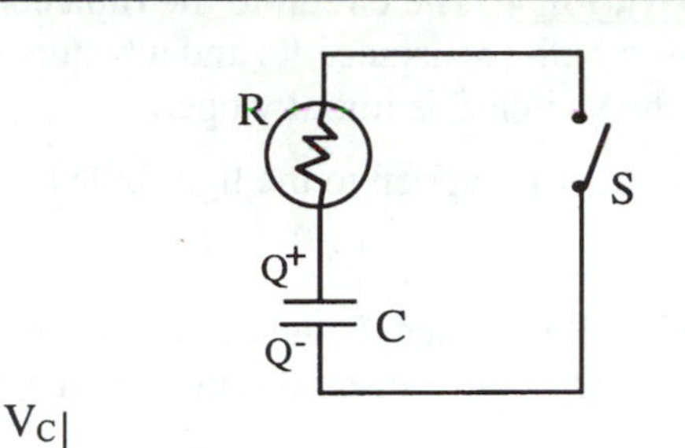

Predict what will happen to the bulb after switch S is closed.

Sketch on the top axes to the right the *voltage across the capacitor* V_c vs. time after the switch S is closed.

V_C

time

Sketch on the bottom axes to the right the *voltage across the bulb* V_b vs. time after the switch S is closed.

V_b

time

Demonstration 2: (Refer to the circuit in Demonstration 1)

- *Just before* switch S is closed, is the voltage across the bulb zero or not zero?
- *Just after* switch S is closed, is the voltage across the bulb zero or not zero?
- *Just after* switch S is closed, what is the voltage V across the bulb in terms of Q and C?
- *Just after* switch S is closed, what is the current through the circuit in terms of Q, C, and R?
- What happens to the charge Q on the capacitor after switch S is closed? Does it *increase, decrease, or stay the same*? (circle one)
- Sketch on the axes to the right the magnitude of the *current*, *I* in the circuit vs. time after S is closed.

I

time

Demonstration 3: The circuit on the right is exactly the same as the one in Demonstration 1 except that the resistance of the bulb is about twice as large. The capacitor is initially charged to the same initial voltage as in Demonstration 1. Sketch on the same axes as for Demonstrations 1 and 2 (above) using dashed lines the voltage across the capacitor and the current in the circuit vs. time after the switch S in the circuit on the right is closed.

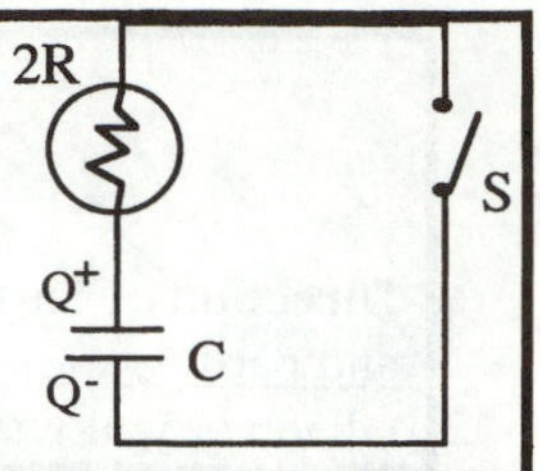

2R S Q^+ C Q^-

Demonstration 4: The circuit to the right consists of an uncharged capacitor, a bulb (resistance R) and a battery of voltage V connected in series. The switch S is initially open.

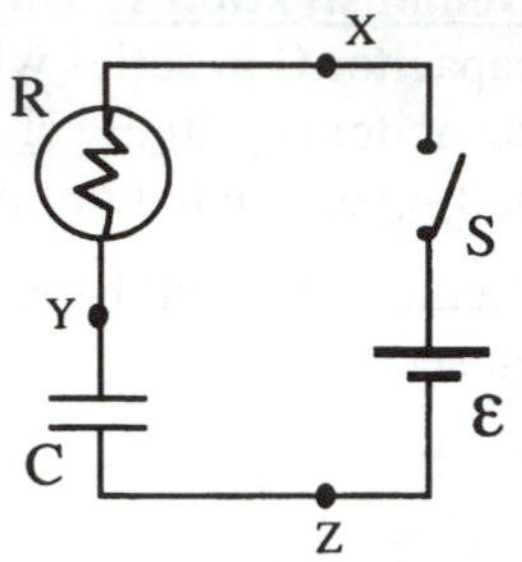

X R S Y ε C Z

Predict what will happen to the light bulb after switch S is closed.

After switch S is closed, is the current through point **X** *greater than, less than, or equal to* the current through point **Y**? (circle one)

After switch S is closed, is the current through point **X** *greater than, less than, or equal to* the current through point **Z**? (circle one)

Sketch on the axes to the right the *voltage across the capacitor*, V_c vs. time after switch S is closed.

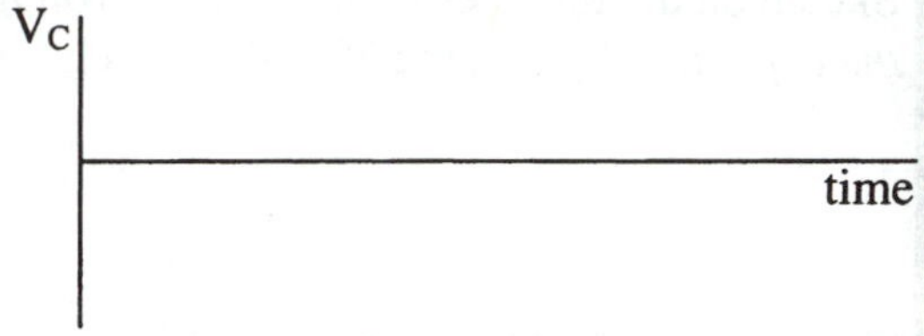

V_C time

Sketch on the axes to the right the *voltage across the bulb*, V_b vs. time after switch S is closed.

V_b time

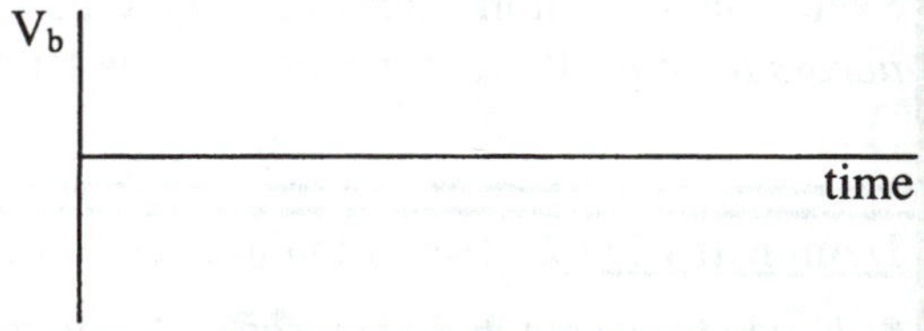

Demonstration 5: (Refer to the circuit in Demonstration 4)

- *Just after* the switch is closed, is the voltage across the capacitor zero or not zero?
- *Just after* the switch is closed, is the voltage across the bulb zero or not zero?
- *Just after* the switch is closed, is the current through the circuit zero or not zero?
- What happens to the amount of charge on the capacitor after switch S is closed? Does it *increase, decrease, or stay the same*? (circle one)

Sketch on the axes to the right the magnitude of the *current*, *I* in the circuit after switch S is closed.

I time

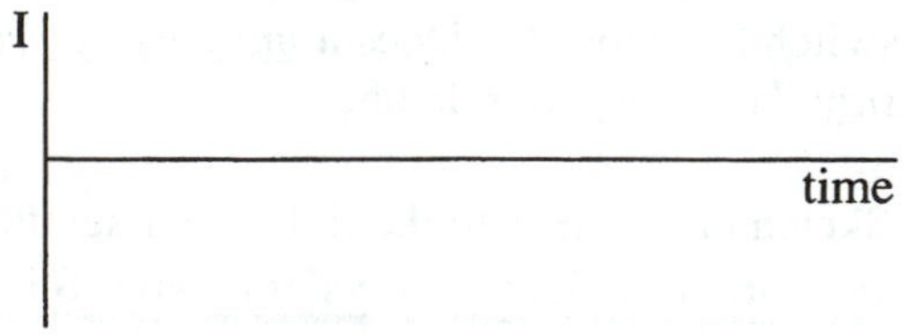

Keep this sheet

INTERACTIVE LECTURE DEMONSTRATIONS
RESULTS SHEET—**RC CIRCUITS**

You may write whatever you wish on this sheet and take it with you.

Demonstration 1: The circuit on the right consists of capacitor C in series with a bulb of resistance R. The capacitor is initially charged with +Q on the top plate and -Q on the bottom plate.

Predict what will happen to the bulb after switch S is closed.

R
S
Q^+
C
Q^-

V_C

time

Sketch on the top axes to the right the *voltage across the capacitor* V_c vs. time after the switch S is closed.

V_b

time

Sketch on the bottom axes to the right the *voltage across the bulb* V_b vs. time after the switch S is closed.

Demonstration 2: (Refer to the circuit in Demonstration 1)

- *Just before* switch S is closed, is the voltage across the bulb zero or not zero?
- *Just after* switch S is closed, is the voltage across the bulb zero or not zero?
- *Just after* switch S is closed, what is the voltage V across the bulb in terms of Q and C?
- *Just after* switch S is closed, what is the current through the circuit in terms of Q, C, and R?

I

time

- What happens to the charge Q on the capacitor after switch S is closed? Does it *increase, decrease, or stay the same*? (circle one)
- Sketch on the axes to the right the magnitude of the *current*, *I* in the circuit vs. time after S is closed.

Demonstration 3: The circuit on the right is exactly the same as the one in Demonstration 1 except that <u>the resistance of the bulb is about twice as large</u>. The capacitor is initially charged to the same initial voltage as in Demonstration 1. Sketch on the same axes as for Demonstrations 1 and 2 (above) <u>using dashed lines</u> the voltage across the capacitor and the current in the circuit vs. time after the switch S in the circuit on the right is closed.

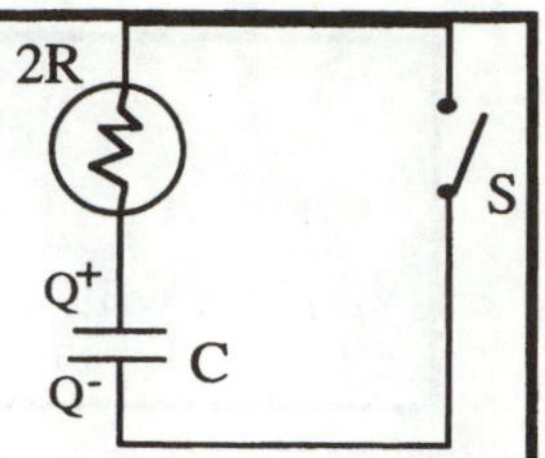

2R S Q^+ C Q^-

Demonstration 4: The circuit to the right consists of an <u>uncharged</u> capacitor, a bulb (resistance R) and a battery of voltage V connected in series. The switch S is initially open.

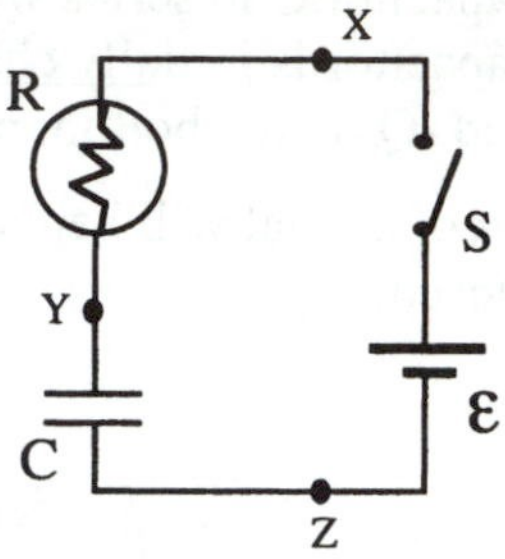

Predict what will happen to the light bulb after switch S is closed.

X R S Y ε C Z

After switch S is closed, is the current through point **X** *greater than, less than, or equal to* the current through point **Y**? (circle one)

After switch S is closed, is the current through point **X** *greater than, less than, or equal to* the current through point **Z**? (circle one)

Sketch on the axes to the right the *voltage across the capacitor*, V_c vs. time after switch S is closed.

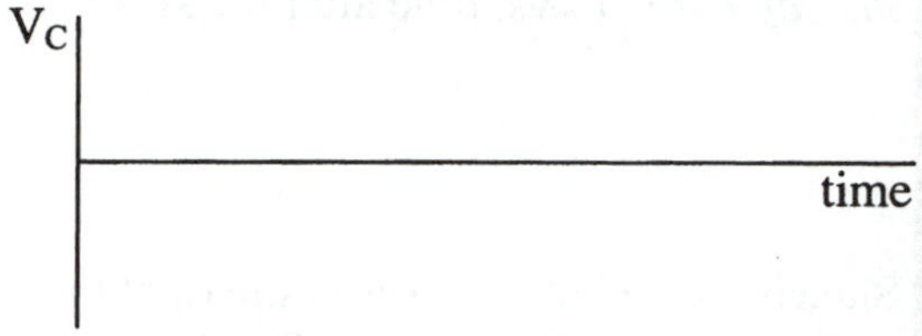

V_C time

Sketch on the axes to the right the *voltage across the bulb*, V_b vs. time after switch S is closed.

V_b time

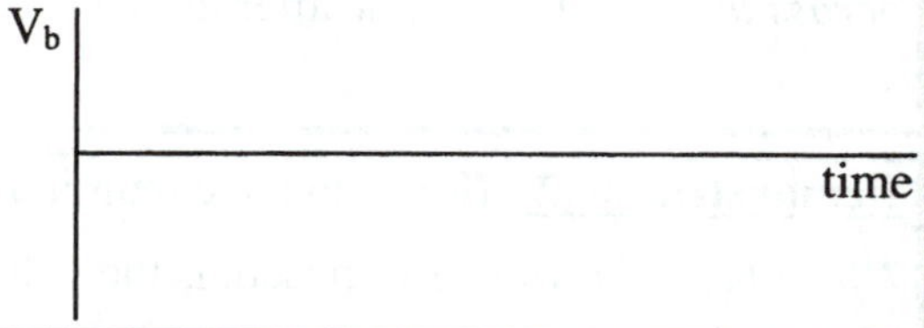

Demonstration 5: (Refer to the circuit in Demonstration 4)

- *Just after* the switch is closed, is the voltage across the capacitor zero or not zero?
- *Just after* the switch is closed, is the voltage across the bulb zero or not zero?
- *Just after* the switch is closed, is the current through the circuit zero or not zero?
- What happens to the amount of charge on the capacitor after switch S is closed? Does it *increase, decrease, or stay the same*? (circle one)

Sketch on the axes to the right the magnitude of the *current*, *I* in the circuit after switch S is closed.

I time

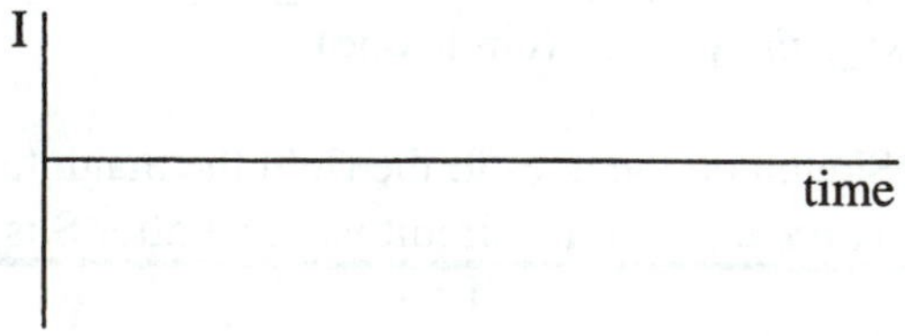

RC CIRCUITS (RCC)
TEACHER'S GUIDE

Prerequisites:

ILD sequences *Introduction to DC Circuits* and *Series and Parallel Circuits* are prerequisite to this sequence. A qualitative understanding of Kirchhoff's rules will help. Students should also have been introduced to capacitors, and the process of charging as a flow of charge from one plate to the other through the connecting wires.

Equipment:

computer-based laboratory system
current probe (See below.)
2 voltage probes (See below.)
ILD experiment configuration files
circuit board with capacitor (see below), bulb socket, bulb and switch (See below.)
circuit board with battery, capacitor (see below), bulb socket, bulb and switch (See below.)
light bulb with approximately twice the resistance that will screw into the same socket
6 V lantern battery

General Notes on Preparation and Equipment:

Current and Voltage Probes:
Current and Voltage probes are available from Vernier Software and Technology (www.vernier.com) and from PASCO Scientific (www.pasco.com). The Vernier probes are the Current Probe (DCP-BTA) and the Differential Voltage Probe (DVP-BTA). The PASCO probes are the Current Sensor (CI-6556) and the Voltage Sensor (CI-6503).

Capacitor:
To get a long, visible decay of the light bulb, a large capacitor is needed. 25,000 μF, 0.47 F and 1.0 F capacitors are readily available, for example from PASCO (SE-8626, EM-8632).

Circuit Boards:
The size of the circuit boards should be appropriate to the size of the lecture class. They can be fabricated from poster board or foam board. Circuit diagrams identical to those on the Prediction and Results Sheets should be drawn on the board with a thick marker, and the circuit elements should be mounted over the diagram in the appropriate places. The circuits should be wired in front, where students can see the wires, and provision should be made for connecting the probes. The switch can just be two alligator clips that can be connected together. The same board may be used for Demonstrations 1-3 and for Demonstrations 4-5 if provision is made to break the circuit and connect a battery. In this case, it should also be possible to add the image of a battery to the circuit board.

Demonstrations and Sample Graphs:

Demonstration 1: Discharge of a capacitor through a bulb—brightness and voltage. First have the students only make predictions for what happens to the bulb. Charge the capacitor by connecting wires from the battery to the terminals of the capacitor. Be sure that the polarity is as shown in the diagram.

Only after displaying and discussing these observations should students be asked to predict the voltages across the capacitor and bulb. (Use experiment configuration file **RCCD1**. This will display two sets of voltage vs. time axes, with the voltage across the capacitor on the top and the voltage across the bulb on the bottom.) Be sure to connect the correct voltage probe across each circuit element: Voltage Probe 1 across the bulb and Voltage Probe 2 across the capacitor. The positive lead of each probe should be connected to the top of each element as seen in the circuit diagram. Typical graphs are shown in Figure V-5.

Discussion after the capacitor discharges: Ask students to describe what happened to the bulb after the switch was closed. It may be best to hold off on a discussion of why the brightness decayed until after displaying the results for the voltages.

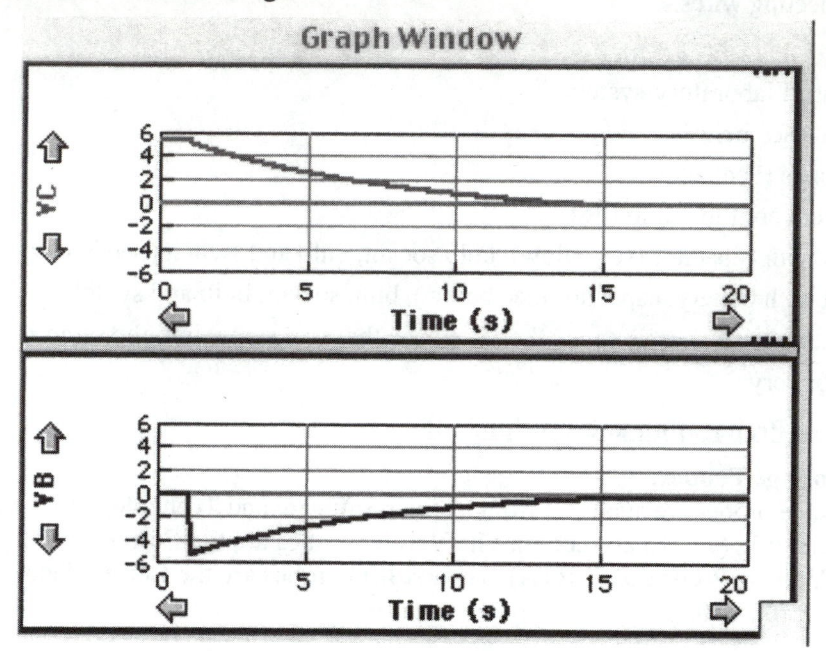

Figure V-5: Voltage B and Voltage C vs. time in Demonstration 1.

Discussion after the voltages are displayed: Ask students to describe the Voltage C graph. Why does the voltage across the capacitor decay after the switch is closed? What is happening to the charge on the capacitor? Ask students to describe the Voltage B graph. Why must the voltage across the bulb decrease in the same way as that across the capacitor? How does this explain the decay in brightness of the bulb? Why is the polarity of the bulb reversed from the polarity of the capacitor?

<u>Demonstration 2:</u> Discharge of a capacitor through a bulb—current. First keep the graphs from Demonstration 1 displayed as the students make their predictions.

Disconnect the voltage probe across the bulb, and replace it with a current probe connected to measure the current in the circuit. (Use experiment configuration file **RCCD2**. Two sets of axes will open—Voltage C and Current vs. time.) After all predictions, discussion and display of predictions are completed, recharge the capacitor, begin graphing and close the switch. Figure V-6 shows typical graphs. (Keep the graphs displayed persistently on the screen for comparison in Demonstration 3.)

Discussion after the graphs are displayed: Ask students to describe the current graph. Ask for volunteer(s) to explain this graph in terms of their answers to the four prediction questions, and their observation of the brightness of the bulb. How does the decay of current correspond to the behavior of the bulb?

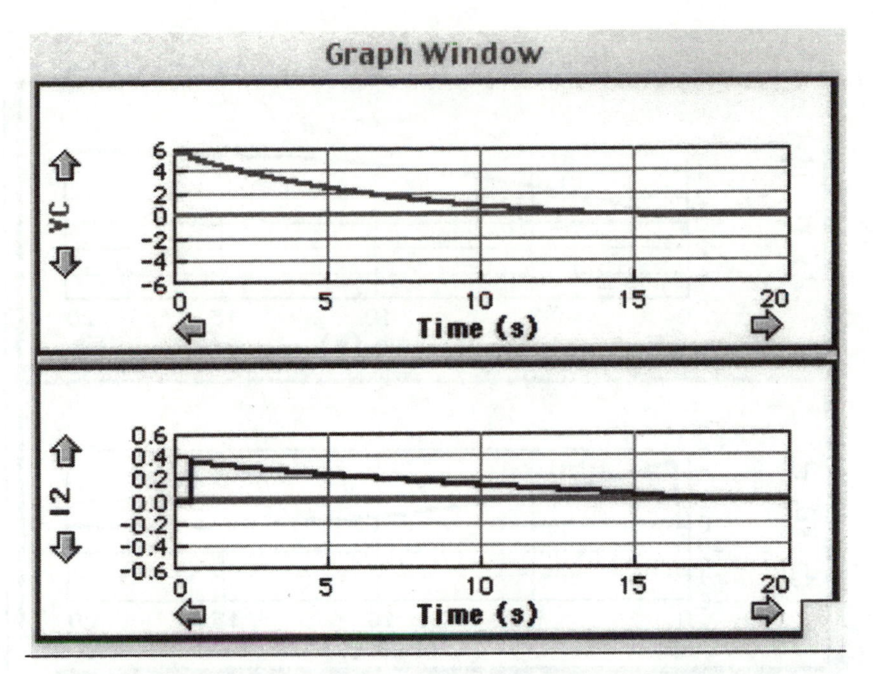

Figure V-6: Voltage across the capacitor and current in the circuit for Demonstration 2.

Demonstration 3: Discharge of a capacitor through twice the resistance. (Use the same experiment configuration file as in Demonstration 2. The graphs from Demonstration 2 should still be displayed persistently on the screen.) Replace the light bulb with the one that has about two times the resistance. (While it is true that the resistance of a light bulb is not constant, it is possible to find a bulb that will behave nearly as if its resistance is twice as large.) Observing the decay of brightness may not be a useful activity, since it is difficult to make even a semi-quantitative comparison of decay times. Therefore, the focus is on the Voltage C and Current graphs.

Discussion after the graphs are displayed: Ask volunteer(s) to describe the graphs, and compare them to Demonstration 2. With the larger resistance, was the decay faster, slower or the same? Was the initial current larger, smaller or the same? Why might the decay be slower through a larger resistor? How does the time constant depend on the resistance and capacitance? How does the initial current depend on the resistance? (See Demonstration 2.)

Demonstration 4: Charging a capacitor through a bulb—brightness and voltage. (Use experiment configuration file **RCCD4**. Two sets of axes will open—Voltage C and Voltage B vs. time.) Break the circuit, and insert the battery. Be sure that the capacitor is discharged to start with. As in Demonstration 1, do predictions and observations of the brightness of the bulb before looking at the graphs. Be sure to connect the correct voltage probe across each circuit element: Voltage Probe 1 across the bulb and Voltage Probe 2 across the capacitor. The positive lead of each probe should be connected to the top of each element as seen in the circuit diagram. Typical graphs are shown in Figure V-7.

Discussion after the capacitor charges: Ask students to describe what happened to the bulb after the switch was closed. It may be best to hold off on a discussion of why the brightness decayed until after displaying the results for the voltages.

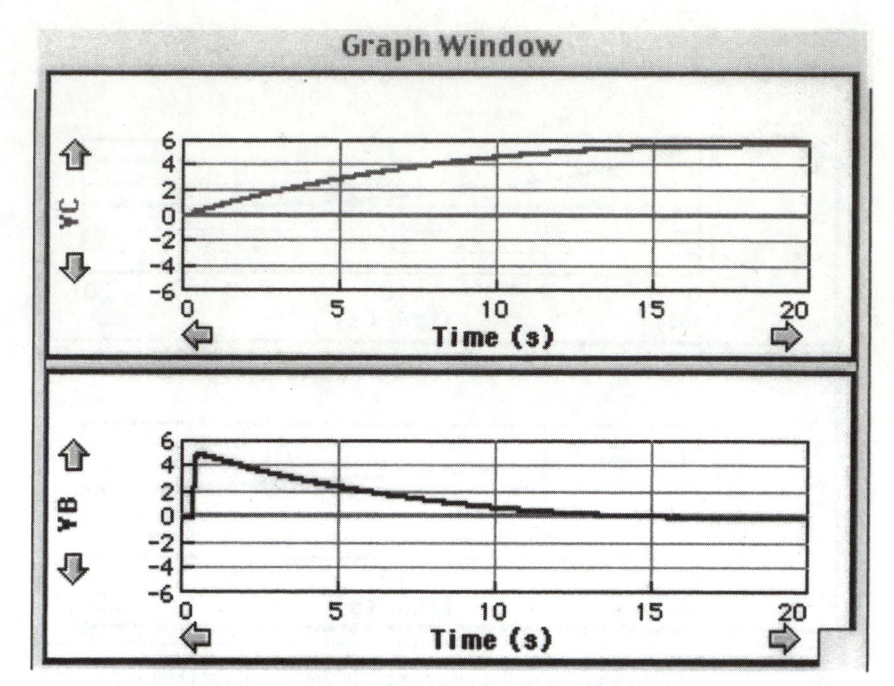

Figure V-7: Voltage across the capacitor and voltage across the bulb vs. time in Demonstrations 4.

Discussion after the voltages are displayed: Ask students to describe the Voltage C graph. Why does the voltage across the capacitor build up slowly after the switch is closed? What is happening to the charge on the capacitor? Ask students to describe the Voltage B graph. Why does the voltage across the bulb decay this time as the voltage across the capacitor grows? Can this be related to Kirchhoff's loop rule? How does this explain the decay in brightness of the bulb?

__Demonstration 5:__ **Charging a capacitor through a bulb—current.** First keep the graphs from Demonstration 4 displayed as the students make their predictions.

Disconnect the voltage probe across the bulb, and replace it with a current probe connected to measure the current in the circuit. (Use experiment configuration file **RCCD5**. Two sets of axes will open—Voltage C and Current vs. time.) After all predictions, discussion and display of predictions are completed, recharge the capacitor, begin graphing and close the switch.

Discussion after the graphs are displayed: Ask students to describe the current graph. Ask for volunteer(s) to explain this graph in terms of their answers to the four prediction questions, and their observation of the brightness of the bulb. How does the decay of current correspond to the behavior of the bulb?

RC CIRCUITS (RCC)
TEACHER PRESENTATION NOTES

Demonstration 1: Discharge of a capacitor through a bulb—brightness and voltage. First have the students only make predictions for what happens to the bulb. Charge the capacitor with polarity shown.

- Ask students to describe what happened to the bulb after the switch was closed.

Use experiment configuration file **RCCD1**. Be sure to connect the correct voltage probe across each circuit element. The positive lead of each probe should be connected to the top of each circuit element.

- Ask students to describe the Voltage C graph. Why does the voltage across the capacitor decay? What is happening to the charge on the capacitor?
- Ask students to describe the Voltage B graph. Why must the voltage across the bulb decrease in the same way as that across the capacitor? How does this explain the decay in brightness of the bulb? Why is the polarity of the bulb reversed from the polarity of the capacitor?

Demonstration 2: Discharge of a capacitor through a bulb—current. Keep the graphs from Demonstration 1 displayed as the students make their predictions. Then open experiment configuration file **RCCD2**. Recharge the capacitor, begin graphing and close the switch. (Keep the graphs displayed persistently for comparison in Demonstration 3.)

- Ask students to describe the current graph. Ask for volunteer(s) to explain this graph in terms of their answers to the four prediction questions, and the brightness of the bulb.
- How does the decay of current correspond to the behavior of the bulb?

Demonstration 3: Demonstration 3: Discharge of a capacitor through twice the resistance. Use the same experiment configuration file as in Demonstration 2. Graphs from Demonstration 2 should remain persistently displayed. Replace the light bulb with one that has about two times the resistance.

- Ask volunteer(s) to describe the graphs, and compare them to Demonstration 2. With larger resistance was decay faster, slower or the same? Was initial current larger, smaller or the same?
- Why is decay slower through a larger resistor? How does time constant depend on resistance and capacitance? How does initial current depend on the resistance? (See Demonstration 2.)

Demonstration 4: Charging a capacitor through a bulb—brightness and voltage. As in Demonstration 1, do predictions and observations of the brightness of the bulb before looking at graphs. Be sure that capacitor is discharged to start with. Be sure polarity is as shown in the diagram.

- Ask students to describe what happened to the bulb after the switch was closed.

Use experiment configuration file **RCCD4**. The positive lead of each probe should be connected to the top of each element as shown in the circuit diagram.

- Ask students to describe the Voltage C graph. Why does the voltage across the capacitor build up slowly after the switch is closed? What is happening to the charge on the capacitor?
- Ask students to describe the Voltage B graph. Why does the voltage across the bulb decay this time as the voltage across the capacitor grows? Can this be related to Kirchhoff's loop rule? How does this explain the decay in brightness of the bulb?

Demonstration 5: Charging a capacitor through a bulb—current. Keep the graphs from Demonstration 4 displayed as the students make their predictions.

Then open experiment configuration file **RCCD5**. Two sets of axes will open—Voltage C and Current vs. time. Replace Voltage Probe B with a current probe connected to measure the current in the circuit. Recharge the capacitor, begin graphing and close the switch.

- Ask students to describe the current graph. Ask for volunteer(s) to explain this graph in terms of their answers to the four prediction questions, and the brightness of the bulb.
- How does the decay of current correspond to the behavior of the bulb?

MAGNETISM (MAG)

Hand in this sheet **Name**__________________________

INTERACTIVE LECTURE DEMONSTRATIONS
PREDICTION SHEET—**MAGNETISM**

Directions: This sheet will be collected. Write your name at the top to record your presence and participation in these demonstrations. Follow your instructor's directions. You may write whatever you wish on the attached Results Sheet and take it with you.

Demonstration 1: A Van de Graaff generator is turned on, and it becomes positively charged. An *uncharged* piece of Styrofoam, wrapped with aluminum foil, is brought near the generator. Predict whether the aluminum-covered Styrofoam will be attracted to the generator, repulsed by the generator, or unaffected.

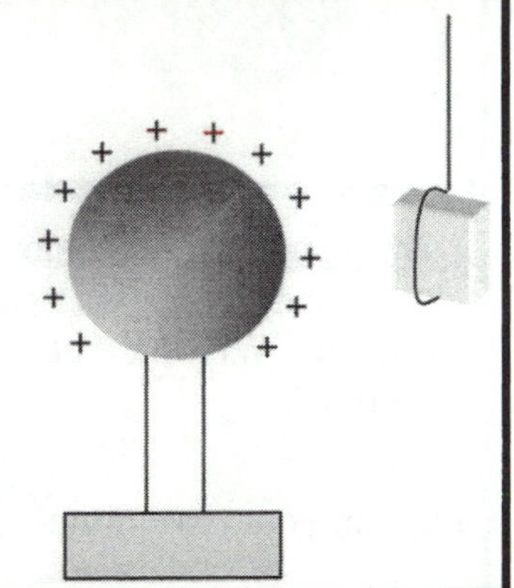

Explain your reasoning:

Demonstration 2: The uncharged, aluminum-covered Styrofoam is now brought near a magnet from a direction perpendicular to the north pole magnet, as shown to the right. Predict whether it will be attracted to the magnet, repulsed by the magnet, or unaffected.

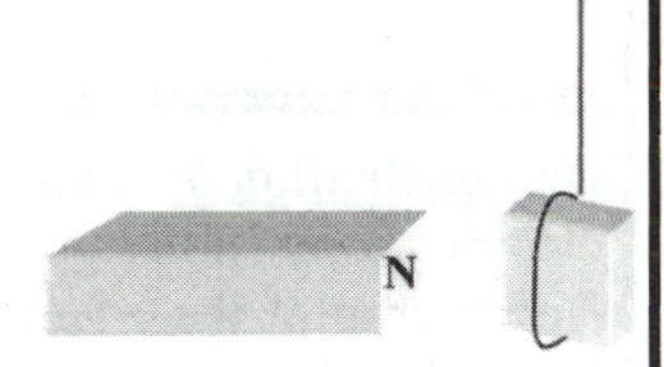

Explain your reasoning:

Demonstration 3: Demonstration 2 is repeated, only this time the aluminum-covered Styrofoam has a positive charge on it. It is brought near the magnet from a direction perpendicular to the north pole magnet, as shown to the right. Predict whether it will be attracted to the magnet, repulsed by the magnet, or unaffected.

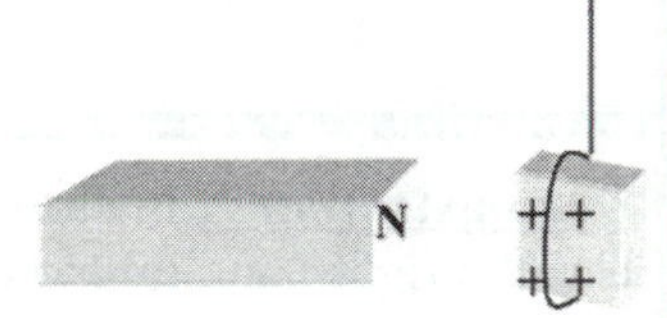

Explain your reasoning:

Demonstration 4: An insulated wire is near a magnet as shown to the right, and there is initially no electric current flowing through the wire. If an electric current begins to move straight up through the wire as shown, will there be a force on the wire? If so, in what direction?

Explain your reasoning:

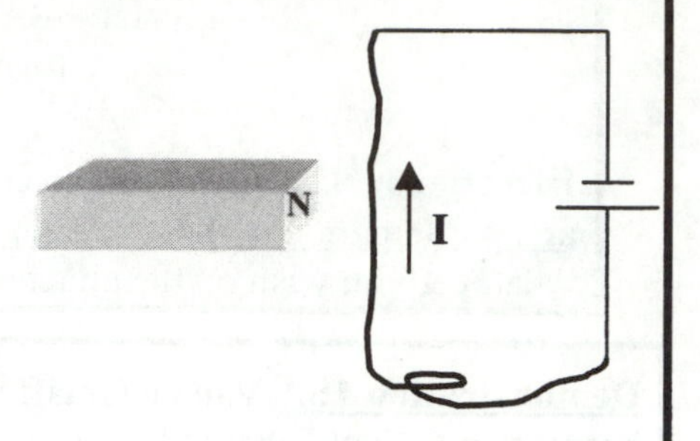

Demonstration 5: Demonstration 4 is repeated, but now an electric current begins to move straight *down* through the wire, as shown. Will there be a force on the wire? If so, in what direction?

Explain your reasoning:

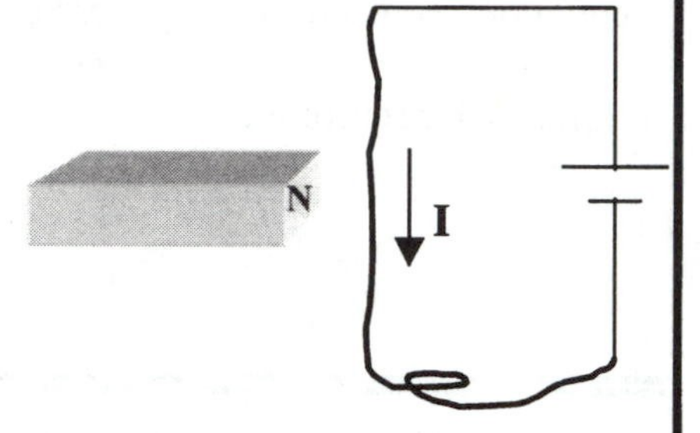

Demonstration 6: Demonstration 5 is repeated, with the electric current still moving straight down through the wire. However, now the orientation of the magnet is changed so that its south pole points straight towards the wire, as shown. Will there be a force on the wire? If so, in what direction?

Explain your reasoning:

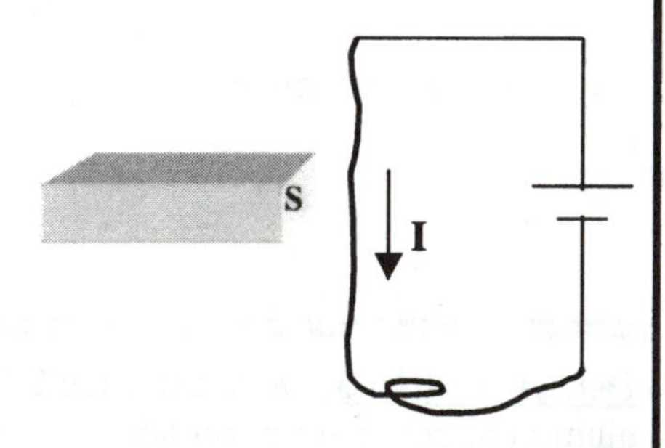

Demonstration 7: In each of the situations shown, a *positive* charge is moving in a magnetic field. Use the right-hand rule to predict the direction of the force or magnetic field (whichever is missing).

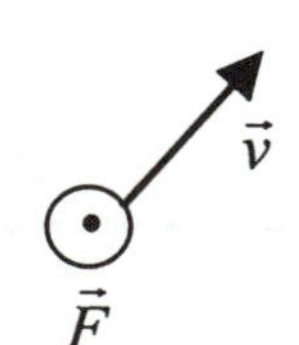

Keep this sheet

INTERACTIVE LECTURE DEMONSTRATIONS

RESULTS SHEET—**MAGNETISM**

You may write whatever you wish on this sheet and take it with you.

Demonstration 1: A Van de Graaff generator is turned on, and it becomes positively charged. An *uncharged* piece of Styrofoam, wrapped with aluminum foil, is brought near the generator. Predict whether the aluminum-covered Styrofoam will be attracted to the generator, repulsed by the generator, or unaffected.

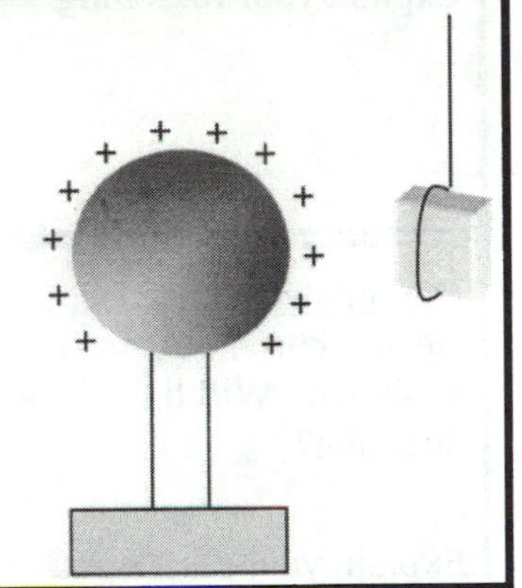

Explain your reasoning:

Demonstration 2: The uncharged, aluminum-covered Styrofoam is now brought near a magnet from a direction perpendicular to the north pole magnet, as shown to the right. Predict whether it will be attracted to the magnet, repulsed by the magnet, or unaffected.

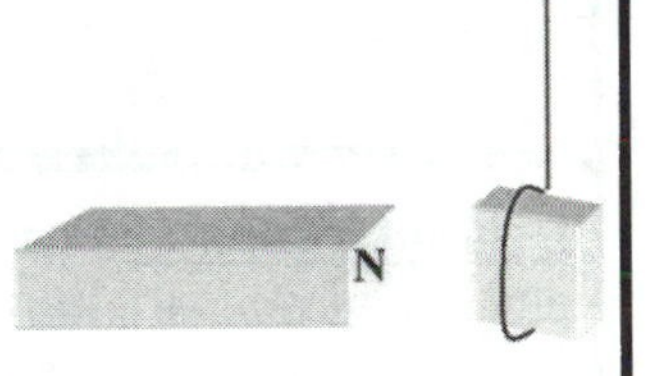

Explain your reasoning:

Demonstration 3: Demonstration 2 is repeated, only this time the aluminum-covered Styrofoam has a positive charge on it. It is brought near the magnet from a direction perpendicular to the north pole magnet, as shown to the right. Predict whether it will be attracted to the magnet, repulsed by the magnet, or unaffected.

Explain your reasoning:

Demonstration 4: An insulated wire is near a magnet as shown to the right, and there is initially no electric current flowing through the wire. If an electric current begins to move straight up through the wire as shown, will there be a force on the wire? If so, in what direction?

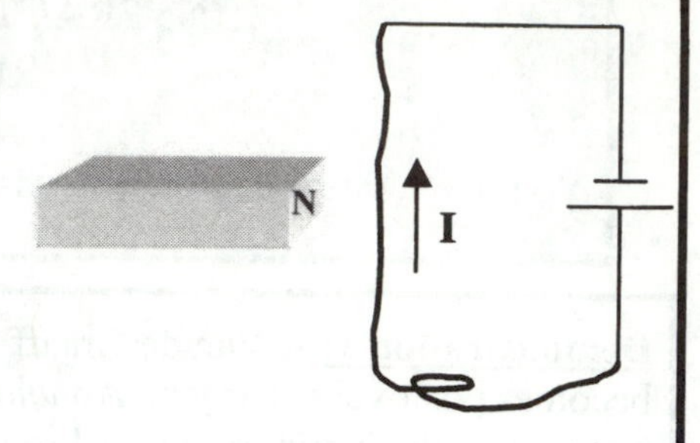

Explain your reasoning:

Demonstration 5: Demonstration 4 is repeated, but now an electric current begins to move straight *down* through the wire, as shown. Will there be a force on the wire? If so, in what direction?

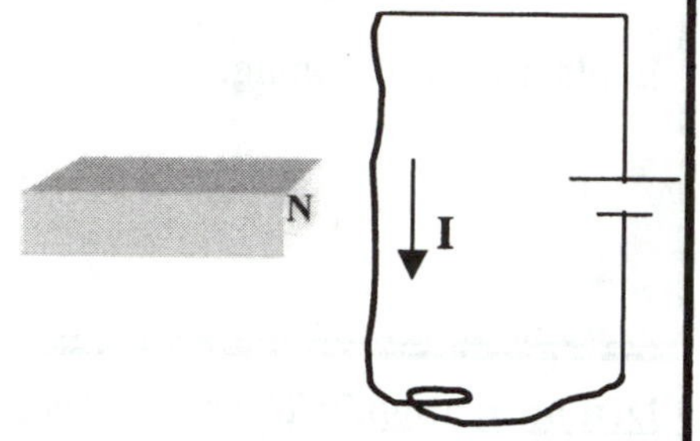

Explain your reasoning:

Demonstration 6: Demonstration 5 is repeated, with the electric current still moving straight down through the wire. However, now the orientation of the magnet is changed so that its south pole points straight towards the wire, as shown. Will there be a force on the wire? If so, in what direction?

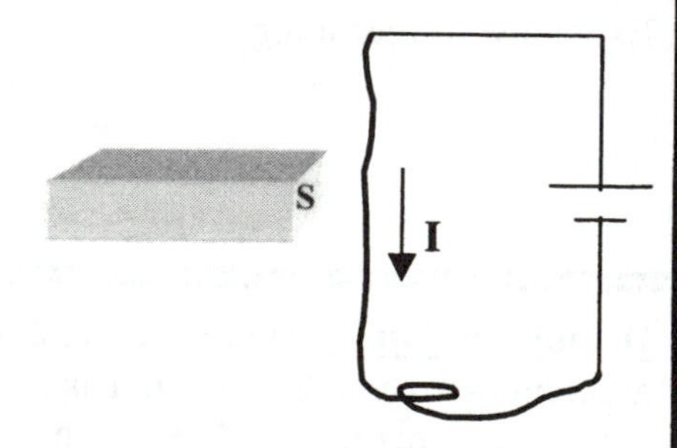

Explain your reasoning:

Demonstration 7: In each of the situations shown, a *positive* charge is moving in a magnetic field. Use the right-hand rule to predict the direction of the force or magnetic field (whichever is missing).

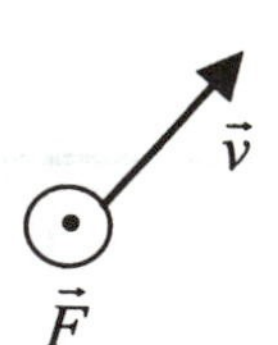

MAGNETISM (MAG)
TEACHER'S GUIDE

Prerequisites:

This ILD sequence can be used as an introduction to magnetic fields and forces. It can also be used as a review after these topics have been introduced in lecture or through text readings. It will be useful for students to know about the right-hand rule for finding magnetic forces.

Equipment:

- Van de Graaff generator
- small piece of Styrofoam coated with conducting paint or wrapped in aluminum foil
- string
- electrostatic rods, fur and cloth to charge the Styrofoam
- relatively strong bar magnet
- insulated wire
- 6 V lantern battery or DC power supply

General Notes on Preparation and Equipment:

The apparatus for these demonstrations is standard and straightforward. There are a variety of ways that the demonstrations can be set up.

Demonstrations and Sample Results:

Demonstration 1: Uncharged Styrofoam and Van de Graaff generator: After the prediction and discussion steps, bring the Styrofoam suspended from the string near the sphere of the operating generator. The Styrofoam will be attracted, since an opposite charge to the one on the sphere will be induced on it, and unlike charges attract.

Discussion after observing the results: Ask students to describe the result. Why should there be an attraction here when there was no charge on the Styrofoam? What is charging by induction? What sign of charge will be induced on the side of the Styrofoam nearest the generator sphere?

Demonstration 2: Uncharged Styrofoam near magnet. After the prediction and discussion steps, bring the Styrofoam suspended from the string near the North pole of the magnet. The Styrofoam will not be attracted, since a magnetic field cannot induce charge on the Styrofoam, and in any case, a magnetic field only exerts a force on moving charges.

Discussion after observing the results: Ask students to describe the (non-)result. Is there a force on the Styrofoam? Why not? Can a steady magnetic field induce charges on a conductor at rest? Can a magnetic field cause a force on charges at rest?

Demonstration 3: Charged Styrofoam near magnet. After the prediction and discussion steps, bring the Styrofoam suspended from the string near the North pole of the magnet. The Styrofoam will not be attracted, since a magnetic field only exerts a force on moving charges.

Discussion after observing the results: Ask students to describe the (non-)result. Is there a force on the Styrofoam? Why not? Can a magnetic field cause a force on charges at rest?

Demonstration 4: **Current-carrying wire near North pole of magnet.** After the prediction and discussion steps, bring the North pole of the magnet near the wire, and turn on the current in the wire. The wire will feel a force in a direction into the paper.

Discussion after observing the results: Ask students to describe the result. What is an electric current? What is different about the charges in the wire when a current is flowing compared to the charges on the Styrofoam? Does a magnetic field exert a force on *moving* charges? How can the right-hand rule be used to explain the direction of the force on the wire?

Demonstration 5: **Current-carrying wire near North pole of magnet with current in opposite direction.** After the prediction and discussion steps, bring the North pole of the magnet near the wire, and turn on the current in the wire. The wire will feel a force in a direction out of the paper.

Discussion after observing the results: Ask students to describe the result. What is different from Demonstration 4? How can the right-hand rule be used to explain the direction of the force on the wire?

Demonstration 6: **Current-carrying wire near South pole of magnet with current in opposite direction.** After the prediction and discussion steps, bring the South pole of the magnet near the wire, and turn on the current in the wire. The wire will feel a force in a direction into the paper.

Discussion after observing the results: Ask students to describe the result. What is different from Demonstration 5? How can the right-hand rule be used to explain the direction of the force on the wire?

Demonstration 7: **Practice with the right-hand rule.** No actual demonstrations will be done for this demonstration. After the prediction and discussion steps, ask students for their predictions for each case. Show them the correct situations. In the first, the force is toward the right. In the second, the force is into the paper. In the last, the magnetic field must be in the plane of the paper, making a counterclockwise angle less than 180° with the velocity.

Discussion after observing the results: Ask students to describe how the right-hand rule can be used in each of these cases.

MAGNETISM (MAG)
TEACHER PRESENTATION NOTES

Demonstration 1: Uncharged Styrofoam and Van de Graaff generator: Bring the Styrofoam near the sphere of the generator. The Styrofoam will be attracted because of induced charge.

- Why should there be an attraction if there was no charge on the Styrofoam? What is charging by induction? What sign of charge will be induced on the side of the Styrofoam nearest the sphere?

Demonstration 2: Uncharged Styrofoam near magnet. Bring the Styrofoam near the North pole of the magnet. The Styrofoam will not be attracted, since a magnetic field cannot induce charge and only exerts a force on moving charges.

- Is there a force on the Styrofoam? Why not? Can a magnetic field induce charges on a conductor at rest? Can a magnetic field cause a force on charges at rest?

Demonstration 3: Charged Styrofoam near magnet. Bring the Styrofoam near the North pole of the magnet. The Styrofoam will not be attracted, since a magnetic field only exerts a force on moving charges.

- Is there a force on the Styrofoam? Why not? Can a magnetic field exert a force on charges at rest?

Demonstration 4: Current-carrying wire near North pole of magnet. Bring the North pole of the magnet near the wire, and turn on the current. The wire will feel a force into the paper.

- What is an electric current? What is different about the charges in the wire when a current is flowing compared to the charges on the Styrofoam?
- Does a magnetic field exert a force on *moving* charges?
- How can the right-hand rule be used to explain the direction of the force on the wire?

Demonstration 5: Current-carrying wire near North pole of magnet with current in opposite direction. Bring the North pole of the magnet near the wire, and turn on the current. The wire will feel a force out of the paper.

- What is different from Demonstration 4? How can the right-hand rule be used to explain the direction of the force on the wire?

Demonstration 6: Current-carrying wire near South pole of magnet with current in opposite direction. Bring the South pole of the magnet near the wire, and turn on the current. The wire will feel a force into the paper.

- What is different from Demonstration 5? How can the right-hand rule be used to explain the direction of the force on the wire?

Demonstration 7: Practice with the right-hand rule. Ask students their predictions for each case. Show them the correct situations.

- Ask students to describe how the right-hand rule can be used in each of these cases?

Electromagnetic Induction (EMIN)

Hand in this sheet **Name**__________________________

INTERACTIVE LECTURE DEMONSTRATIONS

PREDICTION SHEET—**ELECTROMAGNETIC INDUCTION**

Directions: This sheet will be collected. Write your name at the top to record your presence and participation in these demonstrations. Follow your instructor's directions. You may write whatever you wish on the attached Results Sheet and take it with you.

Demonstration 1: A magnet is near a coil of wire. The switch, S_1, is initially open. When the switch is closed, will there be a current induced in the coil? Explain.

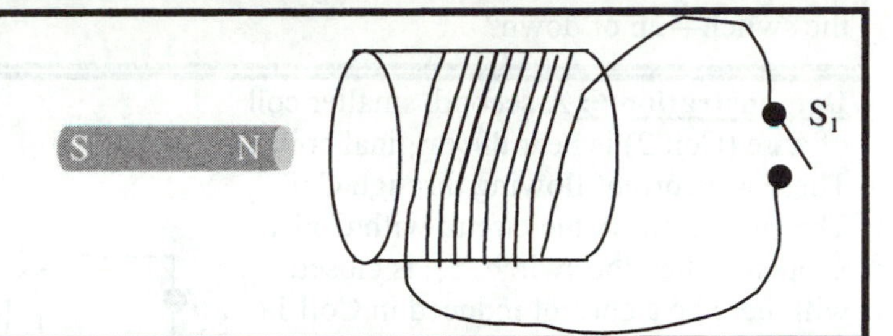

Demonstration 2: Switch S_1 is now closed and remains closed. The magnet is moved away from the coil of wire. Will there be a current induced in the coil? Explain.

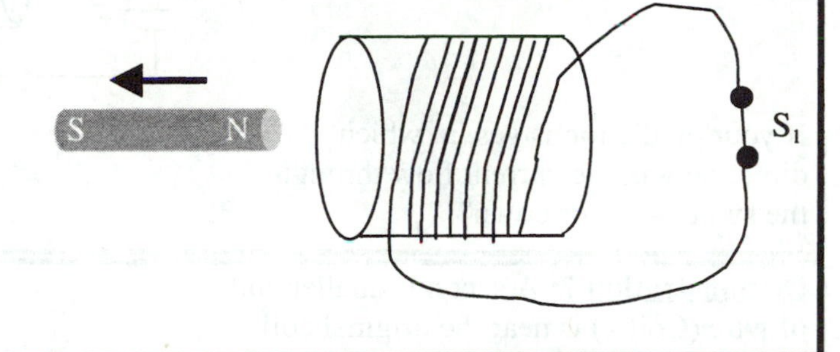

If your prediction is yes, in which direction will the current flow through the switch—up or down?

Demonstration 3: Switch S_1 remains closed. The magnet is now moved toward the coil of wire. Will there be a current induced in the coil? Explain.

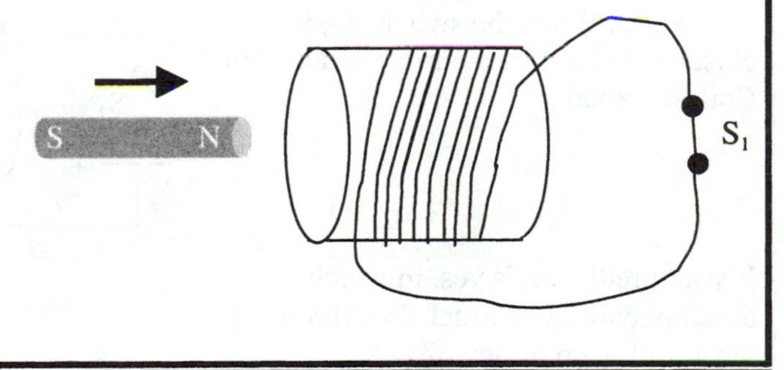

If your prediction is yes, in which direction will the current flow through the switch—up or down?

Demonstration 4: Switch S_1 remains closed. The coil is moved away from the magnet. Will there be a current induced in the coil? Explain.

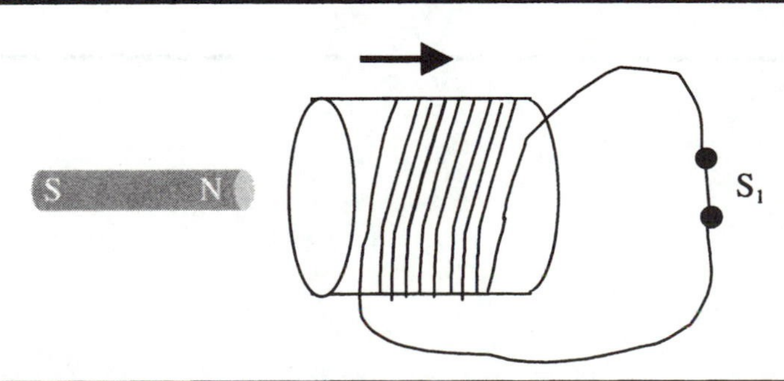

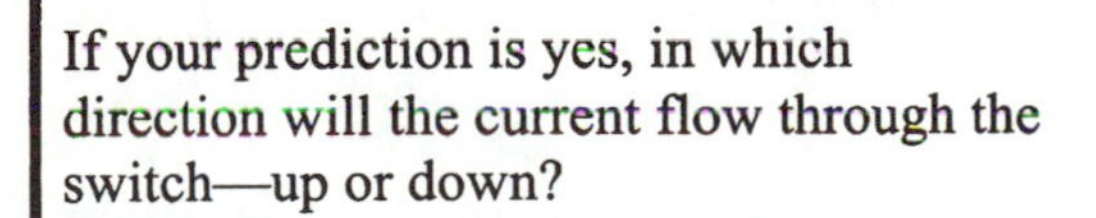

If your prediction is yes, in which direction will the current flow through the switch—up or down?

Demonstration 5: Switch S_1 remains closed. The coil is moved towards the magnet. Will there be a current induced in the coil? Explain.

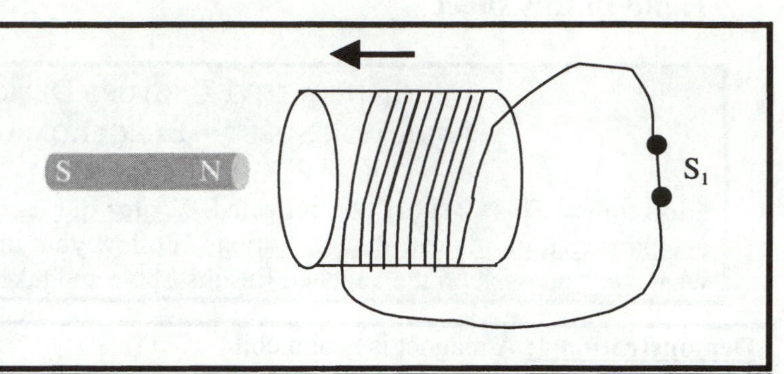

If your prediction is yes, in which direction will the current flow through the switch—up or down?

Demonstration 6: A second, smaller coil of wire (Coil 2) is near the original coil. There is a current flowing through Coil 2. The switch, S_1, in the circuit with Coil 1 is open. When the switch, S_1, is closed will there be a current induced in Coil 1? Explain.

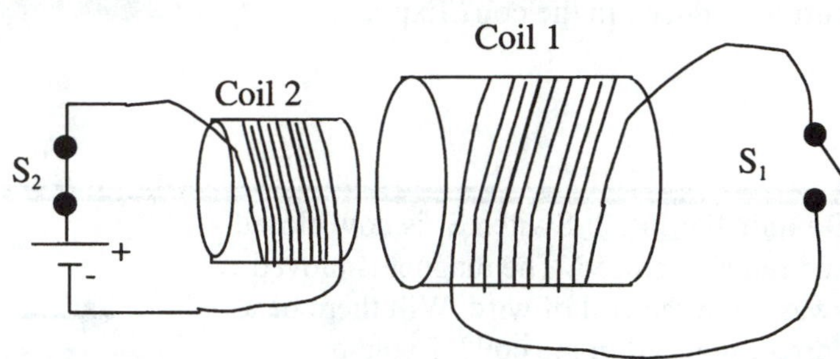

If your prediction is yes, in which direction will the current flow through the switch—up or down?

Demonstration 7: A second, smaller coil of wire (Coil 2) is near the original coil. There is no current flowing through Coil 2. The switch, S_1, in the circuit with Coil 1 is closed. When the switch, S_2, is closed will there be a current induced in Coil 1? Explain.

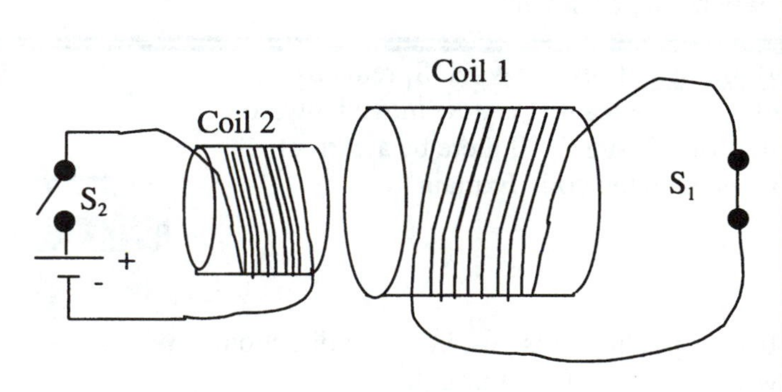

If your prediction is yes, in which direction will the current flow through the switch—up or down?

Keep this sheet

INTERACTIVE LECTURE DEMONSTRATION
RESULTS SHEET—**ELECTROMAGNETIC INDUCTION**

You may write whatever you wish on this sheet and take it with you.

Demonstration 1: A magnet is near a coil of wire. The switch, S_1, is initially open. When the switch is closed, will there be a current induced in the coil? Explain.

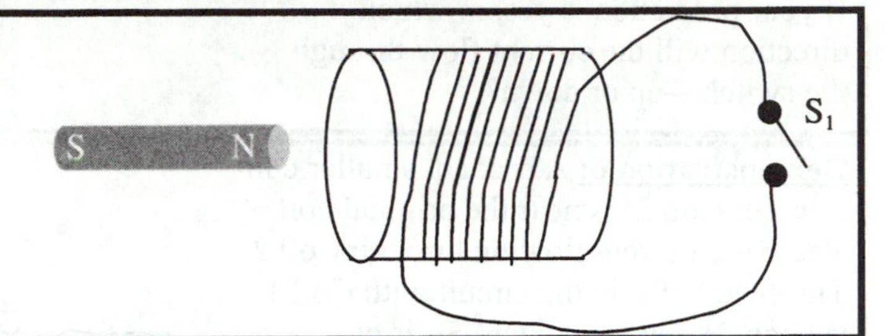

Demonstration 2: Switch S_1 is now closed and remains closed. The magnet is moved away from the coil of wire. Will there be a current induced in the coil? Explain.

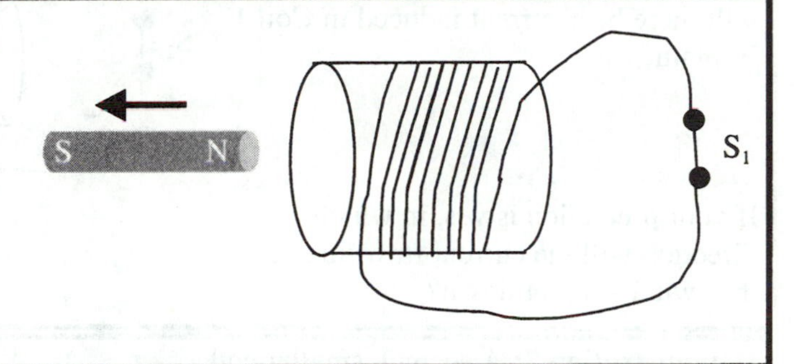

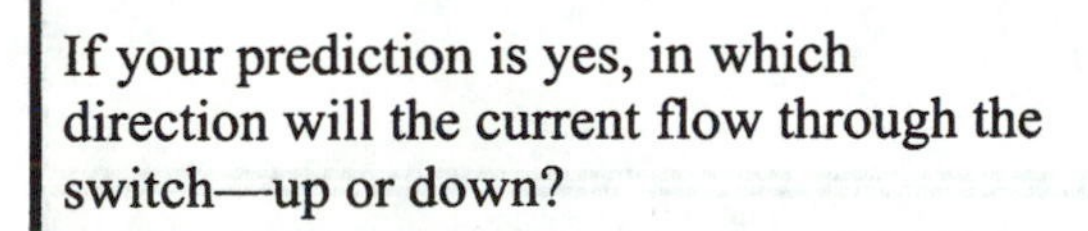

If your prediction is yes, in which direction will the current flow through the switch—up or down?

Demonstration 3: Switch S_1 remains closed. The magnet is now moved toward the coil of wire. Will there be a current induced in the coil? Explain.

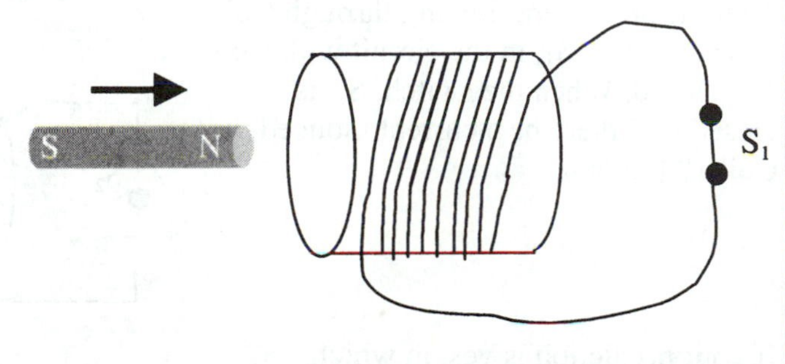

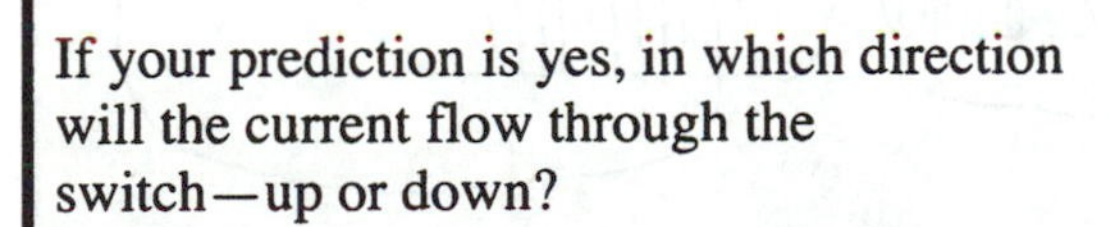

If your prediction is yes, in which direction will the current flow through the switch—up or down?

Demonstration 4: Switch S_1 remains closed. The coil is moved away from the magnet. Will there be a current induced in the coil? Explain.

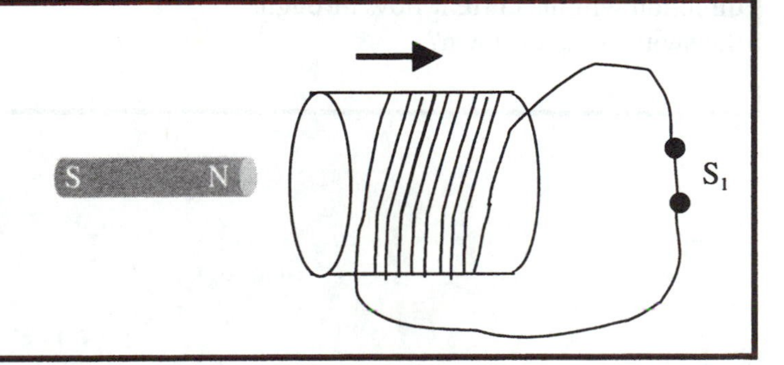

If your prediction is yes, in which direction will the current flow through the switch—up or down?

Demonstration 5: Switch S_1 remains closed. The coil is moved towards the magnet. Will there be a current induced in the coil? Explain.

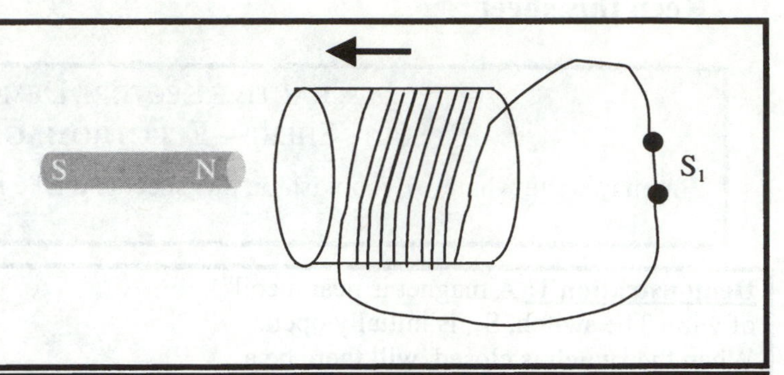

If your prediction is yes, in which direction will the current flow through the switch—up or down?

Demonstration 6: A second, smaller coil of wire (Coil 2) is near the original coil. There is a current flowing through Coil 2. The switch, S_1, in the circuit with Coil 1 is open. When the switch, S_1, is closed will there be a current induced in Coil 1? Explain.

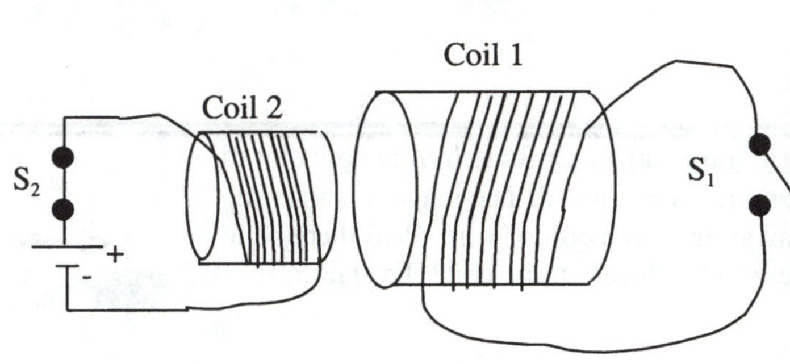

If your prediction is yes, in which direction will the current flow through the switch—up or down?

Demonstration 7: A second, smaller coil of wire (Coil 2) is near the original coil. There is no current flowing through Coil 2. The switch, S_1, in the circuit with Coil 1 is closed. When the switch, S_2, is closed will there be a current induced in Coil 1? Explain.

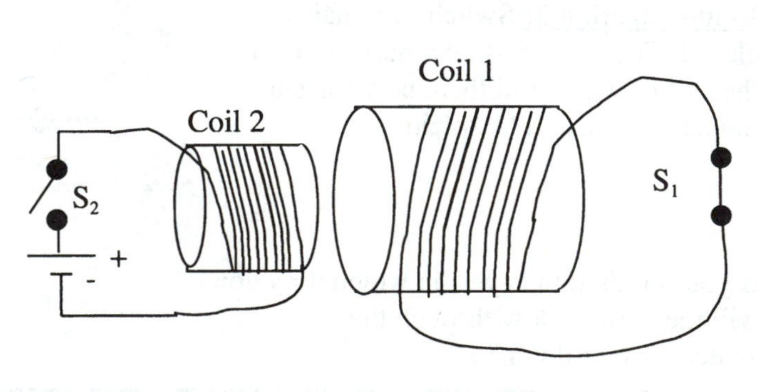

If your prediction is yes, in which direction will the current flow through the switch—up or down?

ELECTROMAGNETIC INDUCTION (EMIN)
TEACHER'S GUIDE

Prerequisites:

This *ILD* sequence can be used as an introduction to electromagnetic induction and Lenz's law. It can also be used as a review after these topics have been introduced in lecture or through text readings. It is best if students have been introduced to the terms magnetic field, magnetic flux, solenoid, induced voltage, induced current, Faraday's law, Lenz's law.

Equipment:

relatively strong bar magnet (See below.)
2 coils of wire with large number of closely wound turns (See below.)
2 knife switches
sensitive projection analog ammeter or galvanometer (See below.)
connecting wires
6 V lantern battery or DC power supply

General Notes on Preparation and Equipment:

Magnet and coils:
The magnet and one coil should be chosen so that a significant current (deflection of the meter needle) is observed in Demonstrations 2-5. An appropriate coil can be found in most physics departments. If you have one of the old Berkeley Physics Laboratory solenoids with 3400 turns of wire, that will work very well with a reasonably strong magnet. Coils from a transformer set like the PASCO (www.pasco.com) SF-8617 will work well. In particular the 3200 turn coil (SF-8613) or the 1600 turn coil (SF-8612) work well. The other coil should be chosen so that it gives a significant deflection of the meter in Demonstration 7.

Meter:
You will probably need a projection meter if you have even a relatively large lecture class. Again choose a meter in conjunction with the coils and magnet that gives significant deflections for the demonstrations. If your department doesn't have a projection galvanometer, you could also use an ordinary meter and a video camera projected on a screen.

Demonstrations and Sample Results:

Demonstration 1: Magnet and coil—both at rest. Be sure that the coil windings are as shown in the diagram (clockwise around the coil as viewed from the right. The meter is connected in series with the coil so that it is possible to tell from the deflection whether the current is flowing up or down through the switch. For example, if the + terminal of the meter is just below the switch, then a positive current would be a current flowing *down* through the switch. After all prediction and discussion steps have been completed, with the magnet close to the coil and both at rest, close and open the switch several times.

Discussion after observing the results: Ask students to describe the (non) result. What is the definition of magnetic flux? Is there a magnetic flux through the coil? Does closing the switch change the flux? What does Faraday's law say about an induced emf and current?

Demonstration 2: Magnet moving away from the coil. Close the switch, and keep it closed. Start with the magnet close to the coil and with the North pole pointing toward the coil. Move the magnet quickly away from the coil. There should be a deflection of the meter while the magnet is moving, that stops when the magnet comes to rest. Be sure to observe the polarity of the deflection, and determine the direction of current through the switch.

Discussion after observing the results: Ask students to describe the result. Does moving the magnet change the flux through the coil? Is the flux increasing or decreasing? Does the observed induced current agree with Faraday's law? What does Lenz's law say about the magnetic field that much be produced by the induced current in the coil? In which direction must the current in the coil flow to produce this field?

Demonstration 3: Magnet moving towards the coil. Keep the switch closed. Start with the magnet a little away from the coil and with the North pole pointing toward the coil. Move the magnet quickly toward the coil. There should be a deflection of the meter while the magnet is moving, that stops when the magnet comes to rest. Be sure to observe the polarity of the deflection, and determine the direction of current through the switch. The direction should be opposite to that in Demonstration 2.

Discussion after observing the results: Ask students to describe the result. Does moving the magnet change the flux through the coil? Is the flux increasing or decreasing? Does the observed induced current agree with Faraday's law? What does Lenz's law say about the magnetic field that much be produced by the current in the coil? In which direction must the current in the coil flow to produce this field? Why is the direction opposite to that in Demonstration 2?

Demonstration 4: Coil moving away from the magnet. Keep the switch closed. Start with the magnet close to the coil and with the North pole pointing toward the coil. Move the coil quickly away from the magnet. There should be a deflection of the meter while the coil is moving, that stops when the coil comes to rest. Be sure to observe the polarity of the deflection, and determine the direction of current through the switch. The direction should be the same as in Demonstration 2.

Discussion after observing the results: Ask students to describe the result. Does moving the coil change the flux through it? Is the flux increasing or decreasing? Does the observed induced current agree with Faraday's law? Why is the direction the same as Demonstration 2 and opposite to Demonstration 3?

Demonstration 5: Coil moving toward the magnet. Keep the switch closed. Start with the magnet a little away from the coil and with the North pole pointing toward the coil. Move the coil quickly toward the coil. There should be a deflection of the meter while the coil is moving, that stops when the coil comes to rest. Be sure to observe the polarity of the deflection, and determine the direction of current through the switch. The direction should be the same as in Demonstration 3, and opposite to Demonstration 4.

Discussion after observing the results: Ask students to describe the result. Does moving the coil change the flux through it? Is the flux increasing or decreasing? Does the observed induced current agree with Faraday's law? Why is the direction the same as Demonstration 3 and opposite to Demonstrations 2 and 4?

Demonstration 6: Coil with steady current through it. Replace the magnet with the other coil. Coil 2 is connected in series with the battery (power supply) and a second switch. Be sure that the windings of the Coil 2 are as shown in the diagram (counterclockwise around the coil as viewed from the right), and that the polarity of the battery is as shown. Have the two coils close together. Coil 1 should be as in Demonstration 1, with the meter in series with it and S_1 open. Close S_2. Then close S_1, and observe if there is any deflection of the meter. There should be none.

Discussion after observing the results: Ask students to describe the (non) result. Does closing S_1 change the flux through Coil 1? What does Faraday's law say?

Demonstration 7: Coil with changing current through it. This time start with S_1 closed and S_2 open. Close S_2. Be sure to observe the polarity of the deflection, and determine the direction of current through the switch. The direction should be the same as in Demonstrations 3 and 5, since the second coil is an electromagnet with its North pole pointing towards the right.

Discussion after observing the results: Ask students to describe the result. Does closing S_2 change the flux through Coil 1? What does Faraday's law say? What does Lenzs' law say?

ELECTROMAGNETIC INDUCTION (EMIN)
TEACHER PRESENTATION NOTES

Demonstration 1: Magnet and coil—both at rest. Be sure that the coil windings are as shown in the diagram (clockwise around the coil as viewed from the right.

- What is the definition of magnetic flux? Is there a magnetic flux through the coil? Does closing the switch change the flux?
- What does Faraday's law say about an induced emf and current?

Demonstration 2: Magnet moving away from the coil. Start with the magnet close to the coil and with the North pole pointing toward the coil. Move the magnet quickly away from the coil.

- Does moving the magnet change the flux through the coil? Is the flux increasing or decreasing? Does the observed induced current agree with Faraday's law?
- What does Lenz's law say about the magnetic field that much be produced by the current in the coil? In which direction must the current in the coil flow to produce this field?

Demonstration 3: Magnet moving towards the coil. Start with the magnet a little away from the coil and with the North pole pointing toward the coil. Move the magnet quickly toward the coil.

- Does moving the magnet change the flux through the coil? Is the flux increasing or decreasing? Does the observed induced current agree with Faraday's law?
- What does Lenz's law say about the magnetic field that much be produced by the current in the coil? In which direction must the current in the coil flow to produce this field? Why is the direction opposite to that in Demonstration 2?

Demonstration 4: Coil moving away from the magnet. Start with the magnet close the coil and with the North pole pointing toward the coil. Move the coil quickly away from the magnet.

- Does moving the coil change the flux through it? Is the flux increasing or decreasing? Does the observed induced current agree with Faraday's law?
- Why is the direction the same as Demonstration 2 and opposite to Demonstration 3?

Demonstration 5: Coil moving toward the magnet. Start with the magnet a little away from the coil and with the North pole pointing toward the coil. Move the coil quickly toward the magnet.

- Does moving the coil change the flux through it? Is the flux increasing or decreasing? Does the observed induced current agree with Faraday's law?
- Why is the direction the same as Demonstration 3 and opposite to Demonstrations 2 and 4?

Demonstration 6: Coil with steady current through it. Close S_1, and observe if there is any deflection of the meter. There should be none.

- Does closing S_1 change the flux through Coil 1? What does Faraday's law say?

Demonstration 7: Coil with changing current through it. This time start with S_1 closed and S_2 open. Close S_2. Observe polarity of the deflection, and determine the direction of current through the switch.

- Does closing S_2 change the flux through Coil 1? What does Faraday's law say?
- What does Lenzs' law say?

AC CIRCUITS (ACC)

Hand in this sheet **Name**___________________________

INTERACTIVE LECTURE DEMONSTRATIONS
PREDICTION SHEET—**AC CIRCUITS**

Directions: This sheet will be collected. Write your name at the top to record your presence and participation in these demonstrations. Follow your instructor's directions. You may write whatever you wish on the attached Results Sheet and take it with you.

Demonstration 1:

In the circuit on the right, a resistor is connected to an AC source of potential difference (voltage). A voltage vs. time graph for the source is shown. Sketch your prediction of the current vs. time graph for this circuit. Pay special attention to any phase difference between the voltage and current.

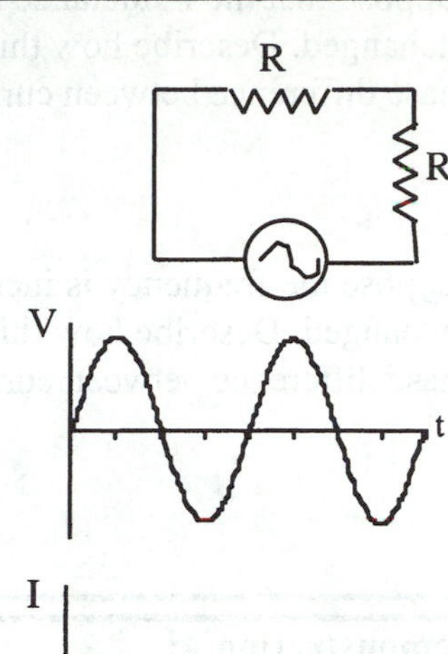

Suppose that the resistance R' is increased. Describe how this would affect the peak current.

Suppose the frequency is increased. Describe how this would affect the peak current and the phase difference between current and voltage.

Demonstration 2:

In the circuit on the right, a resistor and a capacitor are connected in series to an AC source of potential difference (voltage). A voltage vs. time graph for the source is shown. Sketch your prediction of the current vs. time graph for this circuit. Pay special attention to any phase difference between the voltage and current.

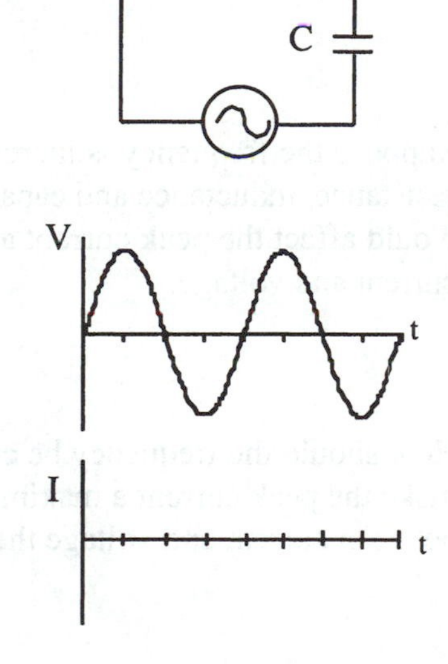

Suppose that the capacitance is increased, with the resistance unchanged. Describe how this would affect the peak current and the phase difference between current and voltage.

Suppose the frequency is increased, with the resistance and capacitance unchanged. Describe how this would affect the peak current and the phase difference between current and voltage.

Demonstration 3:
In the circuit on the right, a resistor and an inductor are connected in series with an AC source of potential difference (voltage). A voltage vs. time graph for the source is shown. Sketch your prediction of the current vs. time graph for this circuit. Pay special attention to any phase difference between the voltage and current.

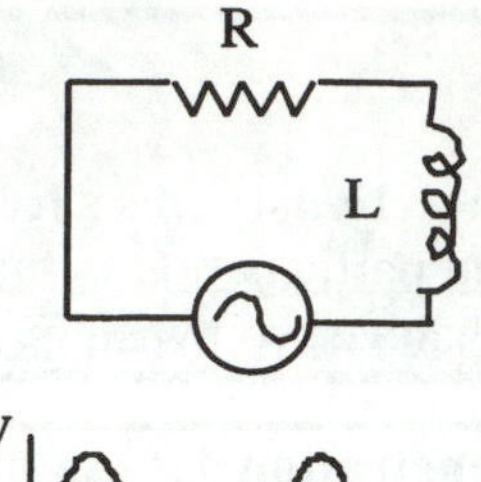

Suppose that the inductance is increased, with the resistance unchanged. Describe how this would affect the peak current and the phase difference between current and voltage.

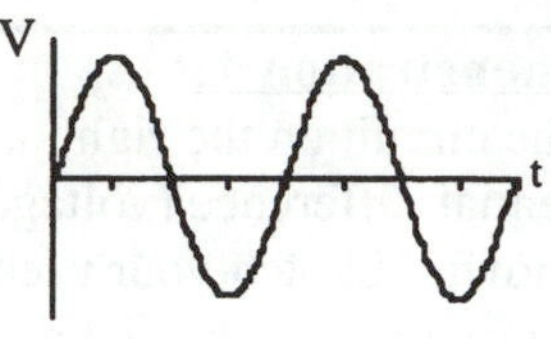

Suppose the frequency is increased, with the resistance and inductance unchanged. Describe how this would affect the peak current and the phase difference between current and voltage.

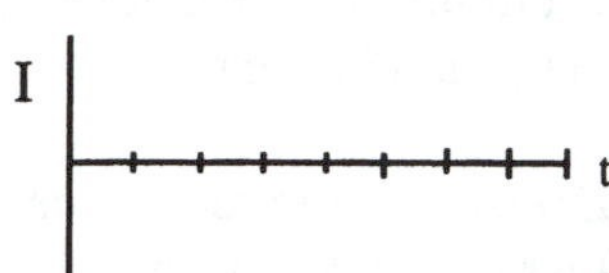

Demonstration 4:
In the circuit on the right, a resistor, an inductor and a capacitor are connected in series with an AC source of potential difference (voltage). Voltage vs. time and current vs. time graphs for the circuit are shown.

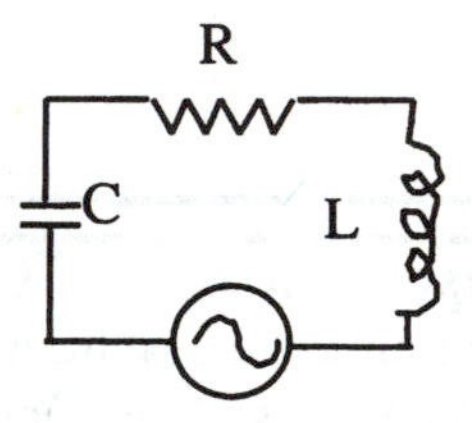

Suppose that the inductance is increased a small amount, with the resistance and capacitance unchanged. Describe how this would affect the peak current and the phase difference between current and voltage.

Suppose the frequency is increased a small amount, with the resistance, inductance and capacitance unchanged. Describe how this would affect the peak current and the phase difference between current and voltage.

How should the frequency be changed (increased or decreased) to make the peak current a maximum? What will the phase difference between current and voltage then be?

Keep this sheet

INTERACTIVE LECTURE DEMONSTRATIONS
RESULTS SHEET—AC CIRCUITS

You may write whatever you wish on this sheet and take it with you.

Demonstration 1:

In the circuit on the right, a resistor is connected to an AC source of potential difference (voltage). A voltage vs. time graph for the source is shown. Sketch your prediction of the current vs. time graph for this circuit. Pay special attention to any phase difference between the voltage and current.

R

R'

Suppose that the resistance R' is increased. Describe how this would affect the peak current.

V

t

Suppose the frequency is increased. Describe how this would affect the peak current and the phase difference between current and voltage.

I

t

Demonstration 2:

In the circuit on the right, a resistor and a capacitor are connected in series to an AC source of potential difference (voltage). A voltage vs. time graph for the source is shown. Sketch your prediction of the current vs. time graph for this circuit. Pay special attention to any phase difference between the voltage and current.

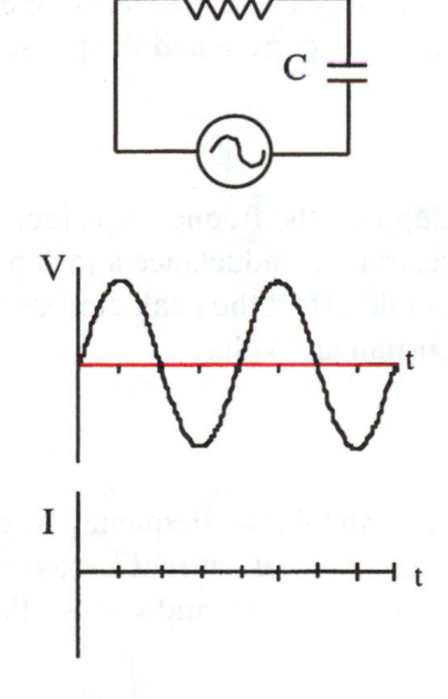

Suppose that the capacitance is increased, with the resistance unchanged. Describe how this would affect the peak current and the phase difference between current and voltage.

Suppose the frequency is increased, with the resistance and capacitance unchanged. Describe how this would affect the peak current and the phase difference between current and voltage.

Demonstration 3:

In the circuit on the right, a resistor and an inductor are connected in series with an AC source of potential difference (voltage). A voltage vs. time graph for the source is shown. Sketch your prediction of the current vs. time graph for this circuit. Pay special attention to any phase difference between the voltage and current.

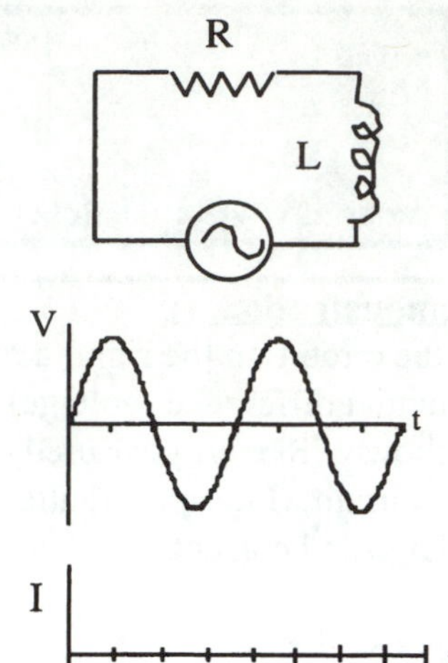

Suppose that the inductance is increased, with the resistance unchanged. Describe how this would affect the peak current and the phase difference between current and voltage.

Suppose the frequency is increased, with the resistance and inductance unchanged. Describe how this would affect the peak current and the phase difference between current and voltage.

Demonstration 4:

In the circuit on the right, a resistor, an inductor and a capacitor are connected in series with an AC source of potential difference (voltage). Voltage vs. time and current vs. time graphs for the circuit are shown.

Suppose that the inductance is increased a small amount, with the resistance and capacitance unchanged. Describe how this would affect the peak current and the phase difference between current and voltage.

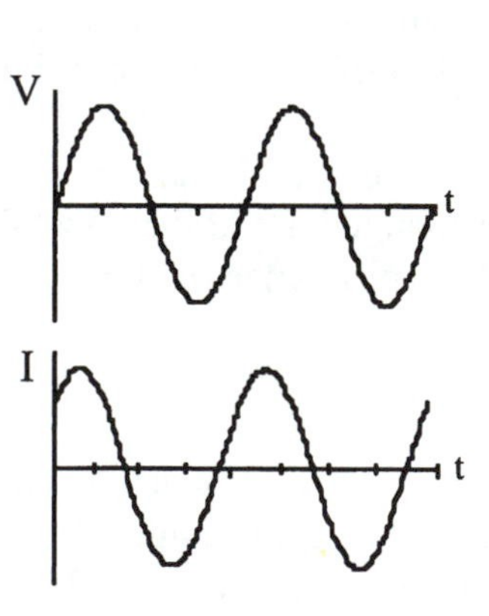

Suppose the frequency is increased a small amount, with the resistance, inductance and capacitance unchanged. Describe how this would affect the peak current and the phase difference between current and voltage.

How should the frequency be changed (increased or decreased) to make the peak current a maximum? What will the phase difference between current and voltage then be?

AC CIRCUITS (ACC)
TEACHER'S GUIDE

Prerequisites:

This *ILD* sequence can be used as an introduction to AC Circuits, or as a review of this topic from lectures and/or text readings. It is designed to give students a semi-quantitative understanding of the behavior of circuits with resistors, capacitors and/or inductors driven by an AC voltage. Students should have studied basic DC circuits, capacitors and inductors, and should also be familiar with the following terms: impedance, capacitative reactance, inductive reactance, peak current and voltage and phase.

Equipment:

oscilloscope with dual input and output to video projector or large monitor (See below.)

signal generator with output of 10 V from 30 to 6000 Hz

2 capacitors with capacitance of several μF

2 10 Ω resistors and a 5 Ω resistor

inductor with inductance around 10 mH with a relatively low resistance (See below.)

iron core for the inductor (optional) (See below.)

connecting wires

Measurements of current and voltage with the oscilloscope:

The voltage across a resistor is proportional to the current through it, and has the same phase as the current. Therefore, the two inputs to the oscilloscope should always be the applied voltage and the voltage across the resistor R. An alternative to an oscilloscope is to use current and voltage probes and a computer. Current and Voltage probes are available from Vernier Software and Technology (www.vernier.com) and from PASCO Scientific (www.pasco.com). The Vernier probes are the Current Probe (DCP-BTA) and the Differential Voltage Probe (DVP-BTA). The PASCO probes are the Current Sensor (CI-6556) and the Voltage Sensor (CI-6503). If you use these probes, you will need to use circuit elements with values close to R=100 5 Ω, L = 800 mH and C = 47 μF.

Inductor:

A good source of a 10 mH inductor is a coil from a transformer set such as the PASCO SF-8616 or SF-8617. A 500 turn coil from an older transformer set used in these demonstrations had a nominal inductance of 10 mH and resistance of 2.5 Ω . One advantage of using such a coil is that an iron core can be placed in its center, increasing the inductance. For the coil described, the iron core increased its inductance to 45 mH.

Demonstrations and Sample Results:

Demonstration 1: AC circuit with two resistors. Begin with R =R' =10 Ω, with the applied voltage displayed. After the prediction and discussion steps are completed, connect the second probe to display the voltage across the resistor R which is proportional to the current through the resistor. You can change the resistance R' by connecting the 5 Ω resistor either in series or in parallel with the 10 Ω one. Also increase the frequency.

Discussion after observing the results: Is the current in phase with the applied voltage? How do you know? Ask students to describe what happens to the peak current (voltage across R) when R' is increased. Why does the peak current decrease? What happens to the peak current and phase when the frequency of the applied voltage is increased? Are the current and applied voltage always in phase for any frequency?

<u>Demonstration 2:</u> **AC circuit with a resistor and a capacitor (RC).** Begin with the applied voltage displayed. After the prediction and discussion steps are completed, connect the second probe to display the voltage across the resistor R which is proportional to the current through the resistor. You can increase the capacitance by connecting the second capacitor in parallel with the first. You can also increase the frequency of the applied signal. Figure V-7 shows a typical oscilloscope trace for the current and voltage in this circuit at a low frequency (around 100 Hz) and a high frequency (around 6000 Hz). In the former, the current leads the applied voltage by close to 90°, while in the latter, the current and voltage are much closer to being in phase because the capacitative reactance ($X_C=1/2\pi fC$) is much smaller for higher frequency.

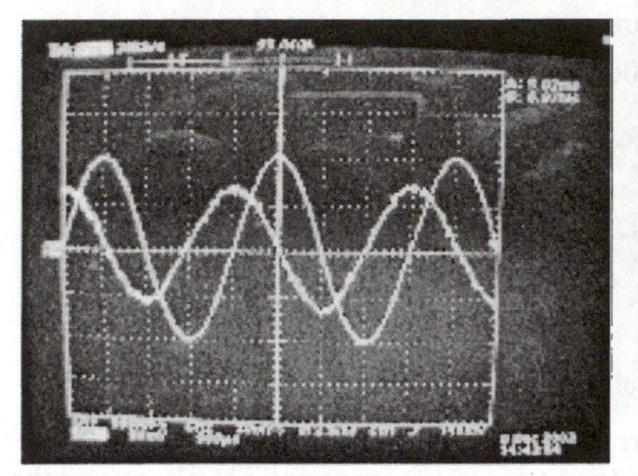

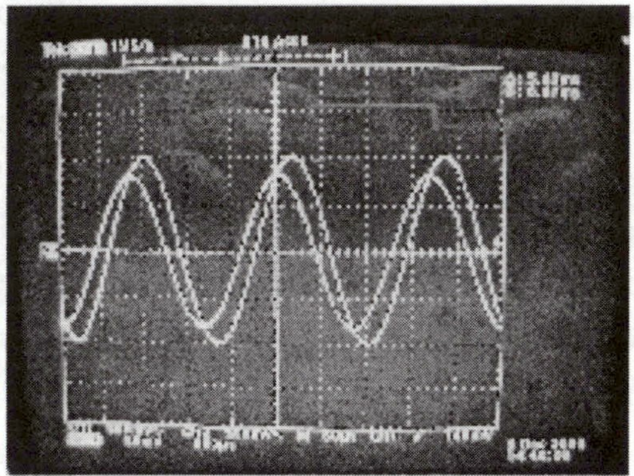

Figure II-8: Applied voltage and current in a circuit with resistor and capacitor at two different frequencies. The left figure is at a low frequency, and shows the current (smaller signal) leading the applied voltage by nearly 90°. The right figure is at a high frequency, and shows the current still leading the applied voltage, but by a much smaller angle.

Discussion after observing the results: Is the current in phase with the applied voltage? How do you know? Can you think of any reason why the current might lead the voltage in a circuit with a capacitor? What has to happen to the capacitor before the voltage across it increases? Ask students to describe what happens to the peak current when C is increased. Why does the peak current decrease? What happens to the peak current and phase when the frequency of the applied voltage is increased? How does capacitative reactance depend on frequency? What happens to the capacitative reactance in this circuit when the frequency is increased? Why would the phase angle get smaller in this case?

<u>Demonstration 3:</u> **AC circuit with a resistor and an inductor (RL).** Begin with the applied voltage displayed. After the prediction and discussion steps are completed, connect the second probe to display the voltage across the resistor R which is proportional to the current through the resistor. You can increase the inductance easily if you are using a transformer coil and have an iron core that can be inserted in it. Also increase the frequency of the applied signal, as in Demonstration 2. Figure V-9 shows a typical oscilloscope trace for the current and voltage in this circuit at a relatively high frequency (around 1000 Hz). The applied voltage leads the current by close to 90°. Remember that the inductive reactance is $X_L=2\pi fL$, and this should become more important as the frequency increases.

Discussion after observing the results: Is the current in phase with the applied voltage? How do you know? Can you think of any reason why the voltage might lead the current in a circuit with an inductor? Does the voltage induced across an inductor oppose the increase of current through it? Ask students to describe what happens to the peak current when L is increased. What happens to the peak current and phase when the frequency of the applied voltage is increased? How does inductive reactance depend on frequency? What happens to the inductive reactance in this circuit when the frequency is increased? Why would the phase angle get larger in this case?

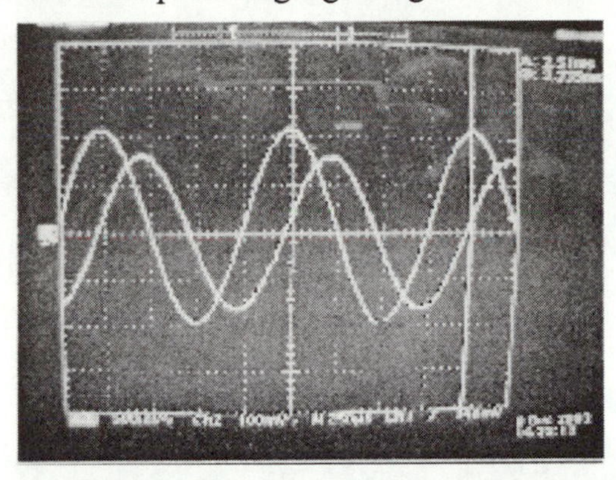

Figure II-9: Applied voltage and current in a circuit with resistor and inductor at a relatively high frequency. The current (larger signal) leads the applied voltage by nearly 90°.

Demonstration 4: AC circuit with a resistor, inductor and capacitor (RLC). Begin with the applied voltage displayed. After the prediction and discussion steps are completed, connect the second probe to display the voltage across the resistor R which is proportional to the current through the resistor. You can increase the inductance by inserting the iron core. You can easily increase the frequency of the applied signal, as in Demonstrations 2 and 3. Figure V-10 shows a series of oscilloscope traces for the current and voltage in this circuit at frequencies below the resonance frequency (current leads the voltage), at the resonance frequency (current and voltage in phase) and above the resonance frequency (voltage leads current).

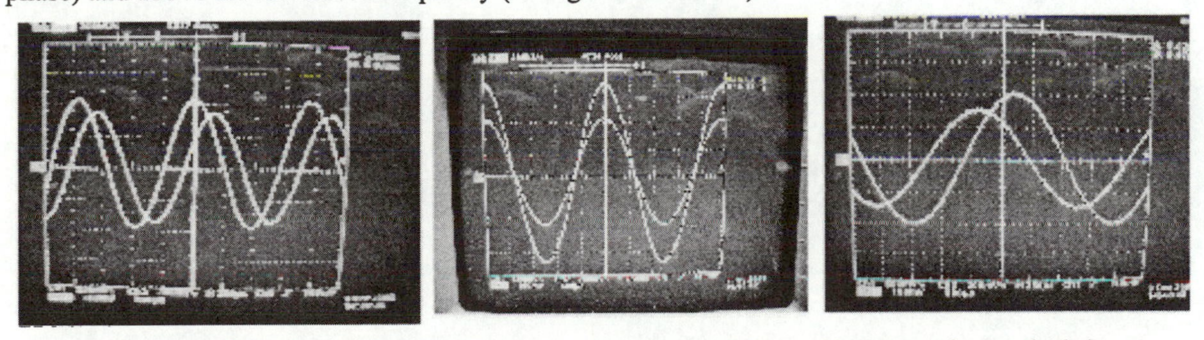

Figure V-10: Applied voltage and current in a circuit with resistor, capacitor and inductor at frequencies below, at and above the resonance frequency. (The current is the larger trace in each picture.)

Discussion after observing the results: Does the current lead the applied voltage? How do you know? Can you think of any reason why the voltage might lead the current for an RLC circuit at low frequency? Which circuit element has the larger reactance at low frequencies? If you increase the inductance what happens to the inductive reactance? As you increase the frequency, what happens to the inductive reactance? To the capacitative reactance? What happens at the resonance frequency? Is this where the impedance has its minimum value and the peak current has its maximum value? What is the phase between the applied voltage and the current at resonance?

AC CIRCUITS (ILDEM07)
TEACHER PRESENTATION NOTES

Demonstration 1: **AC circuit with two resistors.** Begin with R =R' =10 Ω, with the applied voltage displayed. Connect the second probe to display the current. Increase the resistance R'. Increase the frequency.

- Is the current in phase with the applied voltage? How do you know? What happens to the peak current (voltage across R) when R' is increased. Why does the peak current decrease?
- What happens to the peak current and phase when the frequency is increased? Are the current and applied voltage always in phase for any frequency?

Demonstration 2: **AC circuit with a resistor and a capacitor (RC).** Begin with the applied voltage displayed. After the prediction and discussion steps, connect the second probe to display current. Increase the capacitance by connecting the second capacitor in parallel with the first. Increase the frequency of the applied signal. Recall that the capacitative reactance is $X_C=1/2\pi fC$.

- Is the current in phase with the applied voltage? How do you know? Can you think of any reason why the current might lead the voltage in a circuit with a capacitor? What has to happen to the capacitor before the voltage across it increases?
- What happens to the peak current when C is increased? Why does the peak current decrease?
- What happens to the peak current and phase when the frequency is increased? How does capacitative reactance depend on frequency? What happens to the capacitative reactance in this circuit when the frequency is increased? Why would the phase angle get smaller?

Demonstration 3: **AC circuit with a resistor and an inductor (RL).** Begin with the applied voltage displayed. After the prediction and discussion steps, connect the second probe to display current. Increase the inductance by inserting the iron core. Increase the frequency of the applied signal. Remember that the inductive reactance is $X_L=2\pi fL$.

- Is the current in phase with the applied voltage? How do you know? Why does the voltage lead the current in a circuit with an inductor? Does the voltage induced across an inductor oppose the increase of current through it?
- What happens to the peak current when L is increased.
- What happens to the peak current and phase when the frequency is increased? How does inductive reactance depend on frequency? What happens to the inductive reactance in this circuit when the frequency is increased? Why would the phase angle get larger ?

Demonstration 4: **AC circuit with a resistor, inductor and capacitor (RLC).** Begin with the applied voltage displayed. After the prediction and discussion steps, connect the second probe to display current. You can increase the inductance by inserting the iron core. Increase the frequency of the applied signal.

- Does the current lead the applied voltage? How do you know? Why does the voltage lead the current for an RLC circuit at low frequency? Which circuit element has the larger reactance at low frequencies?
- If you increase the inductance what happens to the inductive reactance?
- As you increase the frequency, what happens to the inductive reactance? To the capacitative reactance? What happens at the resonance frequency? Does the impedance have its minimum value and the peak current its maximum value? What is the phase difference at resonance?

Section VI: Interactive Lecture Demonstrations in Light and Optics

REFRACTION AND REFRACTION OF LIGHT (RRLT)

Hand in this sheet **Name**______________________

INTERACTIVE LECTURE DEMONSTRATIONS
PREDICTION SHEET—**RELECTION AND REFRACTION OF LIGHT**

Directions: This sheet will be collected. Write your name at the top to record your presence and participation in these demonstrations. Follow your instructor's directions. You may write whatever you wish on the attached Results Sheet and take it with you.

Demonstration 1: Light is incident as shown on a plane mirror, like the one in your bathroom. The light ray is in the plane of this paper.

Sketch the normal to the surface of the mirror at the point where the light ray hits the mirror.

Predict the direction of the reflected ray, and sketch it on the diagram. Must the reflected ray be in the plane of the paper?

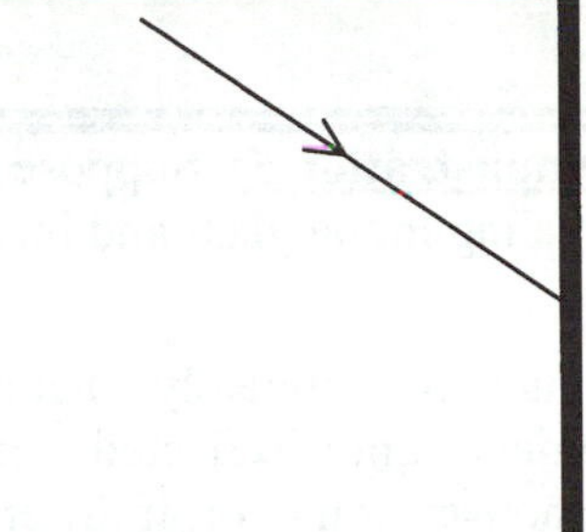

Demonstration 2: You are standing fairly close to the front of the mirror in your bathroom, and you see your image in the mirror. Sketch a stick figure prediction of your image on the diagram. Be sure to carefully show

- The position of your image
- The direction your image is facing
- The height of your image
- Mark with arrows on your image about how much of your body you will actually be able to see

Demonstration 3: A light ray is incident on the surface of a slab of glass.

Which has a larger index of refraction—air or glass?

Sketch predictions of the reflected ray and transmitted ray on the diagram.

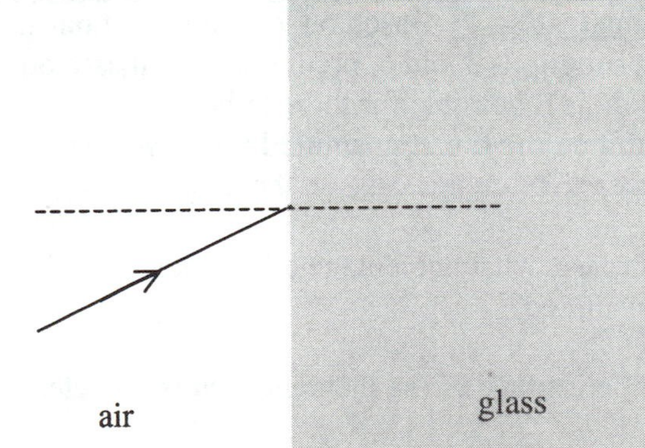

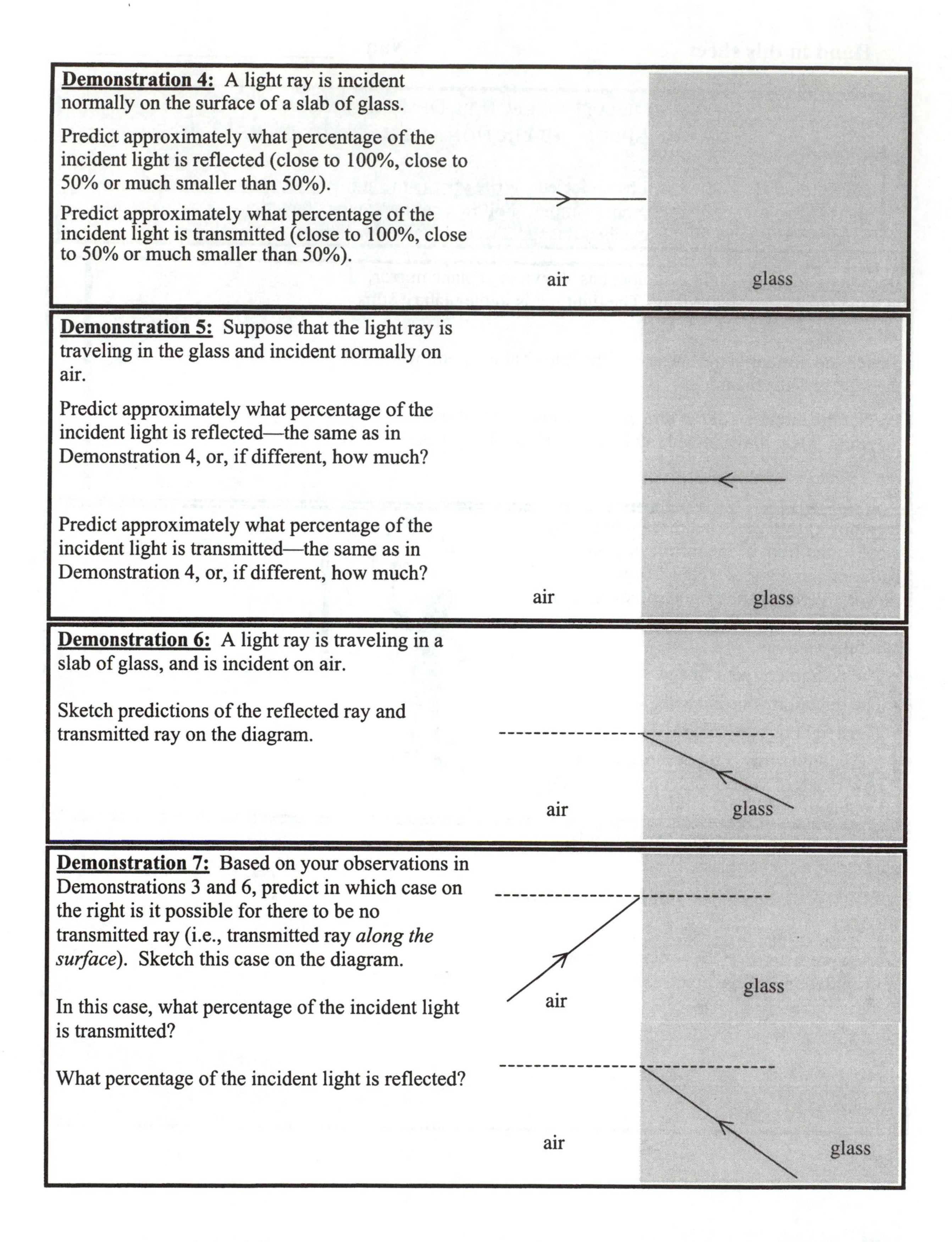

Demonstration 4: A light ray is incident normally on the surface of a slab of glass.

Predict approximately what percentage of the incident light is reflected (close to 100%, close to 50% or much smaller than 50%).

Predict approximately what percentage of the incident light is transmitted (close to 100%, close to 50% or much smaller than 50%).

Demonstration 5: Suppose that the light ray is traveling in the glass and incident normally on air.

Predict approximately what percentage of the incident light is reflected—the same as in Demonstration 4, or, if different, how much?

Predict approximately what percentage of the incident light is transmitted—the same as in Demonstration 4, or, if different, how much?

Demonstration 6: A light ray is traveling in a slab of glass, and is incident on air.

Sketch predictions of the reflected ray and transmitted ray on the diagram.

Demonstration 7: Based on your observations in Demonstrations 3 and 6, predict in which case on the right is it possible for there to be no transmitted ray (i.e., transmitted ray *along the surface*). Sketch this case on the diagram.

In this case, what percentage of the incident light is transmitted?

What percentage of the incident light is reflected?

Keep this sheet

INTERACTIVE LECTURE DEMONSTRATIONS
RESULTS SHEET—REFLECTION AND REFRACTION OF LIGHT

You may write whatever you wish on this sheet and take it with you.

Demonstration 1: Light is incident as shown on a plane mirror, like the one in your bathroom. The light ray is in the plane of this paper.

Sketch the normal to the surface of the mirror at the point where the light ray hits the mirror.

Predict the direction of the reflected ray, and sketch it on the diagram. Must the reflected ray be in the plane of the paper?

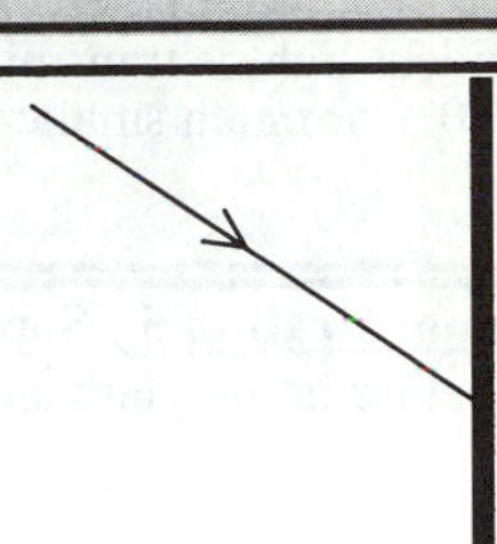

Demonstration 2: You are standing fairly close to the front of the mirror in your bathroom, and you see your image in the mirror. Sketch a stick figure prediction of your image on the diagram. Be sure to carefully show

- The position of your image
- The direction your image is facing
- The height of your image
- Mark with arrows on your image about how much of your body you will actually be able to see

Demonstration 3: A light ray is incident on the surface of a slab of glass.

Which has a larger index of refraction—air or glass?

Sketch predictions of the reflected ray and transmitted ray on the diagram.

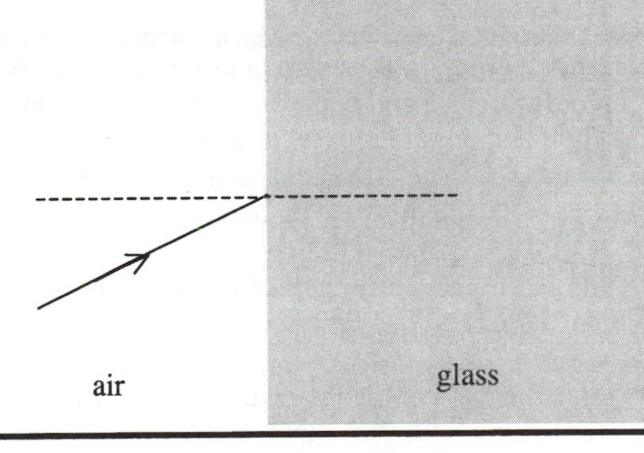

Demonstration 4: A light ray is incident normally on the surface of a slab of glass.

Predict approximately what percentage of the incident light is reflected (close to 100%, close to 50% or much smaller than 50%).

Predict approximately what percentage of the incident light is transmitted (close to 100%, close to 50% or much smaller than 50%).

air glass

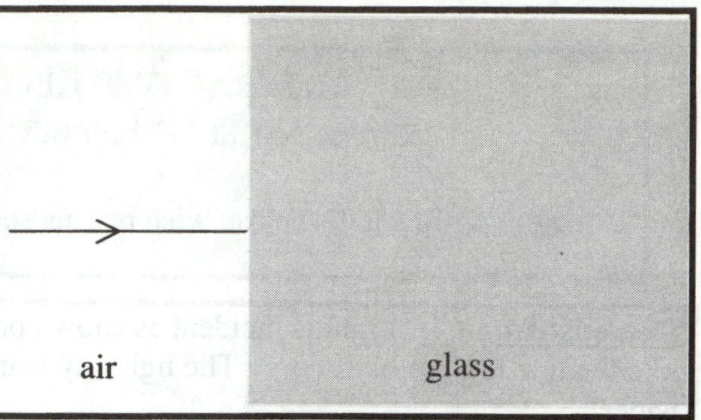

Demonstration 5: Suppose that the light ray is traveling in the glass and incident normally on air.

Predict approximately what percentage of the incident light is reflected—the same as in Demonstration 4, or, if different, how much?

Predict approximately what percentage of the incident light is transmitted—the same as in Demonstration 4, or, if different, how much?

air glass

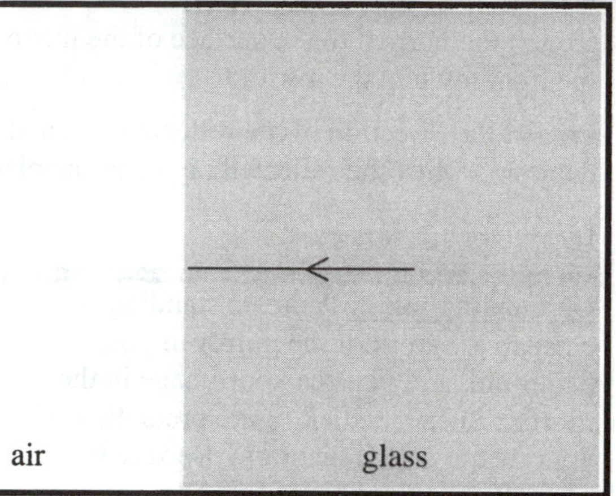

Demonstration 6: A light ray is traveling in a slab of glass, and is incident on air.

Sketch predictions of the reflected ray and transmitted ray on the diagram.

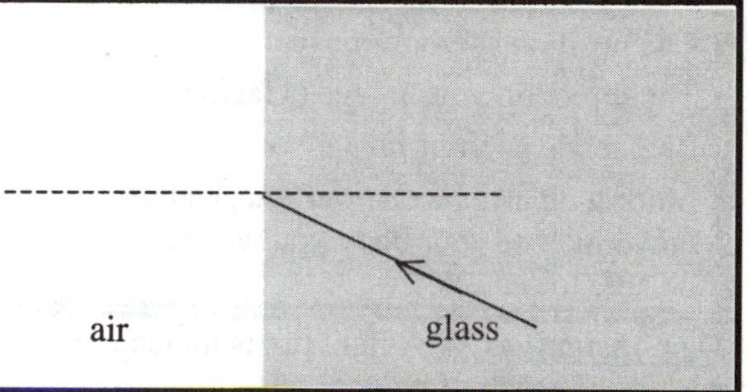

air glass

Demonstration 7: Based on your observations in Demonstrations 3 and 6, predict in which case on the right is it possible for there to be no transmitted ray (i.e., transmitted ray *along the surface*). Sketch this case on the diagram.

In this case, what percentage of the incident light is transmitted?

What percentage of the incident light is reflected?

air glass

air glass

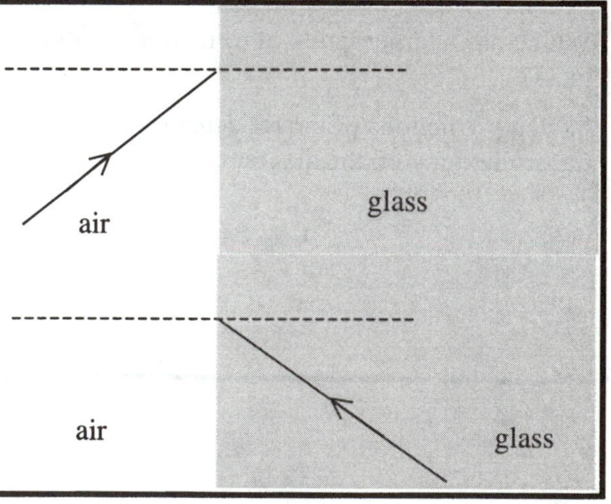

REFLECTION AND REFRACTION OF LIGHT (RRLT)
TEACHER'S GUIDE

Prerequisites:

This *ILD* sequence can be used as an introduction to reflection and refraction. It can also be used as a review after these topics have been introduced in lecture or through text readings. It will be helpful if students have been introduced to the terms index of refraction, normal, object, image and ray.

Equipment:

large plane mirror with horizontal mount (See below.)
large sheet of glass or Lucite (See below.)
cloth to cover mirror
laser (See below.)
blackboard eraser and chalk dust
cylindrical Lucite lens (semi-circular) (See below.)
rotating mount for the Lucite lens (See below.)
ruler

General Notes on Preparation and Equipment:

Plane mirror:
The mirror should be the size of a bathroom mirror. It should be mounted vertically facing the class, with the bottom at about waste level. It should be covered by a cloth at first, so that the students cannot see it.

Sheet of Lucite:
The size of the sheet of Lucite should be appropriate to the size of the class. A sheet about 0.6 m x 0.6 m should be adequate.

Laser:
The laser doesn't need to be too powerful. A laser pointer will work fine.

Cylindrical Lucite lens (semi-circular) and mount:
The Lucite lens should be large enough for the class to see—at least 10 cm radius. Its cross-section should be semi-circular, and it should be at least 5 cm thick. (See the photograph in Figure VI-1 in the Teacher's Guide for *Image Formation with Lenses*.) If you have a blackboard optics kit (for example PASCO (www.pasco.com) SE-9193 or SE-9194), it probably includes a lens like this. You can attach the lens to a black or whiteboard. It is most convenient to mount the lens so that it can rotate about an axis perpendicular to its cross-section. Some blackboard optics kits include such a mount for the lens, or you can construct one.

Demonstrations and Sample Results:

Demonstration 1: Law of reflection. After the prediction and discussion steps, shine the laser at the flat side of the cylindrical lens, so that it is incident on the surface as shown in the diagram on the student sheets. Use chalk dust to view the paths of the incident and reflected rays. The ruler may be used to show the normal to the surface.

Discussion after observing the results: Ask students to describe the direction of the reflected ray. Is the angle of the reflected ray measured from the normal (the angle of reflection) the same as the angle of the incident ray with the normal (the angle of incidence)? Are the incident ray, the normal and the reflected ray all in the same plane? Have a volunteer summarize the law of reflection.

Demonstration 2: Image in a plane mirror. The large mirror should be covered and facing the class. Stand in front of the covered mirror—as shown in the diagram—and ask students to make their predictions. After the prediction and discussion steps, uncover the mirror. For a large class, you will probably need to describe what you can see. For a small class, at least some students could come up and view their images in the mirror. Also let seated students view their images in the mirror, since all but the last prediction can be analyzed based on these observations.

Discussion after observing the results: Ask students to describe their images in the mirror. Where does the image appear to be? Is it upright or inverted? Does it appear to be larger, smaller or the same size as the object? Can you see most of your body, or just the part equal to the height of the mirror? Ask students to explain why you can see your whole body (or most of it). It will be helpful to sketch several rays leaving the character in the diagram and reflected from the surface of the mirror.

Demonstration 3: Refraction at the surface of a more dense medium. After the prediction and discussion steps, shine the laser at an angle with the flat face of the Lucite lens, as in the diagram. This time focus attention on the ray transmitted through the surface and coming out through the curved surface. Chalk dust will again be helpful. If the laser beam is incident at the center of the flat face, it will be refracted at this surface, but not at the curved surface. (The refracted ray will be along a radius and perpendicular to the curved surface.)

Discussion after observing the results: Ask students to describe the direction of the refracted ray. Is the ray bent at the surface? Is the angle of refraction greater than or less than the angle of incidence (both measured with the normal to the surface)? Are the incident and refracted rays and the normal in the same plane? Ask for volunteer(s) to state the law of refraction qualitatively.

Demonstration 4: Percent reflection at the surface of a more dense medium. After the prediction and discussion steps, shine the laser at a small incident angle with the flat surface of the Lucite lens. Use chalk dust to show the incident and reflected rays. It will also be helpful to hold up the large sheet of Lucite, and let the students observe their reflections in it.

Discussion after observing the results: Ask students to compare the incident and reflected intensities (brightness of the beams). If you used the Lucite sheet, ask students to describe how bright their image was. Was very much light reflected from the Lucite to form an image? (Actually, only about 4% of the incident light is reflected by one surface of Lucite or glass.) Why does the mirror in your bathroom reflect more than this?

Demonstration 5: Percent reflection at the surface of a less dense medium. After the prediction and discussion steps, shine the laser through the curved surface of the Lucite lens along a radius so that it strikes the flat surface at the center. The light was moving in the more dense medium, and is now reflected by the less dense medium (air). It is probably easier to use chalk dust to look at the transmitted ray, and compare its brightness to the ray incident on the curved surface. This is not an easy comparison to make.

Discussion after observing the results: Ask students to compare the incident and reflected intensities (brightness of the beams). Is there any reason why the percentage reflected should be different because of the different direction of the incident beam? Actually, the same percentage—only about 4% of the incident light—is reflected by one surface of Lucite or glass in this case as well.

Demonstration 6: Refraction at the surface of a less dense medium. After the prediction and discussion steps, shine the laser through the curved surface of the Lucite lens, along a radius so that it strikes the flat surface at its center. Focus attention on the ray transmitted through the flat surface. Use chalk dust to make the transmitted ray visible.

Discussion after observing the results: Ask students to describe the direction of the refracted ray. Is the ray bent at the surface? Is the angle of refraction greater than or less than the angle of incidence (both measured with the normal to he surface)? How does this result differ from Demonstration 3?

Demonstration 7: Total internal reflection. After the prediction and discussion steps are finished, shine the laser first as in Demonstration 3, rotating the lens to demonstrate that there is always a transmitted ray. Then shine the laser as in Demonstration 6, rotating the lens until there is no transmitted ray and all of the light is reflected back into the Lucite.

Discussion after observing the results: Ask students if there is an angle at which there is no transmitted ray in the top diagram? In the bottom diagram? Ask for volunteer(s) to describe what total internal reflection is.

REFLECTION AND REFRACTION OF LIGHT (RRLT)
TEACHER PRESENTATION NOTES

Demonstration 1: Law of reflection. Shine the laser at the flat side of the cylindrical lens. Use chalk dust to view the paths of the incident and reflected rays.

- Is the angle of reflection equal to the angle of incidence? Are the incident ray, the normal and the reflected ray all in the same plane?
- Have a volunteer summarize the law of reflection

Demonstration 2: Image in a plane mirror. The large mirror should be covered and facing the class. After the prediction steps, uncover the mirror.

- Where does the image appear to be? Is it upright or inverted? Does it appear to be larger, smaller or the same size as the object? Can you see most of your body?
- Ask students to explain why you can see your whole body (or most of it). Sketch several rays leaving the character in the diagram and reflected from the surface of the mirror.

Demonstration 3: Refraction at the surface of a more dense medium. Shine the laser at an angle with the flat face of the Lucite lens. Use chalk dust to focus attention on the transmitted ray coming out through the curved surface. (Be sure that the laser is incident at the center of the flat face.)

- Is the transmitted ray bent at the surface? Is the angle of refraction greater than or less than the angle of incidence (both measured with the normal to he surface)? Are the incident and refracted rays and the normal in the same plane?
- Ask for volunteer(s) to state the law of refraction qualitatively.

Demonstration 4: Percent reflection at the surface of a more dense medium. Shine the laser at a small incident angle with the flat surface of the Lucite lens. Use chalk dust. Hold up the large sheet of Lucite, and let the students observe their reflections in it.

- Ask students to compare the incident and reflected intensities (brightness of the beams).
- Ask students how bright their image was in the Lucite. Was very much light reflected?

Demonstration 5: Percent reflection at the surface of a less dense medium. Shine the laser through the curved surface of the Lucite lens along a radius. The light is now reflected by the less dense medium (air). Use chalk dust. Compare incident and transmitted brightness.

- Ask students to compare the incident and reflected intensities (brightness of the beams).
- Is there any reason why the percentage reflected should be different than in Demonstration 4?

Demonstration 6: Refraction at the surface of a less dense medium. Shine the laser through the curved surface of the Lucite lens, along a radius to the center of the flat surface. Use chalk dust.

- Is the ray bent at the surface? Is the angle of refraction greater than or less than the angle of incidence? How does this result differ from Demonstration 3?

Demonstration 7: Total internal reflection. Shine the laser first as in Demonstration 3 and then as in Demonstration 6, rotating the lens to different angles.

- Ask students if there is an angle at which there is no transmitted ray in the top diagram? In the bottom diagram?
- Ask for volunteer(s) to describe what total internal reflection is.

IMAGE FORMATION WITH LENSES (IMFL)

Hand in this sheet **Name**______________________

INTERACTIVE LECTURE DEMONSTRATIONS
PREDICTION SHEET—**IMAGE FORMATION WITH LENSES**

Directions: This sheet will be collected. Write your name at the top to record your presence and participation in these demonstrations. Follow your instructor's directions. You may write whatever you wish on the attached Results Sheet and take it with you.

Demonstration 1: You have a converging lens. An object in the shape of an arrow is positioned a distance larger than the focal length to the left of the lens, as shown in the diagram on the right. Draw several rays from the head of the arrow and several rays from the foot of the arrow to show how the image of the arrow is formed by the lens.

Is this a real or a virtual image?

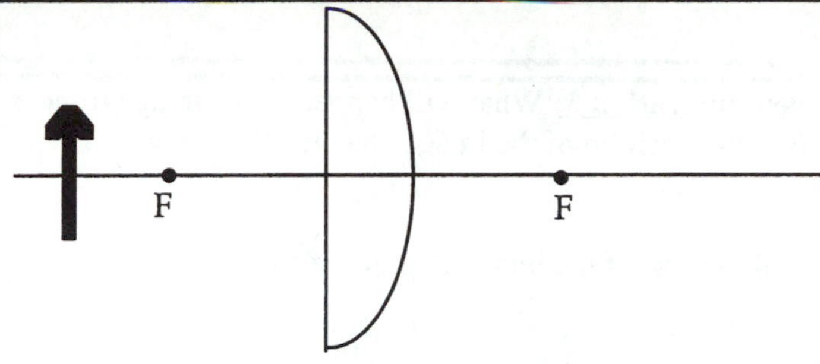

Demonstration 2: What will happen to the image if you block the top half of the *lens* with a card? Answer in words and show what happens on the diagram on the right by making any changes needed in the rays you drew in Demonstration 1.

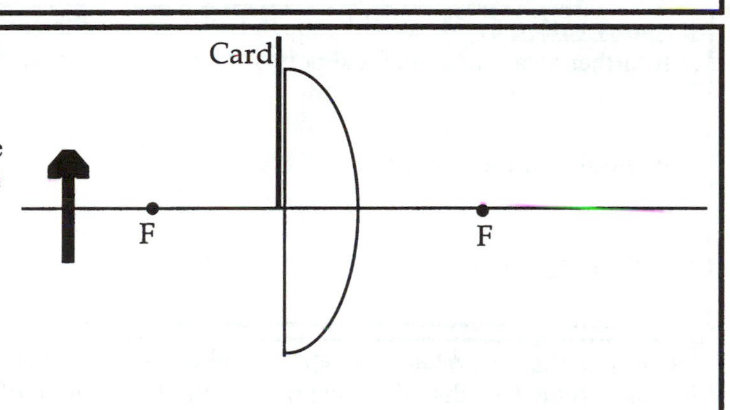

Demonstration 3: What will happen to the image if you block the top half of the *object* with a card? Answer in words and show what happens on the diagram on the right by making any changes needed in the rays you drew above for Demonstration 1.

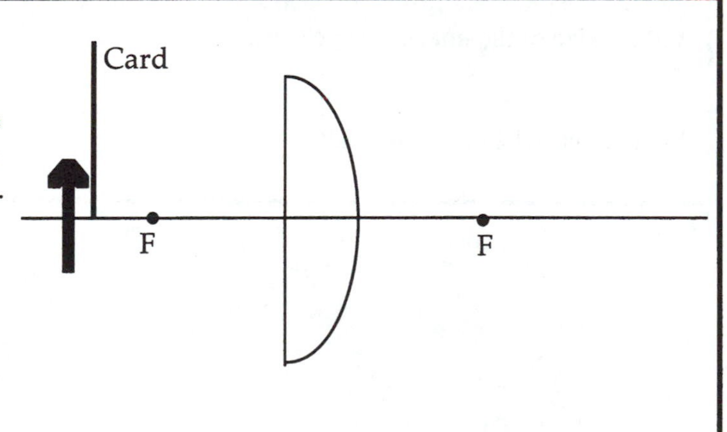

Demonstration 4: What will happen to the image if you remove the lens? Answer in words and show what happens on the diagram on the right by making any changes needed in the rays you drew above for Demonstration 1.

Demonstration 5: What will happen to the image if the object is moved further away from the lens? Will the position of the image change? If so, how?

Will the size of the image change? If so, how?

Will the image be real or virtual?

Demonstration 6: What will happen to the image if the object is moved closer to the lens (but is still further away than the focal point)? Will the position of the image change? If so, how?

Will the size of the image change? If so, how?

Will the image be real or virtual?

Demonstration 7: What will happen to the image if the object is moved closer to the lens so that it is closer to the lens than the focal point? Will the position of the image change? If so, how?

Will the size of the image change? If so, how?

Will the image be real or virtual?

Keep this sheet

INTERACTIVE LECTURE DEMONSTRATIONS
RESULTS SHEET—**IMAGE FORMATION WITH LENSES**

You may write whatever you wish on this sheet and take it with you.

Demonstration 1: You have a converging lens. An object in the shape of an arrow is positioned a distance larger than the focal length to the left of the lens, as shown in the diagram on the right. Draw several rays from the head of the arrow and several rays from the foot of the arrow to show how the image of the arrow is formed by the lens.

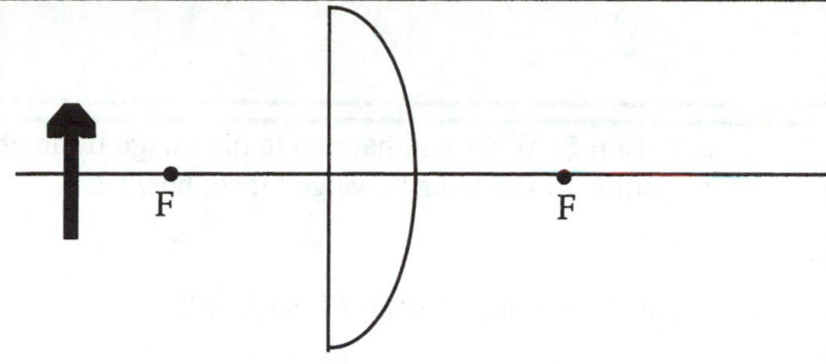

Is this a real or a virtual image?

Demonstration 2: What will happen to the image if you block the top half of the *lens* with a card? Answer in words and show what happens on the diagram on the right by making any changes needed in the rays you drew in Demonstration 1.

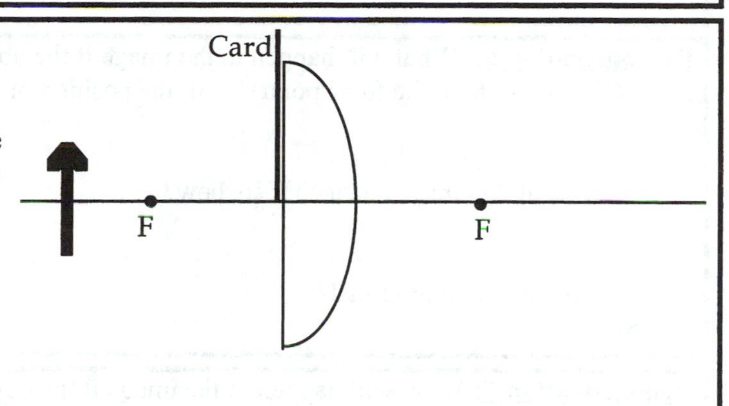

Demonstration 3: What will happen to the image if you block the top half of the *object* with a card? Answer in words and show what happens on the diagram on the right by making any changes needed in the rays you drew above for Demonstration 1.

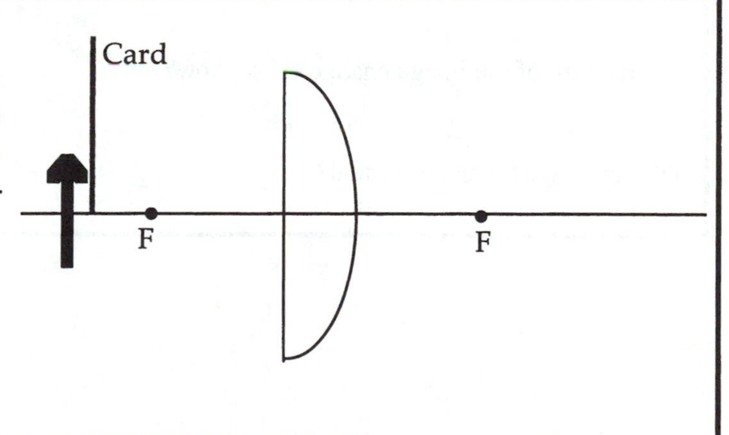

Demonstration 4: What will happen to the image if you remove the lens? Answer in words and show what happens on the diagram on the right by making any changes needed in the rays you drew above for Demonstration 1.

Demonstration 5: What will happen to the image if the object is moved further away from the lens? Will the position of the image change? If so, how?

Will the size of the image change? If so, how?

Will the image be real or virtual?

Demonstration 6: What will happen to the image if the object is moved closer to the lens (but is still further away than the focal point)? Will the position of the image change? If so, how?

Will the size of the image change? If so, how?

Will the image be real or virtual?

Demonstration 7: What will happen to the image if the object is moved closer to the lens so that it is closer to the lens than the focal point? Will the position of the image change? If so, how?

Will the size of the image change? If so, how?

Will the image be real or virtual?

IMAGE FORMATION WITH LENSES (IMFL)
TEACHER'S GUIDE

Prerequisites:

This *ILD* sequence can be used as an introduction to image formation with lenses. It can also be used as a review after this topic has been introduced in lecture or through text readings. It will be helpful if the students have been introduced to the terms ray, lens, object, image, real, virtual, upright and inverted.

Equipment:

cylindrical Lucite lens (semi-circular) (See below.)
mount for the Lucite lens (See below.)
arrow about half the diameter of the lens, either drawn on the white (black) board, or on a card
two small, bright light bulbs, sockets and connecting leads (See below.)
mounts for bulbs
lantern battery or power supply
green filter
comb
index card

General Notes on Preparation and Equipment:

Cylindrical Lucite lens (semi-circular) and mount:
The Lucite lens should be large enough for the class to see—at least 10 cm radius. Its cross-section should be semi-circular, and it should be at least 5 cm thick. If you have a blackboard optics kit (for example PASCO (www.pasco.com) SE-9193 or SE-9194), it probably includes a lens like this. You can attach the lens to a black or whiteboard. Some blackboard optics kits include a magnetic or suction mount for the lens, or you can construct one.

Light bulbs:
The bulbs should be fairly bright, but any 6 V flashlight bulbs and sockets should work. The sockets should be wired in parallel with enough wire to reach the battery or power supply. If you glue ceramic magnets to the bottoms of the sockets, you can easily mount the bulbs on a white board or black board that has an iron backing.

Demonstrations and Sample Results:

Demonstration 1: Ray diagram. Figure VI-1 shows the setup with the lens, arrow and bulbs. Figure VI-2 shows the setup with the bulbs lighted. Begin by showing students the setup with the bulbs off. These first predictions will obviously be easier for students who have already studied image formation in lecture, and who have had some practice drawing ray diagrams. (This does not mean that they will necessarilydraw the diagram correctly!) The idea behind this sequence of demonstrations is that students will better understand the function of a lens in forming images if they see what happens to *all* the light rays from a point on the object incident on the lens.

After the prediction and discussion steps, tell the students that the two bulbs represent point sources of light at the head and foot of the arrow. Turn on both bulbs. Then put the green filter in front of the top bulb. Put the comb in front of the two bulbs to give the illusion of multiple rays.

Discussion after observing the results: Ask students to describe where the image points are for the top and bottom bulbs. What does the lens do to all of the rays from a point on the object that are incident on the lens? Where are the image points of each bulb? What does image formation mean? Is this a *real* or *virtual* image? Have students define each of these. Is the image upright or inverted?

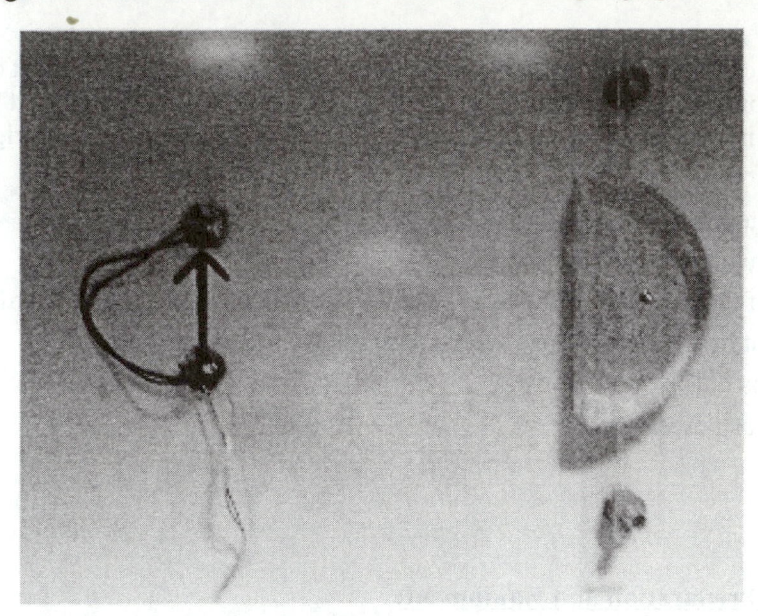

Figure VI-1: Setup for observing image formation in these demonstrations.

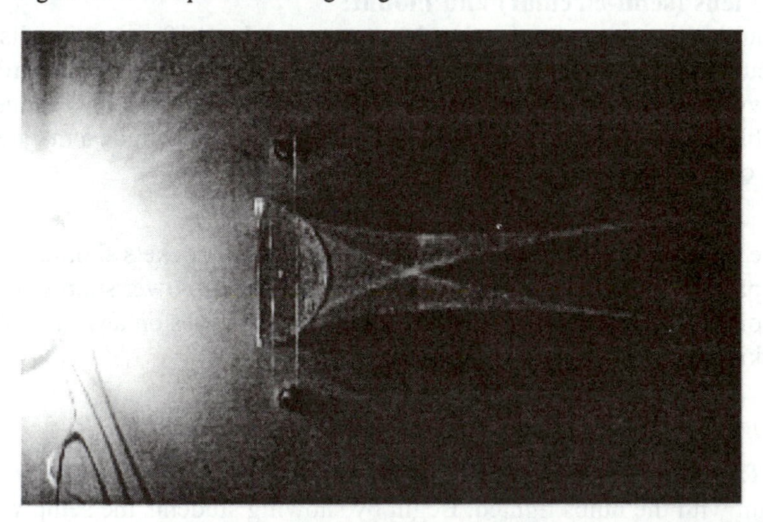

Figure VI-2: Actual observations of image formation with both point source bulbs lighted.

Demonstration 2: Blocking top half of the lens. After the prediction and discussion steps, keep both bulbs lighted, and place the card so that it blocks the top half of the lens. Be sure that students can see clearly that light from both bulbs incident on the bottom half of the lens is still focused to two image points. (The image is whole, at the same location and dimmer.)

Discussion after observing the results: Would the whole image of the arrow still be seen? Would it be in the same location? How would it be changed?

<u>Demonstration 3:</u> Blocking half the object. After the prediction and discussion steps, keep both bulbs lighted and place the card so that it blocks the top bulb (part of the top half of the object). Be sure that students can see clearly that light from the top bulb no longer reaches the lens, and is not focused to an image point. (Only the bottom half of the arrow is now imaged by the lens.)

Discussion after observing the results: Would the whole image of the arrow still be there? Why or why not?

<u>Demonstration 4:</u> Removing the lens. After the prediction and discussion steps, keep both bulbs lighted and remove the lens. Be sure that students can see that light rays diverge from both bulbs, and there is no focusing element to focus them to image points.

Discussion after observing the results: Would there still be an image? Why or why not? What is an image?

<u>Demonstration 5:</u> Moving lens further away. After the prediction and discussion steps, keep both bulbs lighted and move the lens further away from the object. Be sure that students can see that there are still two image points that are now closer to the lens and closer together. (The image is closer to the lens and smaller than before.) The rays from each bulb are not diverging as much when they hit the lens, and are focused more effectively by the lens.

Discussion after observing the results: Ask students to describe where the image is now formed. Is it closer or further away from the lens? Is it larger or smaller than before? How do you know?

<u>Demonstration 6:</u> Moving lens closer. After the prediction and discussion steps, keep both bulbs lighted and move the lens closer to the object (but still further away than the focal point). Be sure that students can see that there are still two image points that are now further away from the lens and further apart. (The image is further from the lens and larger than before.) The rays from each bulb are diverging more when they hit the lens, and are focused less effectively by the lens.

Discussion after observing the results: Ask students to describe where the image is now formed. Is it closer or further away from the lens? Is it larger or smaller than before? How do you know?

<u>Demonstration 7:</u> Virtual image. After the prediction and discussion steps, keep both bulbs lighted and move the lens still closer to the object, this time closer than the focal point. Be sure that students can see that light rays from each bulb are now still diverging when they leave the lens. (There are no *real* image points formed beyond the lens.) The rays from each bulb are diverging still more when they hit the lens, and the lens isn't strong enough to focus them to image points.

Discussion after observing the results: Ask students to describe what they now observe. Are the rays from an object point converged to a real image point beyond the lens? If you were standing to the right of the lens looking back into it, where would rays appear to be diverging from? What kind of image is this, and where is it located?

Image Formation with Lenses (IMFL)
Teacher Presentation Notes

Demonstration 1: Ray diagram. Begin by showing students the setup with the bulbs off. After the prediction and discussion steps, turn on both bulbs. Put the green filter in front of the top bulb. Put the comb in front of the two bulbs.

- Ask students to describe where the image points are for the top and bottom bulbs.
- What does the lens do to all of the rays from a point on the object that are incident on the lens? What does image formation mean?
- Is this a *real* or virtual *image* image? Have students define each of these.

Demonstration 2: Blocking top half of the lens. The card blocks the top half of the lens.

- Would the whole image of the arrow still be there? Would it be in the same location? How would it be changed?

Demonstration 3: Blocking half the object. The card so that it blocks the top bulb.

- Would the whole image of the arrow still be there? Why or why not?

Demonstration 4: Removing the lens. Remove the lens.

- Would there still be an image? Why or why not? What is an image?

Demonstration 5: Moving lens further away. Move the lens further away from the object..

- Ask students to describe where the image is now formed. Is it closer or further away from the lens? Is it larger or smaller than before? How do you know?

Demonstration 6: Moving lens closer. Move the lens closer to the object (but still further away than the focal point).

- Ask students to describe where the image is now formed. Is it closer or further away from the lens? Is it larger or smaller than before? How do you know?

Demonstration 7: Virtual image. Move the lens still closer to the object, this time closer than the focal point.

- Ask students to describe what they now observe. Are the rays from an object point converged to a real image point beyond the lens?
- If you were standing to the right of the lens looking back into it, where would rays appear to be diverging from? What kind of image is this—real or virtual, and where is it located?

MIRRORS (MIRR)

Hand in this sheet **Name**___________________________

INTERACTIVE LECTURE DEMONSTRATIONS
PREDICTION SHEET—**MIRRORS**

Directions: This sheet will be collected. Write your name at the top to record your presence and participation in these demonstrations. Follow your instructor's directions. You may write whatever you wish on the attached Results Sheet and take it with you.

Demonstration 1: A concave mirror has focal length *f*. Where should an object be placed so that a real image is formed that is larger than the object? Sketch the object arrow on the diagram showing a possible location of the object. (Note: there may be more than one possible location for the object.)

Where will the image be formed? Will it be upright or inverted?

center of curvature
focal point
object arrow

Demonstration 2: For the same concave mirror as in Demonstration 1, where should an object be placed so that a real image is formed that is smaller than the object? Sketch the object arrow on the diagram showing a possible location of the object. (Note: there may be more than one possible location for the object.)

Where will the image be formed? Will it be upright or inverted?

center of curvature
focal point
object arrow

Demonstration 3: For the same concave mirror as in Demonstrations 1 and 2, where should an object be placed so that a virtual image is formed? Sketch the object arrow on the diagram showing a possible location of the object. (Note: there may be more than one possible location for the object.)

Where will the image be formed? Will it be larger or smaller than the object? Upright or inverted?

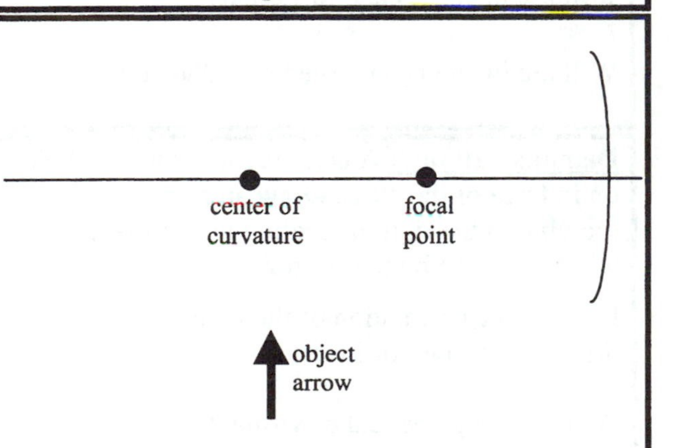

Demonstration 4: A convex mirror has focal length *f*. Where should an object be placed so that a real image is formed? Sketch the object arrow showing a possible location of the object. (Note: there may be more than one possible location for the object.)

Where will the image be formed? Will it be upright or inverted? Larger or smaller than the object?

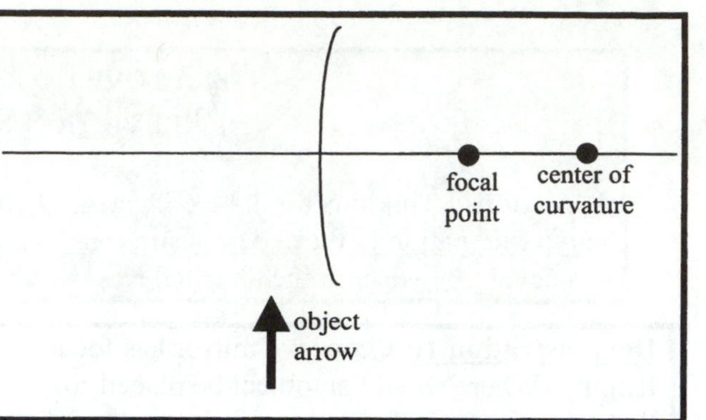

Demonstration 5: A convex mirror has focal length *f*. Where should an object be placed so that a virtual image is formed? Sketch the object arrow showing a possible location of the object. (Note: there may be more than one possible location for the object.)

Where will the image be formed? Will it be upright or inverted? Larger or smaller than the object?

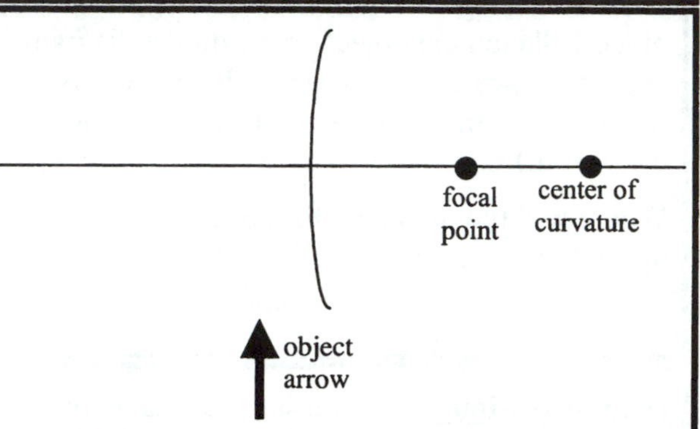

Demonstration 6: A concave mirror of relatively short focal length will be held up in front of the class. Draw an arrow on the drawing to the right indicating where the image will be formed by the mirror.

Indicate the orientation of the image by drawing the arrow upright or inverted.

Will the image be real or virtual?

Will the image be enlarged or reduced in size?

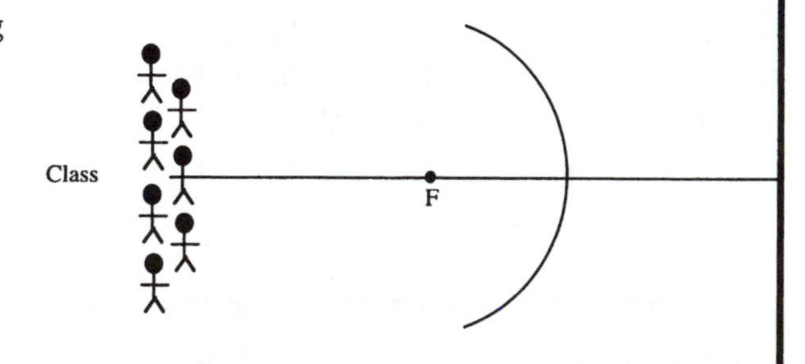

Demonstration 7: A convex mirror will be held up in front of the class. Draw an arrow on the drawing to the right indicating where the image will be formed by the mirror.

Indicate the orientation of the image by drawing the arrow upright or inverted.

Will the image be real or virtual?

Will the image be enlarged or reduced in size?

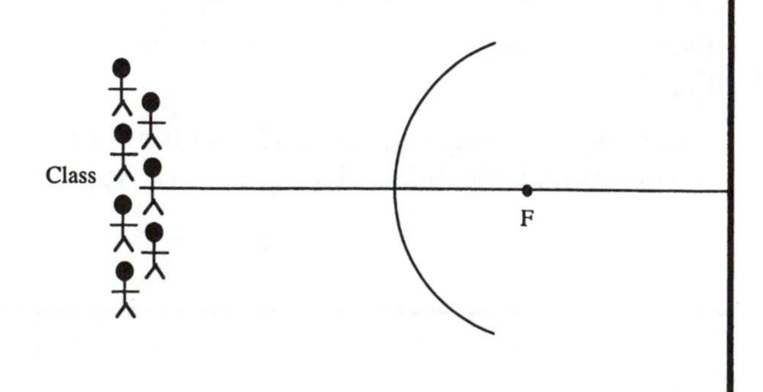

Keep this sheet

INTERACTIVE LECTURE DEMONSTRATIONS
RESULTS SHEET—**MIRRORS**

You may write whatever you wish on this sheet and take it with you.

Demonstration 1: A concave mirror has focal length *f*. Where should an object be placed so that a real image is formed that is larger than the object? Sketch the object arrow on the diagram showing a possible location of the object. (Note: there may be more than one possible location for the object.)

Where will the image be formed? Will it be upright or inverted?

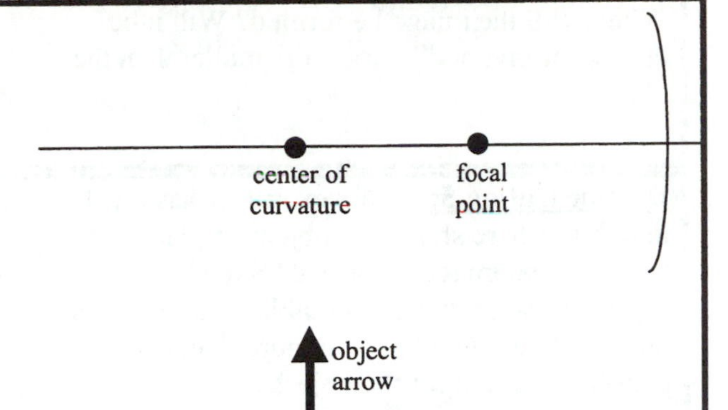

Demonstration 2: For the same concave mirror as in Demonstration 1, where should an object be placed so that a real image is formed that is smaller than the object? Sketch the object arrow on the diagram showing a possible location of the object. (Note: there may be more than one possible location for the object.)

Where will the image be formed? Will it be upright or inverted?

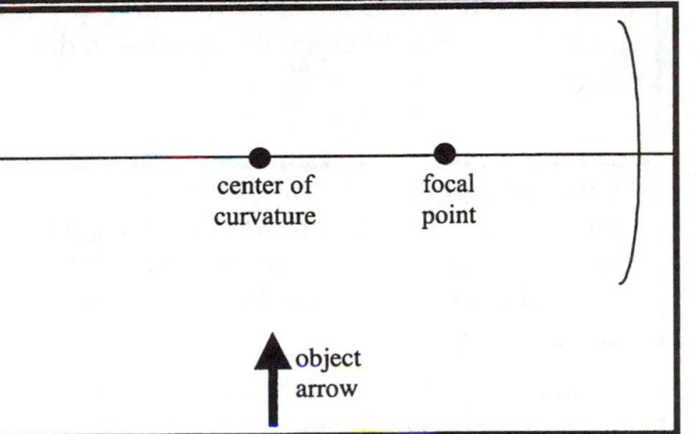

Demonstration 3: For the same concave mirror as in Demonstrations 1 and 2, where should an object be placed so that a virtual image is formed? Sketch the object arrow on the diagram showing a possible location of the object. (Note: there may be more than one possible location for the object.)

Where will the image be formed? Will it be larger or smaller than the object? Upright or inverted?

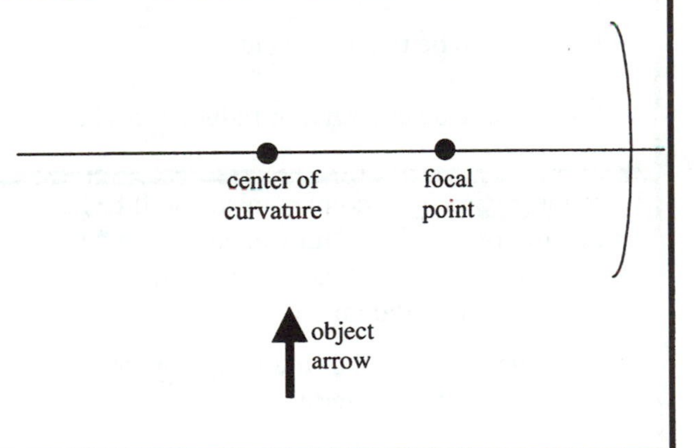

Demonstration 4: A convex mirror has focal length *f*. Where should an object be placed so that a real image is formed? Sketch the object arrow showing a possible location of the object. (Note: there may be more than one possible location for the object.)

Where will the image be formed? Will it be upright or inverted? Larger or smaller than the object?

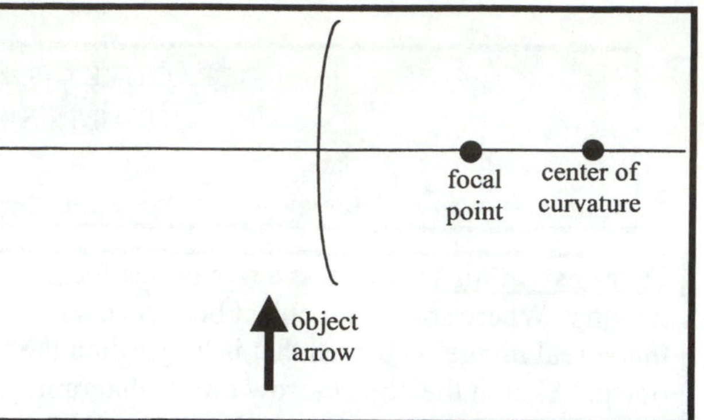

Demonstration 5: A convex mirror has focal length *f*. Where should an object be placed so that a virtual image is formed? Sketch the object arrow showing a possible location of the object. (Note: there may be more than one possible location for the object.)

Where will the image be formed? Will it be upright or inverted? Larger or smaller than the object?

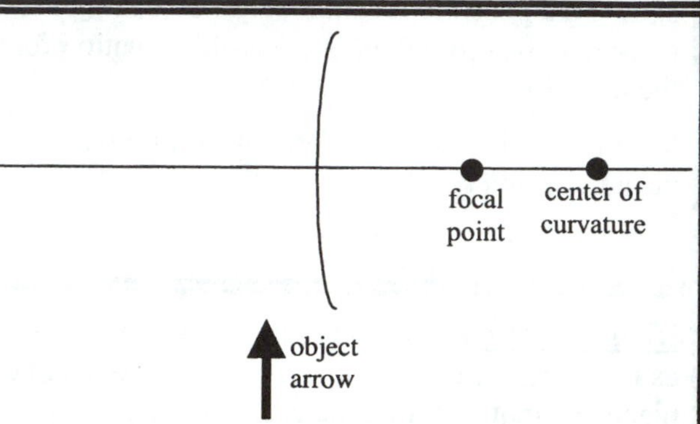

Demonstration 6: A concave mirror of relatively short focal length will be held up in front of the class. Draw an arrow on the drawing to the right indicating where the image will be formed by the mirror.

Indicate the orientation of the image by drawing the arrow upright or inverted.

Will the image be real or virtual?

Will the image be enlarged or reduced in size?

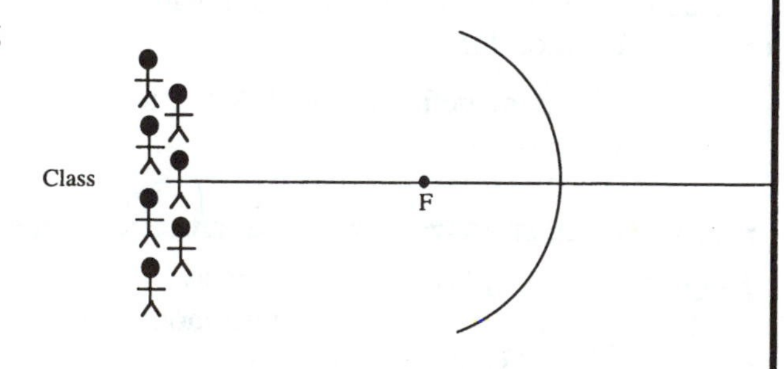

Demonstration 7: A convex mirror will be held up in front of the class. Draw an arrow on the drawing to the right indicating where the image will be formed by the mirror.

Indicate the orientation of the image by drawing the arrow upright or inverted.

Will the image be real or virtual?

Will the image be enlarged or reduced in size?

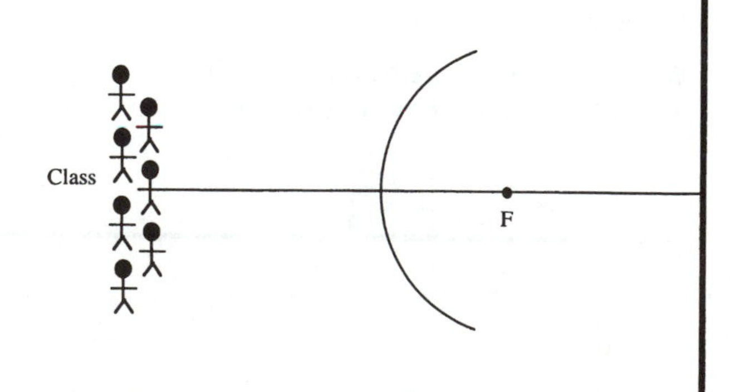

Mirrors (MIRR)
Teacher's Guide

Prerequisites:

This *ILD* sequence can be used as an introduction to image formation with non-plane mirrors. It can also be used as a review after these topics have been introduced in lecture or through text readings. Students should have been introduced to the terms ray, lens, object, image, real, virtual, upright and inverted.

Equipment:

- large concave mirror (See below.)
- lighted object arrow (See below.)
- large convex mirror (See below.)
- optical rail (See below.)
- holder for mirrors
- translucent screen

General Notes on Preparation and Equipment:

Concave and convex mirrors:
Choose mirrors with diameters and focal lengths appropriate to the class size. Most physics departments have a variety of mirrors including some large diameter ones. For example, the Sargeant Welch (www.sargeantwelch.com) CP46312-00 would work for both mirrors, although it is a too small for large lecture classes.

Lighted object arrow:
A rear-lighted flat object will work best. Pick a source that is appropriate in size for your class. Some examples are the Sargeant Welch (www.sargeantwelch.com) CP86027-00, CP85607-00 or CP86038-00 (with a separate light source).

Optical rail:
Either a commercially available or homemade rail will work. Examples are the Sargeant Welch WL3620 or CP85801-01. It is not essential to make precise measurements of image and object distances. The most important considerations are 1) ease of movement of the object, mirror and screen and 2) appropriate size so that the class can see the elements. It is best to set up the rail somewhat diagonally so that the students can see the object, mirror and image. The mirror will need to be tilted, and the screen will need to be off-center so that both the image and object can be seen simultaneously. It is useful to be able to rotate the whole table with the optical rail so that students can also look into the mirror. This is especially useful in looking at virtual images.

Demonstrations and Sample Results:

Demonstration 1: Real *larger* image with concave mirror. After the prediction and discussion steps, place the object outside the focal point, and move the screen until you see a sharp image.

Discussion after observing the results: Ask students to describe the position of the object and image relative to the focal point of the mirror. What does it mean to have a *real* image? Is this image real? How did the mirror form a real image? Is the image upright or inverted? In general, for what range of object distances (locations of the object) will the image be real, inverted and larger than the object?

Demonstration 2: Real *smaller* image with concave mirror. After the prediction and discussion steps, move the object further from the focal point (more than twice the focal length from the mirror), and move the screen until you see a sharp image.

Discussion after observing the results: Ask students to describe the position of the object and image relative to the focal point of the mirror. What is different about the positions of the object and image compared to Demonstration 1? Is this image real? Is the image upright or inverted? In general, for what range of object distances will the image be real, inverted and smaller than the object?

Demonstration 3: Virtual image with concave mirror. After the prediction and discussion steps, place the object closer to the mirror than the focal point, and move the screen to demonstrate that there is no location that gives a sharp image. Ask students to look into the mirror to see the virtual image. It may be helpful to rotate the optical rail.

Discussion after observing the results: Ask students to describe the position of the object relative to the focal point of the mirror. What is a *virtual* image? What is different about the position of the object and image compared to Demonstrations 1 and 2? Why is no real image formed? Where is the virtual image formed? Is it upright or inverted? Is it larger or smaller than the object? In general, for what range of object distances will the image be virtual and upright? (Object closer than focal point.) Can a virtual image produced by a concave mirror ever be smaller than the object? (Never.)

Demonstration 4: *Real* image with convex mirror? Replace the concave mirror with the convex one. After the prediction and discussion steps, place the object somewhere in front of the mirror, and move the screen to demonstrate that there is no location that gives a sharp image. Ask students to look into the mirror to see the virtual image. It may be helpful to rotate the optical rail. Try several positions of the object—further away than the focal point and closer than the focal point.

Discussion after observing the results: Can a real image be formed by a convex mirror? What is different and what is the same about this image and the image in Demonstration 3? What is different about the effect of the convex mirror compared to a concave mirror on rays diverging from object points? Where is the virtual image formed? Is it upright or inverted? Is it larger or smaller than the object? In general, for what range of object distances will the image be virtual and upright? (All!) Can a virtual image produced by a convex mirror ever be larger than the object? (Never!)

Demonstration 5: Virtual image with convex mirror. This demonstration has really been done in Demonstration 4. If you like, you can separate the two demonstrations, and try a variety of object distances here. Otherwise, just skip to Demonstration 6.

Discussion after observing the results: For what range of object distances will the image be virtual and upright? (All!) Can a virtual image produced by a convex mirror be larger than the object? (Never!)

Demonstration 6: Image in the concave side of a spoon. A good analogy is looking at your image in the concave side of a spoon. The focal length of the spoon is small compared to the possible object distances, and the image formed is always real, inverted and smaller, as in Demonstration 2. It is fun to put a handle on a mirror with both concave and convex polished surfaces, so that it looks like a spoon. After the prediction and discussion steps, hold up the concave side of the spoon (or the mirror from Demonstrations 1-3), and let the students observe their images.

Discussion after observing the results: Ask students to describe their images. Is this a real or virtual image? Why is the image inverted? Smaller than the object? Which previous demonstration is similar to this one? Is it possible to get a virtual, upright image of yourself from the concave side of a spoon?

Demonstration 7: **Image in the convex side of a spoon.** After the prediction and discussion steps, hold up the convex side of the spoon (or the mirror from Demonstrations 4-5), and let the students observe their images.

Discussion after observing the results: Ask students to describe their images. Is this a real or virtual image? Why is the image upright? Why is the image smaller than the object? Which previous demonstration is similar to this one? Is it possible to get a real, inverted image of yourself with the convex side of a spoon?

MIRRORS (MIRR)
TEACHER PRESENTATION NOTES

Demonstration 1: Real *larger* image with concave mirror. After the prediction and discussion steps, place the object outside the focal point, and move the screen until you see a sharp image.

- Ask students to describe the position of the object and image relative to the focal point. What does *real* mean? Is this image real? Upright or inverted? How is a real image formed?
- For what range of object distances will the image be real, inverted and larger than the object?

Demonstration 2: Real *smaller* image with concave mirror. After the prediction and discussion steps, move the object more than twice the focal length from the mirror, and find a sharp image.

- Ask students to describe the position of the object and image relative to the focal point. What is different from Demonstration 1? Is this image real? Upright or inverted?
- For what range of object distances will the image be real, inverted and smaller than the object?

Demonstration 3: Virtual image with concave mirror. After the prediction and discussion steps, move the object closer than the focal point, and demonstrate that there is no sharp image. Ask students to look into the mirror to see the virtual image. It may be helpful to rotate the optical rail.

- Ask students to describe the position of the object relative to the focal point. What is a *virtual* image? What is different compared to Demonstrations 1 and 2?
- Why is no real image formed? Where is the image formed? Is it upright or inverted? Is it larger or smaller than the object? For what range of object distances will the image be virtual?
- Can a virtual image produced by a concave mirror ever be smaller than the object?

Demonstration 4: *Real* image with convex mirror? Replace concave mirror with convex one. After the prediction and discussion steps, place the object in front of mirror, and demonstrate that there is no sharp image. Ask students to look into the mirror to see virtual image for several locations of the object.

- Can a real image be formed by a convex mirror? What is different and the same about this image and Demonstration 3? What is different about the effect of a convex mirror?
- Where is the virtual image formed? Is it upright or inverted? Is it larger or smaller than the object? In general, for what range of object distances will the image be virtual and upright?
- Can a virtual image produced by a convex mirror ever be larger than the object?

Demonstration 5: Virtual image with convex mirror. Try a variety of object distances.

- For what range of object distances will the image be virtual and smaller than the object?

Demonstration 6: Image in the concave side of a spoon. After the prediction and discussion steps, hold up the concave side of the spoon and let the students observe their images.

- Is this a real or virtual image? Upright or inverted? Larger or smaller?
- Which demonstration is similar? Can you get a virtual image of yourself in the concave side?

Demonstration 7: Image in the convex side of a spoon. After the prediction and discussion steps, hold up the convex side of the spoon and let the students observe their images.

- Is this a real or virtual image? Upright or inverted? Larger or smaller?
- Which demonstration is similar? Can you get a real image of yourself in the convex side?

POLARIZATION (POL)

Hand in this sheet **Name**_________________________________

INTERACTIVE LECTURE DEMONSTRATIONS
PREDICTION SHEET—**POLARIZED LIGHT**

Directions: This sheet will be collected. Write your name at the top to record your presence and participation in these demonstrations. Follow your instructor's directions. You may write whatever you wish on the attached Results Sheet and take it with you.

Demonstration 1: Consider a vertical object approaching a picket fence that is tilted by some angle θ with respect to that object. Will this object pass through the fence to the other side?

Suppose that the vertical object acts like a vector, and is equivalent to the sum of its components in any coordinate system. In terms of A (the length of the vector) and θ, how much of the vector will make it through the picket fence?

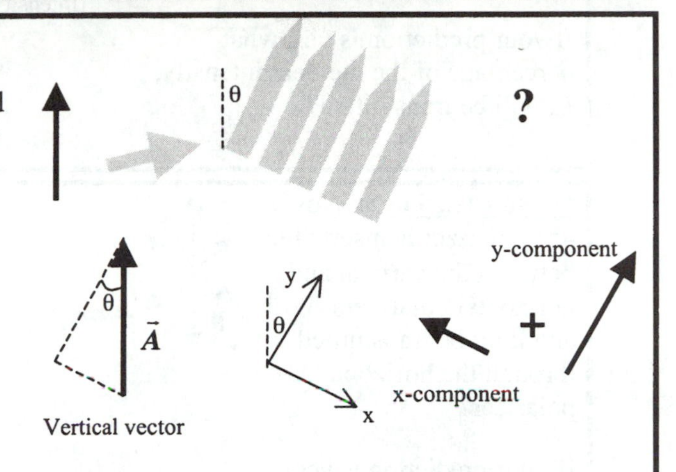

Demonstration 2: Consider unpolarized light, which consists of electromagnetic waves with an electric field vector, $\vec{E}_o$, that oscillates in every transverse direction. A piece of Polaroid (polarizer) works just like the picket fence in Demonstration 1. If unpolarized light of intensity I_o is incident on the polarizer, what is the intensity, I, of the polarized light that passes through?

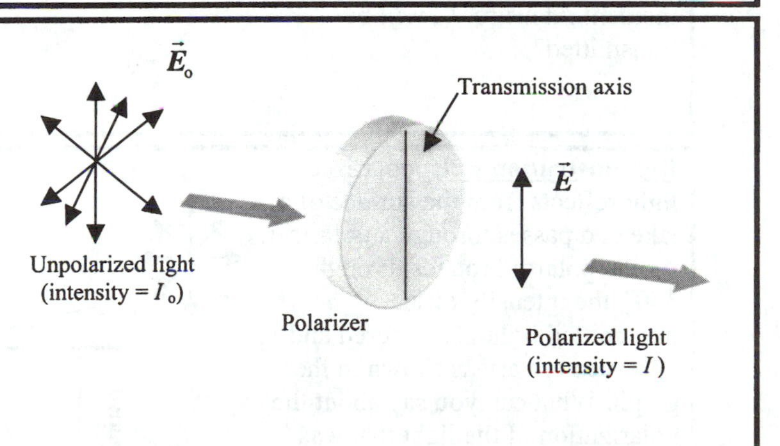

Demonstration 3: Light is polarized vertically and then sent through a polarizer that has its transmission axis oriented horizontally. Will any light be transmitted through the horizontal polarizer?

If your prediction is yes, what percentage of the incident intensity, I_o, will be transmitted?

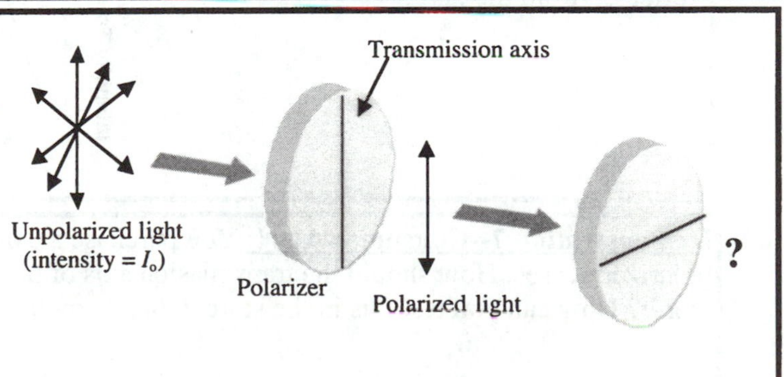

Demonstration 4: Light is polarized vertically and then sent through a polarizer that has its transmission axis rotated 45° to the vertical. Will any light be transmitted through the horizontal polarizer?

If your prediction is yes, what percentage of the incident intensity, I_o, will be transmitted?

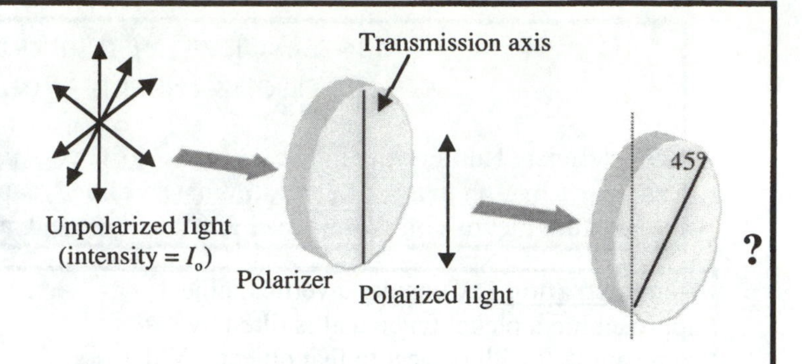

Demonstration 5: Now a 45° polarizer is inserted in between the vertical and horizontal polarizers. Will any light be transmitted through the horizontal polarizer?

If your prediction is yes, what percentage of the incident intensity, I_o, will be transmitted?

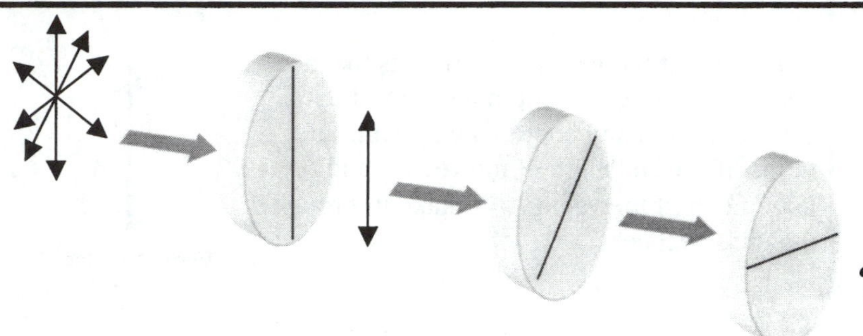

Demonstration 6: Unpolarized light reflects from the surface of a lake and passes through a polarizer. As the polaroid rotates through 360°, the intensity of the transmitted light is measured and observed to vary as shown in the graph. What can you say about the polarization of the light that was reflected from the lake?

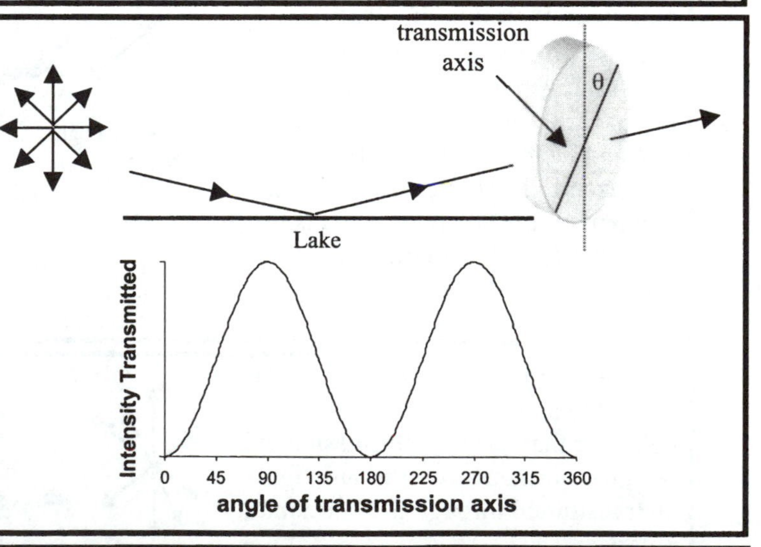

Demonstration 7--Consumer Alert: You purchase a pair of sunglasses advertised as having polarized lenses. How should the transmission axis of the lenses be oriented so you get your money's worth? How can you test this in the store without any fancy equipment?

Keep this sheet

INTERACTIVE LECTURE DEMONSTRATIONS
RESULTS SHEET—**POLARIZED LIGHT**

You may write whatever you wish on this sheet and take it with you.

Demonstration 1: Consider a vertical object approaching a picket fence that is tilted by some angle θ with respect to that object. Will this object pass through the fence to the other side?

Suppose that the vertical object acts like a vector, and is equivalent to the sum of its components in any coordinate system. In terms of A (the length of the vector) and θ, how much of the vector will make it through the picket fence?

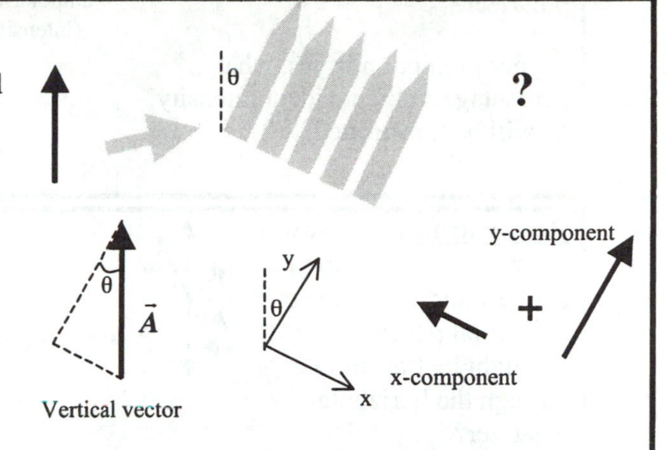

Demonstration 2: Consider un-polarized light, which consists of electromagnetic waves with an electric field vector, $\vec{E}_o$, that oscillates in every transverse direction. A piece of Polaroid (polarizer) works just like the picket fence in Demonstration 1. If un-polarized light of intensity I_0 is incident on the polarizer, what is the intensity, I, of the polarized light that passes through?

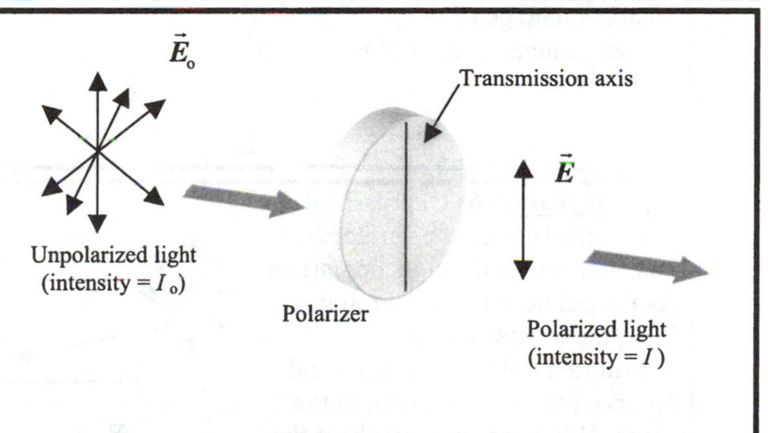

Demonstration 3: Light is polarized vertically and then sent through a polarizer that has its transmission axis oriented horizontally. Will any light be transmitted through the horizontal polarizer?

If your prediction is yes, what percentage of the incident intensity, I_0, will be transmitted?

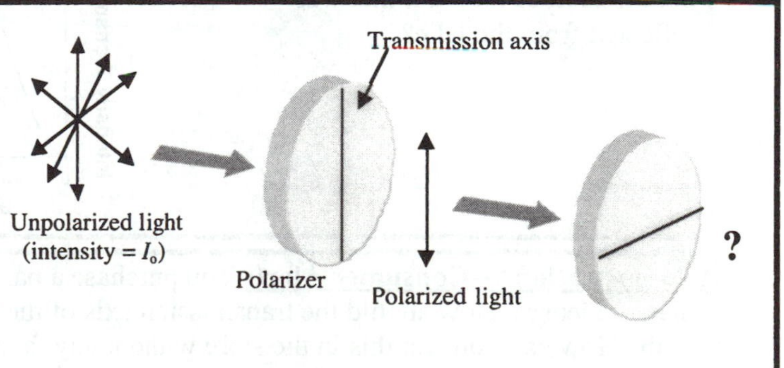

Demonstration 4: Light is polarized vertically and then sent through a polarizer that has its transmission axis rotated 45° to the vertical. Will any light be transmitted through the horizontal polarizer?

If your prediction is yes, what percentage of the incident intensity, I_o, will be transmitted?

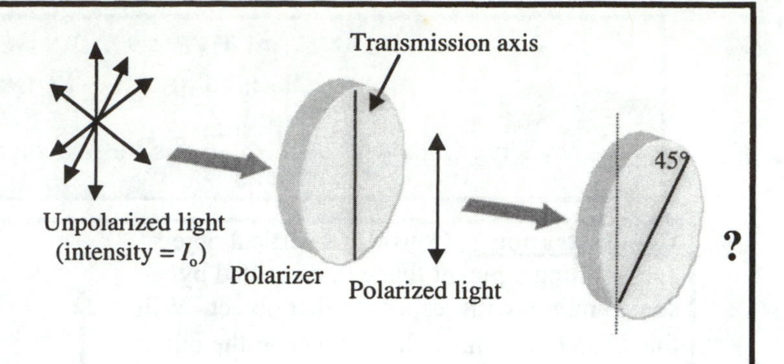

Demonstration 5: Now a 45° polarizer is inserted in between the vertical and horizontal polarizers. Will any light be transmitted through the horizontal polarizer?

If your prediction is yes, what percentage of the incident intensity, I_o, will be transmitted?

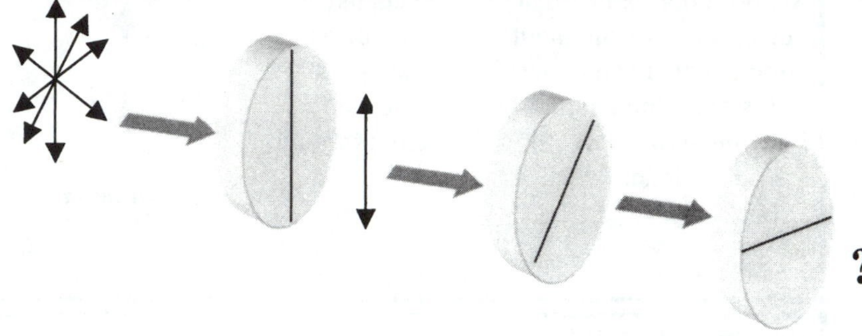

Demonstration 6: Unpolarized light reflects from the surface of a lake and passes through a polarizer. As the polaroid rotates through 360°, the intensity of the transmitted light is measured and observed to vary as shown in the graph. What can you say about the polarization of the light that was reflected from the lake?

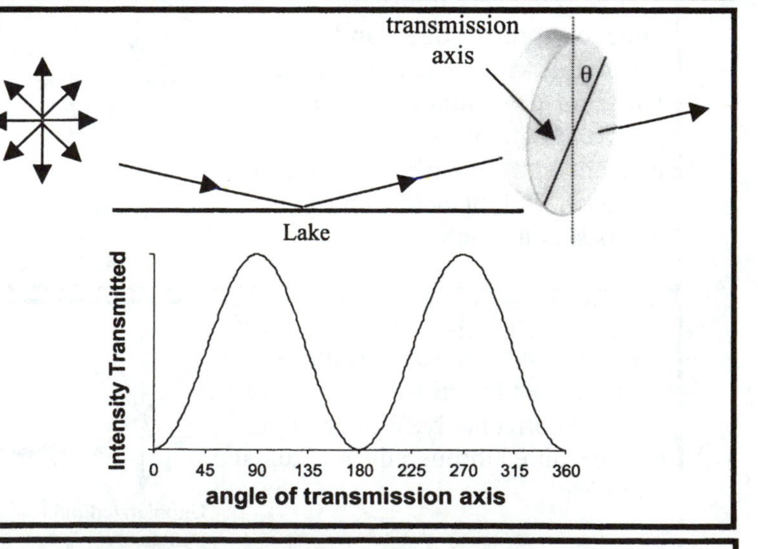

Demonstration 7--Consumer Alert: You purchase a pair of sunglasses advertised as having polarized lenses. How should the transmission axis of the lenses be oriented so you get your money's worth? How can you test this in the store without any fancy equipment?

POLARIZED LIGHT (POL)
TEACHER'S GUIDE

Prerequisites:

This *ILD* sequence can be used as an introduction to polarization. It can also be used as a review after these topics have been introduced in lecture or through text readings. It will be helpful if students have been introduced to the terms intensity, electric field, plane of incidence and axis of polarization.

Equipment:

3 pieces of Polaroid, about 15 cm square, with arrows on them indicating the axis
slide projector, flashlight or other relatively collimated beam of un-polarized light
glass plate
Polaroid sunglasses

Demonstrations and Sample Results:

Demonstration 1: Picket fence. After the prediction and discussion steps, discuss the concepts.

Discussion after observing the results: How would you find the component parallel to the openings in the picket fence?

Demonstration 2: Model for polarized light. After the prediction and discussion steps, shine un-polarized light through a piece of Polaroid with its axis vertical.

Discussion after observing the results: Have a student describe the light transmitted through the Polaroid. About what percentage is transmitted? How does intensity depend on the electric field? Is intensity proportional to the field? Proportional to the field squared? What percentage of the incident un-polarized light should be transmitted?

Demonstration 3: Vertically polarized light through crossed (horizontal) Polaroid. After the prediction and discussion steps, shine the vertically polarized light through the Polaroid with its axis horizontal (perpendicular to the axis of polarization of the light).

Discussion after observing the results: Ask students to describe how much light gets through. How can this be explained in terms of Demonstrations 1 and 2?

Demonstration 4: Polaroid at 45° to polarized light. After the prediction and discussion steps, shine the polarized light through the Polaroid with its axis at 45° to the axis of polarization of the light.

Discussion after observing the results: Ask students to describe how much light gets through. How can this be explained in terms of Demonstrations 1 and 2? If the component of the field $E_o\cos\theta$ is transmitted, what should the transmitted intensity I be? Does this seem to correspond to the observed transmitted intensity?

Demonstration 5: 45° and crossed (horizontal) Polaroids. After the prediction and discussion steps, shine the polarized light through the Polaroid with its axis at 45° to the axis of polarization of the light and then through the crossed (horizontal) Polaroid.

Discussion after observing the results: Ask students to describe how much light gets through. How can this be explained in terms of Demonstrations 2 and 3? What is the axis of polarization of the light that gets through the 45° Polaroid? Does this have a component along the axis of the second Polaroid?

Demonstration 6: Polarization by reflection. After the prediction and discussion steps, shine the beam of un-polarized light so that it is reflecting off the glass plate at an angle of incidence close to Brewster's angle for glass, 56°. Then rotate a Polaroid in front of the reflected beam, and have the students observe the variation in transmitted intensity.

Discussion after observing the results: Have students describe the intensity variation. Does it appear that the un-polarized light became polarized when it was reflected from the glass surface.

Demonstration 7: Polaroid sunglasses. After the prediction and discussion steps, repeat Demonstration 6, and note the direction of the axis of the Polaroid for which the transmission is minimum. Then demonstrate that the Polaroid in sunglasses has its axis oriented this way.

Discussion after observing the results: Does the reflected light appear to have its polarization in one particular direction. Which direction is this—in the plane of incidence or perpendicular to the plane of incidence? Why would Polaroid sunglasses with the axis vertical be most effective in cutting down the glare off the rear windshield of a car or off the surface of a lake?

POLARIZED LIGHT (POL)
TEACHER PRESENTATION NOTES

Demonstration 1: Picket fence. Discuss the concepts.

- How would you find the component parallel to the openings in the picket fence?

Demonstration 2: Model for polarized light. Shine un-polarized light through a piece of Polaroid with its axis vertical.

- About what percentage is transmitted?
- How does intensity depend on the electric field? What percentage should be transmitted?

Demonstration 3: Vertically polarized light through crossed (horizontal) Polaroid.. Shine the polarized light through the Polaroid with its axis perpendicular to the axis of polarization of the light.

- How much light gets through. How can this be explained in terms of Demonstrations 1 and 2?

Demonstration 4: Polaroid at 45° to polarized light. Shine the polarized light through the Polaroid with its axis at 45° to the axis of polarization of the light.

- How much light gets through. How can this be explained in terms of Demonstrations 1 and 2? If the component of the field $E_o\cos\theta$ is transmitted, what should the transmitted intensity I be? Does this seem to correspond to the transmitted intensity?

Demonstration 5: 45° and crossed (horizontal) Polaroids. Shine the polarized light through the Polaroid with its axis at 45° to the axis of polarization of the light and then through the crossed Polaroid.

- How much light gets through. How can this be explained in terms of Demonstrations 2 and 3?
- What is the direction of the axis of polarization of the light that gets through the 45° Polaroid? Does this have a component along the axis of the second Polaroid?

Demonstration 6: Polarization by reflection. Shine the beam of un-polarized light so that it is reflecting off the glass plate at an angle of incidence close to Brewster's angle, 56°. Then rotate a Polaroid in front of the reflected beam, and observe the variation in transmitted intensity.

- Have students describe the intensity variation. Did the light become polarized by reflection?

Demonstration 7: Polaroid sunglasses. After the prediction and discussion steps, repeat Demonstration 6, and note the direction of the axis of the Polaroid for which the transmission is minimum. Then demonstrate that the Polaroid in sunglasses has its axis oriented this way.

- In what direction does the reflected light appear to be polarized—in the plane of incidence or perpendicular to the plane of incidence?
- Why would Polaroid sunglasses with the axis vertical be most effective in cutting down the glare off the rear windshield of a car or off the surface of a lake?

Appendix A: Interactive Lecture Demonstration Experiment Configuration Files

APPENDIX A: INTERACTIVE LECTURE DEMONSTRATION EXPERIMENT CONFIGURATION FILES

Listed below are the settings in the *Experiment Configuration Files* used in these *ILDs*. These files are available from Vernier Software and Technology (www.vernier.com) for *Logger Pro* (Macintosh and Windows) and from PASCO Scientific (www.pasco.com) for *Data Studio* (Macintosh and Windows). With the information below, the user can set up files for any compatible hardware and software package.

Experiment File	Description	Data Collection	Data Handling	Analysis	Display
KIN1D1	Graphs distance vs. time.	20 points/sec Motion sensor	NA	NA	One set of graph axes with line. Distance: 0-2 m Time 0-10 sec
KIN1D2	Graphs velocity vs. time.	20 points/sec Motion sensor	NA	NA	One set of graph axes with line. Velocity: -1 to +1 m/s Time 0-5 sec
KIN1D4	Graphs velocity vs. time. Second window has distance and velocity axes	20 points/sec Motion sensor	NA	NA	One set of graph axes with line. Velocity: -1 to +1 m/s Time 0-15 sec
KIN2D1	Graphs position and velocity vs. time.	20 points/sec Motion sensor	NA	NA	Two sets of graph axes with lines. Position: 0-2 m Velocity: -1 to +1 m/s Time 0-5 sec
KIN2D3	Graphs position, velocityand acceleration vs. time.	20 points/sec Motion sensor	NA	NA	Three sets of graph axes with lines. Position: 0-2 m Velocity: -1 to +1 m/s Accel.: -1 to +1 m/s^2 Time 0-3 sec
KIN2D4	Graphs velocity and acceleration vs. time.	20 points/sec Motion sensor	NA	NA	Two sets of graph axes with lines. Velocity: -1 to +1 m/s Accel.: -2 to +2 m/s^2 Time 0-5 sec
KIN2D6	Graphs velocity and acceleration vs. time.	20 points/sec Motion sensor	NA	NA	Two sets of graph axes with lines. Velocity: -1 to +1 m/s Accel.: -4 to +4 m/s^2 Time 0-5 sec

Experiment File	Description	Data Collection	Data Handling	Analysis	Display
KIN2D8	Graphs velocity and acceleration vs. time.	30 points/sec Motion sensor Distance away is negative	NA	Can use analysis and statistics features to find average acceleration from graph.	Two sets of graph axes with lines. Velocity: -5 to +5 m/s Accel.: -15 to +15 m/s^2 Time 0-2 sec
N1&2D1	Graphs velocity, force and acceleration vs. time.	20 points/sec Motion sensor and force sensor.	NA	NA	Three sets of graph axes with lines. Velocity: -2 to +2 m/s Accel.: -2 to +2 m/s^2 Force: -0.5 to +0.5 N Time 0-3 sec
N1&2D3	Graphs velocity and acceleration vs. time.	20 points/sec Motion sensor.	NA	NA	Two sets of graph axes with lines. Velocity: -2 to +2 m/s Accel.: -2 to +2 m/s^2 Time 0-5 sec
N1&2D4	Graphs velocity, acceleration and force vs. time.	20 points/sec Motion sensor and force sensor.	NA	NA	Three sets of graph axes with lines. Velocity: -1 to +1 m/s Accel.: -2 to +2 m/s^2 Force: -2 to +2 N Time 0-3 sec
N1&2D5	Graphs velocity, acceleration and force vs. time.	20 points/sec Motion sensor and force sensor.	NA	NA	Two sets of graph axes with lines. Velocity: -1 to +1 m/s Force: -0.5 to +0.5 N Accel.: -2 to +2 m/s^2 Time 0-5 sec
N3D1	Graphs two forces vs. time	20 points/sec Two force sensors Sign of one sensor reversed	NA	NA	Both forces graphed on the same set of axes with lines. Forces: -10 to +10 N Time 0-10 sec
N3D4	Graphs two forces vs. time	20 points/sec Two force sensors Sign of one sensor reversed	NA	NA	Both forces graphed on the same set of axes with lines. Forces: -10 to +10 N Time 0-10 sec
N3D5	Graphs two forces vs. time	20 points/sec Two force sensors Sign of one sensor reversed	NA	NA	Both forces graphed on the same set of axes with lines. Forces: -10 to +10 N Time 0-10 sec

Experiment File	Description	Data Collection	Data Handling	Analysis	Display
N3D6	Graphs two forces vs. time	4000 points/sec Two force sensors Sign of one sensor reversed Triggered mode	NA	Use integration routine to find area under force vs. time graphs.	Two sets of graph axes with lines. Forces: -10 to +10 N Time 0-0.2 sec
PROJD1	QuickTime movie of the projectile motion of a tossed tennis ball. (If possible, a movie made live in lecture should be substituted.)	NA	NA	Analysis in *VideoPoint*, *Logger Pro* or other video analysis software.	
ENERD1	Graphs position, velocity and acceleration vs. time. Multiple windows to display different combinations of quantities.	20 points/sec Motion sensor.	Includes calculated columns for kinetic energy, potential energy and total mechanical energy.	NA	Three sets of graph axes with lines. Position: 0 to +1.5 m Velocity: -1 to +1 m/s Accel: -2 to +2 m/s^2 Time 0-5 sec Plus additional axes for *KE*, *PE* and *E*.
FLUSD1	Graphs force vs. time	20 points/sec Force sensor	NA	Use examine feature to measure forces.	One set of graph axes with line. Force: -1 to +1 N Time 0-40 sec
SHMD1	Graphs displacement, velocity, acceleration and force vs. time. Multiple windows to display different combinations of quantities.	20 points/sec Motion sensor and force sensor.	Includes calculated columns for displacement, kinetic energy, potential energy and total mechanical energy. Enter equilibrium position to graph displacement from equilibrium.		Three sets of graph axes with lines. Position: -0.5 to +0.5 m Velocity: -1 to +1 m/s Accel.: -4 to +4 m/s^2 Force: -3 to +3 N Time 0-5 sec Plus additional axes for *KE*, *PE* and *E*.
SNDD1	Graphs sound pressure vs. time.	10,000 points/sec in repeat mode Microphone.		Use examine feature to measure times.	One set of graph axes with line. Sound pressure: -0.5 to +0.5 in arbitrary units Time 0-0.03 sec

Experiment File	Description	Data Collection	Data Handling	Analysis	Display
INHTD1	Graphs two temp. sensors vs. time.	20 points/sec Temp. sensors 1 and 2	Hidden graph of entire cooling process for display after brief collection of data.	NA	One set of graph axes with lines. Temp. 1: 0-100°C Temp. 2: 0-100°C Time: 0-10 min
INHTD3	Graphs two temp. sensors vs. time.	20 points/sec Temp. sensors 1 and 2	NA	NA	One set of graph axes with lines. Temp. 1: 0-100°C Temp. 2: 0-100°C Time: 0-5 min
INHTD4	Graphs one temp. sensor vs. time. Enables heat pulser to transfer heat pulses.	20 points/sec Temp. sensor 1 only Heat pulser enabled, 5 sec pulse length	NA	May use examine feature to read data from graphs after collected	One set of graph axes with line. Temp.: 10-80°C Time: 0-120 sec
INHTD5	Graphs one temp. sensor vs. time. Enables heat pulser to transfer heat pulses.	20 points/sec Temp. sensor 1 only Heat pulser enabled, 5 sec pulse length	NA	May use examine feature to read data from graphs after collected	One set of graph axes with line. Temp.: 20-50°C Time: 0-80 sec
INHTD6	Graphs one temp. sensor vs. time. Enables heat pulser to transfer heat pulses.	20 points/sec Temp. sensor 1 only Heat pulser enabled, 2 sec pulse length	NA	NA	One set of graph axes with line. Temp.: 60-100°C Time: 0-120 sec
SPHTD1	Graphs one temp. sensor vs. time. Enables heat pulser to transfer heat pulses.	20 points/sec Temp. sensor 1 only Heat pulser enabled, 10 sec pulse length	NA	May use examine feature to read data from graphs after collected	One set of graph axes with line. Temp.: 20-100°C Time: 0-120 sec
SPHTD2	Graphs one temp. sensor vs. time. Enables heat pulser to transfer heat pulses.	20 points/sec Temp. sensor 1 only Heat pulser enabled, 10 sec pulse length	NA	May use examine feature to read data from graphs after collected	One set of graph axes with line. Temp.: 20-100°C Time: 0-120 sec
SPHTD3	Graphs one temp. sensor vs. time. Enables heat pulser to transfer heat pulses.	20 points/sec Temp. sensor 1 only Heat pulser enabled, 10 sec pulse length	NA	May use examine feature to read data from graphs after collected	One set of graph axes with line. Temp.: 20-100°C Time: 0-120 sec

Experiment File	Description	Data Collection	Data Handling	Analysis	Display
HTPCD1	Graphs one temp. sensor vs. time. Enables heat pulser to transfer heat pulses.	20 points/sec Temp. sensor 1 only Heat pulser enabled, 10 sec pulse length	NA	NA	One set of graph axes with line. Temp.: 20-100°C Time: 0-120 sec
HTPCD2	Graphs one temp. sensor vs. time. Enables heat pulser to transfer heat pulses.	20 points/sec Temp. sensor 1 only Heat pulser enabled, 10 sec pulse length	NA	May use analysis feature to read data from graphs after collecting	One set of graph axes with line. Temp.: -10 to 30°C Time: 0-5 min
HTPCD3	Graphs one temp. sensor vs. time. Enables heat pulser to transfer heat pulses.	20 points/sec Temp. sensor 1 only Heat pulser enabled, 10 sec pulse length	NA	May use analysis feature to read data from graphs after collected	One set of graph axes with line. Temp.: 70-110°C Time: 0-10 min
HENGD1	Graphs pressure, temperature and volume. Pressure and temperature are continuously measured while volume is entered manually after pressure value is kept.	20 points/sec Pressure sensor 1 and temperature sensor 2 Prompted event mode--keep pressure values only when desired	New columns which display pressure in 10^5 Pa [P(atm) x 1.01325]. and temp. in K [T(°C)+273.15] New prompted manual column to enter volume in cm^3.	Use integration routine to find area enclosed in pressure vs. volume cycle.	One set of graph axes with line. Press.: 0.9-1.1 x 10^5 Pa Volume: 24 - 40 cm^3
INDCD1	Graphs current and voltage vs. time.	25 points/sec Current probe 1 and Voltage probe 2	Change horizontal axis on Current 1 graph to Voltage 2 to display Current vs. Voltage graph.	NA	Two sets of graph axes with lines. Voltage: -6 to +6V Current: -0.6 to +0.6 A Time: 0-10 sec
INDCD3	Digital displays of either two currents.	Digital displays of Current probes 1 and 2	NA	NA	Digital displays
INDCD5	Digital displays of either two currents or two voltages	Digital displays of Current probes 1 and 2 or Voltage probes 1 and 2	NA	NA	Digital displays
SPCD2	Graphs two currents vs. time.	25 points/sec Curent probes 1 and 2	NA	NA	Two sets of graph axes with lines. Current: -.6 to +.6 A Time: 0-10 sec

Experiment File	Description	Data Collection	Data Handling	Analysis	Display
SPCD3	Graphs two voltages vs. time.	25 points/sec Voltage probes 1 and 2	NA	NA	Two sets of graph axes with lines. Voltage: -6 to +6V Time: 0-10 sec
SPCD5	Graphs two currents vs. time.	25 points/sec Curent probes 1 and 2	NA	NA	Two sets of graph axes with lines. Current: -.6 to +.6 A Time: 0-10 sec
SPCD6	Graphs two voltages vs. time.	25 points/sec Voltage probes 1 and 2	NA	NA	Two sets of graph axes with lines. Voltage: -6 to +6V Time: 0-10 sec
RCCD1	Graphs two voltages vs. time.	25 points/sec Voltage probes 1 and 2	NA	NA	Two sets of graph axes with lines. Voltage: -6 to +6V Time: 0-20 sec
RCCD2	Graphs current and voltage vs. time.	25 points/sec Voltage probe 1 and Current probe 2	NA	NA	Two sets of graph axes with lines. Voltage: -6 to +6V Current: -0.6 to +0.6 A Time: 0-20 sec
RCCD4	Graphs two voltages vs. time.	25 points/sec Voltage probes 1 and 2	NA	NA	Two sets of graph axes with lines. Voltage: -6 to +6V Time: 0-20 sec
RCCD5	Graphs voltage and current vs. time.	25 points/sec Voltage probe 1 and Current probe 2	NA	NA	Two sets of graph axes with lines. Voltage: -6 to +6V Current: -0.6 to +0.6 A Time: 0-20 sec